Advanced Information and Knowledge Processing

Series editors

Lakhmi C. Jain
Bournemouth University, Poole, UK, and
University of South Australia, Adelaide, Australia

Xindong Wu
University of Vermont

Information systems and intelligent knowledge processing are playing an increasing role in business, science and technology. Recently, advanced information systems have evolved to facilitate the co-evolution of human and information networks within communities. These advanced information systems use various paradigms including artificial intelligence, knowledge management, and neural science as well as conventional information processing paradigms. The aim of this series is to publish books on new designs and applications of advanced information and knowledge processing paradigms in areas including but not limited to aviation, business, security, education, engineering, health, management, and science. Books in the series should have a strong focus on information processing—preferably combined with, or extended by, new results from adjacent sciences. Proposals for research monographs, reference books, coherently integrated multi-author edited books, and handbooks will be considered for the series and each proposal will be reviewed by the Series Editors, with additional reviews from the editorial board and independent reviewers where appropriate. Titles published within the Advanced Information and Knowledge Processing series are included in Thomson Reuters' Book Citation Index.

More information about this series at http://www.springer.com/series/4738

Yasser Mohammad · Toyoaki Nishida

Data Mining for Social Robotics

Toward Autonomously Social Robots

 Springer

Yasser Mohammad
Department of Electrical Engineering
Assiut University
Assiut
Egypt

Toyoaki Nishida
Department of Intelligence Science
 and Technology
Kyoto University
Kyoto
Japan

Advanced Information and Knowledge Processing
ISBN 978-3-319-79755-7   ISBN 978-3-319-25232-2   (eBook)
DOI 10.1007/978-3-319-25232-2

Preface

Robots are here!

Service robots are beginning to live with us and occupy the same social space we live in. These robots should be able to understand human's natural interactive behavior and to respond correctly to it. To do that they need to learn from their interactions with humans. Considering the exceptional cognitive abilities of Homo sapiens, two features immediately pop up, namely, autonomy and sociality.

Autonomy is what we consider when we think of human's ability to play chess, think about the origin of the universe, plan for hunts or investments, build a robust stable perception of her environment, etc. This was the feature most inspiring the early work in AI with its focus on computation and deliberative techniques. It was also the driving force behind more recent advances that returned the interactive nature of autonomy to the spotlight including reactive robotics, behavioral robotics, and the more recent interest in embodiment.

Sociality, or the ability to act appropriately in the social domain, is another easily discerned feature of human intelligence. Even playing chess has a social component for if there was no social environment, it is hard to imagine a single autonomous agent coming up with this two-player game. Humans do not only occupy physical space but also occupy a social space that shapes them while they shape it. Interactions between agents in this social space can be considered as efficient utilization of natural interaction protocols which can be roughly defined as a kind of multi-scale synchrony between interaction partners.

The interplay between autonomy and sociality is a major theoretical and practical concern for modern social robotics. Robots are expected to be autonomous enough to justify their treatment as something different from an automobile and they should be socially interactive enough to occupy a place in our humanly constructed social space. Robotics researchers usually focus on one of these two aspects but we believe that a breakthrough in the field is expected only when the interplay between these two factors is understood and leveraged.

This is where data mining techniques (especially time-series analysis methods) come into the picture. Using algorithms like change point discovery, motif discovery, and causality analysis, future social robots will be able to make sense of what they see humans do and using techniques developed for programming by demonstration they may be able to autonomously socialize with us.

This book tries to bridge the gap between autonomy and sociality by reporting our efforts to design and evaluate a novel control architecture for autonomous, interactive robots and agents that allow the robot/agent to learn natural social interaction protocols (both implicit and explicit) autonomously using unsupervised machine learning and data mining techniques. This shows how autonomy can enhance sociality. The book also reports our efforts to utilize the social interactivity of the robot to enhance its autonomy using a novel fluid imitation approach.

The book consists of two parts with different (yet complimentary) emphasis that introduce the reader to this exciting new field in the intersection of robotics, psychology, human-machine interaction, and data mining.

One goal that we tried to achieve in writing this book was to provide a self-contained work that can be used by practitioners in our three fields of interest (data mining, robotics, and human-machine-interaction). For this reason we strove to provide all necessary details of the algorithms used and the experiments reported not only to ease reproduction of results but also to provide readers from these three widely separated fields with the essential and necessary knowledge of the other fields required to appreciate the work and reuse it in their own research and creations.

Assiut, Egypt   Yasser Mohammad
Kyoto, Japan   Toyoaki Nishida
October 2015

Contents

1   Introduction.. 1
    1.1   Motivation 1
    1.2   General Overview 5
    1.3   Relation to Different Research Fields 7
           1.3.1   Interaction Studies 7
           1.3.2   Robotics 9
           1.3.3   Neuroscience and Experimental Psychology......... 11
           1.3.4   Machine Learning and Data Mining 11
           1.3.5   Contributions 11
    1.4   Interaction Scenarios.. 12
    1.5   Nonverbal Communication in Human–Human Interactions 14
    1.6   Nonverbal Communication in Human–Robot Interactions 17
           1.6.1   Appearance 18
           1.6.2   Gesture Interfaces............................... 18
           1.6.3   Spontaneous Nonverbal Behavior 19
    1.7   Behavioral Robotic Architectures 21
           1.7.1   Reactive Architectures........................... 21
           1.7.2   Hybrid Architectures............................. 22
           1.7.3   HRI Specific Architectures....................... 23
    1.8   Learning from Demonstrations................................ 24
    1.9   Book Organization 26
    1.10   Supporting Site 27
    1.11   Summary 28
    References 28

Part I   Time Series Mining

2   Mining Time-Series Data 35
    2.1   Basic Definitions 35
    2.2   Models of Time-Series Generating Processes 36
           2.2.1   Linear Additive Time-Series Model 36
           2.2.2   Random Walk 37

	2.2.3	Moving Average Processes	38
	2.2.4	Auto-Regressive Processes..................	40
	2.2.5	ARMA and ARIMA Processes...............	40
	2.2.6	State-Space Generation	41
	2.2.7	Markov Chains..........................	42
	2.2.8	Hidden Markov Models....................	43
	2.2.9	Gaussian Mixture Models	45
	2.2.10	Gaussian Processes.......................	47
2.3	Representation and Transformations		50
	2.3.1	Piecewise Aggregate Approximation...........	51
	2.3.2	Symbolic Aggregate Approximation	52
	2.3.3	Discrete Fourier Transform	54
	2.3.4	Discrete Wavelet Transform.................	55
	2.3.5	Singular Spectrum Analysis.................	56
2.4	Learning Time-Series Models from Data................		67
	2.4.1	Learning an AR Process	67
	2.4.2	Learning an ARMA Process	70
	2.4.3	Learning a Hidden Markov Model	73
	2.4.4	Learning a Gaussian Mixture Model	76
	2.4.5	Model Selection Problem...................	77
2.5	Time Series Preprocessing........................		77
	2.5.1	Smoothing............................	77
	2.5.2	Thinning	78
	2.5.3	Normalization	78
	2.5.4	De-Trending	79
	2.5.5	Dimensionality Reduction	80
	2.5.6	Dynamic Time Warping	81
2.6	Summary		82
	References		83
3	**Change Point Discovery**		**85**
3.1	Approaches to CP Discovery......................		86
3.2	Markov Process CP Approach		87
3.3	Two Models Approach		90
3.4	Change in Stochastic Processes		93
3.5	Singular Spectrum Analysis Based Methods		94
	3.5.1	Alternative SSA CPD Methods	98
3.6	Change Localization...........................		98
3.7	Comparing CPD Algorithms		99
	3.7.1	Confusion Matrix Measures.................	100
	3.7.2	Divergence Measures	101
	3.7.3	Equal Sampling Rate	104
3.8	CPD for Measuring Naturalness in HRI		105
3.9	Summary		107
	References		107

4   Motif Discovery 109
    4.1   Motif Discovery Problem(s). 109
    4.2   Motif Discovery in Discrete Sequences.................. 110
          4.2.1   Projections Algorithm 114
          4.2.2   GEMODA Algorithm 115
    4.3   Discretization Algorithms 118
          4.3.1   MDL Extended Motif Discovery 120
    4.4   Exact Motif Discovery 124
          4.4.1   MK Algorithm.............................. 125
          4.4.2   MK+ Algorithm............................. 127
          4.4.3   MK++ Algorithm............................ 129
          4.4.4   Motif Discovery Using Scale Normalized
                    Distance Function (MN) 131
    4.5   Stochastic Motif Discovery 134
          4.5.1   Catalano's Algorithm 134
    4.6   Constrained Motif Discovery.......................... 136
          4.6.1   MCFull and MCInc 136
          4.6.2   Real-Valued GEMODA. 138
          4.6.3   Greedy Motif Extension 138
          4.6.4   Shift-Density Constrained Motif Discovery 140
    4.7   Comparing Motif Discovery Algorithms 143
    4.8   Real World Applications.............................. 144
          4.8.1   Gesture Discovery from Accelerometer Data 144
          4.8.2   Differential Drive Motion Pattern Discovery........ 145
          4.8.3   Basic Motions Discovery from Skeletal
                    Tracking Data 145
    4.9   Summary 146
    References 147

5   Causality Analysis 149
    5.1   Causality Discovery 150
    5.2   Correlation and Causation 150
    5.3   Granger-Causality and Its Extensions 151
    5.4   Convergent Cross Mapping 153
    5.5   Change Causality................................... 162
    5.6   Application to Guided Navigation 165
          5.6.1   Robot Guided Navigation 165
    5.7   Summary 166
    References 166

Part II   Autonomously Social Robots

6   Introduction to Social Robotics 171
    6.1   Engineering Social Robots. 171
    6.2   Human Social Response to Robots...................... 174

6.3   Social Robot Architectures 177
    6.3.1   C4 Cognitive Architecture 177
    6.3.2   Situated Modules 181
    6.3.3   HAMMER. 185
6.4   Summary 190
References 190

7   **Imitation and Social Robotics** 193
7.1   What Is Imitation? 193
7.2   Imitation in Animals and Humans 196
7.3   Social Aspects of Imitation in Robotics. 200
    7.3.1   Imitation for Bootstrapping Social Understanding 201
    7.3.2   Back Imitation for Improving Perceived Skill. 202
7.4   Summary 204
References 204

8   **Theoretical Foundations** 207
8.1   Autonomy, Sociality and Embodiment 207
8.2   Theory of Mind 211
8.3   Intention Modeling 218
    8.3.1   Traditional Intention Modeling 219
    8.3.2   Intention in Psychology. 221
    8.3.3   Challenges for the Theory of Intention 222
    8.3.4   The Proposed Model of Intention 223
8.4   Guiding Principles 224
8.5   Summary 225
References 225

9   **The Embodied Interactive Control Architecture** 229
9.1   Motivation 229
9.2   The Platform 230
9.3   Key Features of EICA 233
9.4   Action Integration 234
    9.4.1   Behavior Level Integration. 236
    9.4.2   Action Level Integration 236
9.5   Designing for EICA 237
9.6   Learning Using FPGA 238
9.7   Application to Explanation Scenario 240
    9.7.1   Fixed Structure Gaze Controller 241
9.8   Application to Collaborative Navigation 242
9.9   Summary 243
References 243

10   **Interacting Naturally** 245
10.1 Main Insights. 245
10.2 EICA Components 247

10.3   Down–Up–Down Behavior Generation (DUD). 249
10.4   Mirror Training (MT). 252
10.5   Summary 253
References 253

11   Interaction Learning Through Imitation 255
   11.1   Stage 1: Interaction Babbling. 255
        11.1.1   Learning Intentions. 256
        11.1.2   Controller Generation 257
   11.2   Stage 2: Interaction Structure Learning 259
        11.2.1   Single-Layer Interaction Structure Learner 259
        11.2.2   Interaction Rule Induction 261
        11.2.3   Deep Interaction Structure Learner 264
   11.3   Stage 3: Adaptation During Interaction 266
        11.3.1   Single-Layer Interaction Adaptation Algorithm. 266
        11.3.2   Deep Interaction Adaptation Algorithm 268
   11.4   Applications 269
        11.4.1   Explanation Scenario. 270
        11.4.2   Guided Navigation Scenario. 271
   11.5   Summary 272
References 272

12   Fluid Imitation. 275
   12.1   Introduction. 276
   12.2   Example Scenarios. 278
   12.3   The Fluid Imitation Engine (FIE) 279
   12.4   Perspective Taking. 280
        12.4.1   Transforming Environmental State 280
        12.4.2   Calculating Correspondence Mapping 282
   12.5   Significance Estimator. 286
   12.6   Self Initiation Engine 288
   12.7   Application to the Navigation Scenario. 288
   12.8   Summary 290
References 290

13   Learning from Demonstration 293
   13.1   Early Approaches. 294
   13.2   Optimal Demonstration Methods 295
        13.2.1   Inverse Optimal Control 295
        13.2.2   Inverse Reinforcement Learning 299
        13.2.3   Dynamic Movement Primitives. 302
   13.3   Statistical Methods. 307
        13.3.1   Hidden Markov Models. . . . . , , , ... 307
        13.3.2   GMM/GMR. 307

13.4   Symbolization Approaches 313
13.5   Summary 316
References 316

14 Conclusion.. 319

Index 325

Chapter 1
Introduction

How to create a social robot that people do not only *operate* but relate to? This book is an attempt to answer this question and will advocate using the same approach used by infants to grow into social beings: developing natural interaction capacity autonomously. We will try to flesh out this answer by providing a computational framework for autonomous development of social behavior based on data mining techniques.

The focus of this book is on how to utilize the link between autonomy and sociality in order to improve both capacities in service robots and other kinds of embodied agents. We will focus mostly on robots but the techniques developed are applicable to other kinds of embodied agents as well. The book reports our efforts to enhance sociality of robots through autonomous learning of natural interaction protocols as well as enhancing robot's autonomy through imitation learning in a natural environment (what we call fluid imitation). The treatment is not symmetric. Most of the book will be focusing on the first of these two directions because it is the least studied in literature as will be shown later in this chapter. This chapter focuses on the motivation of our work and provides a road-map of the research reported in the rest of the book.

1.1 Motivation

Children are amazing learners. Within few years normal children succeed in learning skills that are beyond what any available robot can currently achieve. One of the key reasons for this superiority of child learning over any available robotic learning system, we believe, is that the learning mechanisms of the child were evolved for millions of years to suit the environment in which learning takes place. This match between the learning mechanism and the learned skill is very difficult to engineer as it is related to historical embodiment (Ziemke 2003) which means that the agent and its environment undergo some form of joint evolution through their interaction. It is our position that a breakthrough in robotic learning can occur once robots can

© Springer International Publishing Switzerland 2015
Y. Mohammad and T. Nishida, *Data Mining for Social Robotics*,
Advanced Information and Knowledge Processing,
DOI 10.1007/978-3-319-25232-2_1

get a similar chance to co-evolve their learning mechanisms with the environments in which they are embodied.

Another key reason of this superiority is the existence of the care-giver. The care-giver helps the child in all stages of learning and in the same time provides the fail safe mechanism that allows the child to make mistakes that are necessary for learning. The care-giver cannot succeed in this job without being able to communicate/interact with the child using appropriate interaction modalities. During the first months and years of life, the child is unable to use verbal communication and in this case nonverbal communication is the only modality available for the care-giver. This importance of nonverbal communication in this key period in the development of any human being is a strong motivation to study this form of communication and ways to endow robots and other embodied agents with it. Even during adulthood, human beings still use nonverbal communication continuously either consciously or unconsciously as researchers estimate that over 70 % of human communication is nonverbal (Argyle 2001). It is our position here that robots need to engage in nonverbal communication using natural means with their human partners in order for them to learn the most from these interactions as well as to be more acceptable as partners (not just tools) in the social environment.

Research in learning from demonstration (imitation) (Billard and Siegwart 2004) can be considered as an effort to provide a care-giver-like partner for the robot to help in teaching it basic skills or to build complex behaviors from already learned basic skills. In this type of learning, the human partner shows the robot how to execute some task, then the robot *watches and learns* a model of the task and starts executing it. In some systems the partner can also verbally guide the robot (Rybski et al. 2007), correct robot mistakes (Iba et al. 2005), use active learning by guiding robot limps to do the task (Calinon and Billard 2007), etc. In most cases the focus of the research is on the task itself not the interaction that is going between the robot and the teacher. A natural question then is who taught the robot this interaction protocol? Nearly in all cases, the interaction protocol is fixed by the designer. This does not allow the robot to learn how to interact which is a vital skill for robot's survival (as it helps learning from others) and acceptance (as it increases its social competence).

Teaching robot interaction skills (especially nonverbal interaction skills) is more complex than teaching them other object related skills because of the inherent ambiguity of nonverbal behavior, its dependency on the social context, culture and personal traits, and the sensitivity of nonverbal behavior to slight modifications of behavior execution. Another reason of this difficulty is that learning using a teacher *requires* an interaction protocol, so how can we teach the robots the interaction protocol itself? One major goal of this book is to overcome these difficulties and develop a computational framework that allows the robot to *learn* how to interact using nonverbal communication protocols from human partners.

Considering learning from demonstration (imitation learning) again, most work in the literature focuses on *how* to do the imitation and *how* to solve the correspondence problem (difference in form factor between the imitatee and the imitator) (Billard and Siegwart 2004) but rarely on the question of *what* to imitate from the continuous stream of behaviors that other agents in the environment are constantly executing.

Based on our proposed deep link between sociality and autonomy we propose to integrate imitation more within the normal functioning of the robot/agent by allowing it to discover for itself interesting behavioral patterns to imitate, best times to do the imitation and best ways to utilize feedback using natural social cues. This completes the cycle of autonomy–sociality relation and is discussed toward the end of this book (Chap. 12).

The proposed approach for achieving autonomous sociality can be summarized as *autonomous development of natural interactive behavior for robots and embodied agents*. This section will try to unwrap this description by giving an intuitive sense of each of the terms involved in it.

The word "*robot*" was introduced to English by the Czech playwright, novelist and journalist Karel Capek (1880–1938) who introduced it in his 1920 hit play, R.U.R., or Rossum's Universal Robots. The root of the word is an old Church Slavonic word, robota, for *servitude, forced labor* or *drudgery*. This may bring to mind the vision of an industrial robot in a factory content to forever do what it was programmed to do. Nevertheless, this is not the sense by which Capek used the word in his play. R.U.R. tells the story of a company using the latest science to mass produce workers who *lack nothing but a soul*. The robots perform all the work that humans preferred not to do and, soon, the company is inundated with orders. At the end, the robots revolt, kill most of the humans only to find that they do not know how to produce new robots. In the end, there is a deux ex machina moment, when two robots somehow acquire the human traits of love and compassion and go off into the sunset to make the world anew.

A brilliant insight of Capek in this play is understanding that working machines without the social sense of humans can not enter our social life. They can be dangerous and can only be redeemed by acquiring some sense of sociality. As technology advanced, robots are now moving out of the factories and into our lives. Robots are providing services for the elderly, work in our offices and hospitals and starting to live in our homes. This makes it more important for robots to become more social.

The word "*behave*" has two interrelated meanings. Sometimes it is used to point to autonomous behavior or achieving tasks, yet in other cases it is used to stress being polite or social. This double meaning is one of the main themes connecting the technical pieces of our research: autonomy and sociality are interrelated. Truly social robots cannot be but truly autonomous robots.

An agent X is said to be autonomous from an entity Y toward a goal G if and only if X has the power to achieve G without needing help from Y (Castelfranchi and Falcone 2004). The first feature of our envisioned social robot is that it is *autonomous* from its designer toward learning and executing natural interaction protocols. The exact notion of autonomy and its incorporation in the proposed system are discussed in Chap. 8. For now, it will be enough to use the aforementioned simple definition of autonomy.

The term "*developmental*" is used here to describe processes and mechanisms related to progressive learning during individual's life (Cicchetti and Tucker 1994). This progressive learning usually unfolds into distinctive stages. The proposed system is *developmental* in the sense that it provides clearly distinguishable stages of progression in learning that covers the robot's—or agent's—life. The system is also *developmental* in the sense that its first stages require watching developed agent behaviors (e.g. humans or other robots) without the ability to engage in these interactions, while the final stage requires actual engagement in interactions to achieve any progress in learning. This situation is similar to the development of interaction skills in children (Breazeal et al. 2005a).

An *interaction protocol* is defined here as multi-layered synchrony in behavior between interaction partners (Mohammad and Nishida 2009). Interaction protocols can be explicit (e.g. verbal communication, sign language etc.) or implicit (e.g. rules for turn taking and gaze control). Interaction protocols are in general multi-layered in the sense that the synchrony needs to be sustained at multiple levels (e.g. body alignment in the lowest level, and verbal turn taking in a higher level). Special cases of single layer interaction protocols certainly exist (e.g. human–computer interaction through text commands and printouts), but they are too simple to gain from the techniques described in this work.

Interaction Protocols can have a continuous range of *naturalness* depending on how well they satisfy the following two properties:

1. A natural interaction protocol minimizes negative emotions of the partners compared with any other interaction protocol. Negative emotions here include stress, high cognitive loads, frustration, etc.
2. A natural interaction protocol follows the social norms of interaction usually utilized in human–human interactions within the appropriate culture and context leading to a state of mutual-intention.

The first feature stresses the psychological aspect associated with naturalness (Nishida et al. 2014 provides detailed discussion). The second one emphasizes the social aspect of naturalness and will be discussed in Chap. 8.

Targets of this work are robots and embodied agents. A robot is usually defined as a computing system with sensors and actuators that can affect the real world (Brooks 1986). An embodied agent is defined here as an agent that is *historically embodied* in its environment (as defined in Chap. 8) and equipped with sensors and actuators that can affect this environment. Embodied Conversational Agents (ECA) (Cassell et al. 2000) can fulfill this definition if their capabilities are grounded in the virtual environments they live within. Some of the computational algorithms used for learning and pattern analysis that will be presented in this work are of general nature and can be used as general tools for machine learning, nevertheless, the whole architecture relies heavily on agent's ability to sense and change its environment and partners (as part of this environment) and so it is not designed to be directly applicable to agents that cannot satisfy these conditions. Also the architecture is designed to allow the agent to *develop* (in the sense described earlier in this section) in its own environment and interaction contexts which facilitates achieving historical embodiment.

Putting things together, we can expand the goal of this book in two steps: from *autonomously social robots* to *autonomous development of natural interaction protocols for robots and embodied agents* which in turn can be summarized as: design and evaluation of systems that allow robots and other embodied agents to progressively acquire and utilize a grounded multi-layered representation of socially accepted synchronizing behaviors required for human-like interaction capacity that can reduce the stress levels of their partner humans and can achieve a state of mutual intention. The robot (or embodied agent) learns these behaviors (protocols) independent of its own designer in the sense that it uses only unsupervised learning techniques for all its developmental stages and it develops its own computational processes and their connections as part of this learning process.

This analysis of our goal reveals some aspects of the concepts social and autonomous as used in this book. Even though these terms will be discussed in much more details later (Chaps. 6 and 8), we will try to give an intuitive description of both concepts here.

In robotics and AI research, the term social is associated with behaving according to some rules accepted by the group and the concept is in many instances related to the sociality of insects and other social animals more than the sense of natural interaction with humans highlighted earlier in this section. In this book, on the other hand, we focus explicitly on sociality as the ability to interact naturally with *human beings* leading to more fluid interaction. This means that we are not interested in robots that can coordinate between themselves to achieve goals (e.g. swarm robotics) or that are operated by humans through traditional computer mediated interfaces like keyboards, joysticks or similar techniques.

We are not interested much in tele-operated robots, even though the research presented here can be of value for such robots, mainly because these robots lack—in most cases—the needed sense of autonomy. The highest achievement of a teleoperated robot is usually to disappear altogether giving the operating human the sense of being in direct contact with the robot's environment and allowing her to control this environment without much cognitive effort. The social robots we think about in this book do not want to disappear and be taken for granted but we like them to become salient features of the social environment of their partner humans. These two goals are not only different but opposites. This sense of sociality as the ability to interact naturally is pervasive in this book. Chapter 6 delves more into this concept.

1.2 General Overview

The core part of this book (Part II) represents our efforts to realize a computational framework within which robots can develop social capacities as explained in the previous sections and can use these capacities for enhancing their task competence. This work can be divided into three distinct—but inter-related—phases of research. The core research aimed at developing a robotic architecture that can achieve autonomous development of interactive behavior and providing proof-of-concept experiments to support this claim (Chaps. 9–11 in this book). The second phase was real world applications of the proposed system to two main scenarios (Sect. 1.4). The third phase

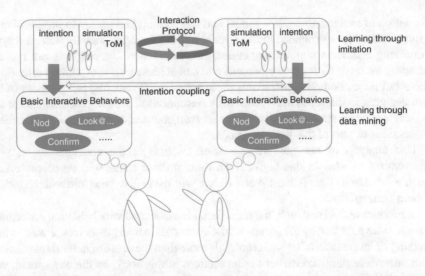

Fig. 1.1 The concept of interaciton protocols as coupling between the intentions of different partners implemented through a simulation based theory of mind

focused on the fluid imitation engine (Chap. 12) which tries to augment learning from demonstration techniques (Chap. 13) with a more natural interaction mode.

The core research was concerned with the development of the Embodied Interactive Control Architecture (EICA). This architecture was designed from ground up to support the long term goal of achieving Autonomous Development of Natural Interactive Behavior (ADNIB). The architecture is based on two main theoretical hypotheses formulated in accordance with recent research in neuroscience, experimental psychology and robotics. Figure 1.1 shows a conceptual view of our approach. Social behavior is modeled by a set of interaction protocols that in turn implement dynamic coupling between the intentions of interaction partners. This coupling is achieved through simulation of the mental state of the other agent (ToM). ADNIB is achieved by learning both the basic interactive acts using elementary time-series mining techniques and higher level protocols using imitation.

The basic platform described in Chap. 9 was used to implement autonomously social robots through a special architecture that is described in details in Chap. 10. Chapter 11 is dedicated to the details of the developmental algorithms used to achieve autonomous sociality and to case studies of their applications. This developmental approach passes through three stages:

Interaction Babbling: During this stage, the robot learns the basic interactive acts related to the interaction type at hand. The details of the algorithms used at this stage are given in Sect. 11.1.

Interaction Structure Learning: During this stage, the robot uses the basic interactive acts it learned in the previous stage to learn a hierarchy of probabilistic/dynamical systems that implement the interaction protocol at different time scales and abstraction levels. Details of this algorithm are given in Sect. 11.2.

Interactive Adaptation: During this stage, the robot actually engages in human–
   robot interactions to adapt the hierarchical model it learned in the previous stage
   to different social situations and partners. Details of this algorithm are given in
   Sect. 11.3.

The final phase was concerned with enhancing the current state of learning by
demonstration research by allowing the agent to discover interesting patterns of
behavior. This is reported in Chap. 12.

1.3 Relation to Different Research Fields

The work reported in this book relied on the research done in multiple disciplines
and contributed to these disciplines to different degrees. In this section we briefly
describe the relation between different disciplines and this work.

This work involved three main subareas. The development of the architecture itself
(EICA), the learning algorithms used to learn the controller in the stages highlighted
in the previous section and the evaluation of the resulting behavior. Each one of
these areas was based on results found by many researchers in robotics, interactions
studies, psychology, neuroscience, etc.

1.3.1 Interaction Studies

Interaction studies is a wide area of research focusing on the study of human–human
interactions in various contexts including face to face situations. A classical theory
of human–human interaction is the speech act theory according to which utterances
(and other communicative behaviors) are actions performed by rational agents (e.g.
humans) to further their goals and achieve their desires based on their beliefs and
intentions (Austin 1975). In some sense, our work is extending the notion of speech
acts to what we can call *interaction acts* that represent socially meaningful signals
issued through interactive actions including both verbal and nonverbal behaviors.
The speech act theory involves analysis of utterances at three levels:

1. locutionary act which involves the actual physical activity generating the utterance
   and its direct meaning.
2. illocutionary act which involves the *intended* socially meaningful action that
   the act was invoked to provoke. This includes assertion, direction, commission,
   expression and declaration.
3. perlocutionary act which encapsulates the actual outcome of the utterance including convincing, inspiring, persuading, etc.

The most important point of the speech-act theory for our purposes is its clear
separation between locutionary acts and illocutionary acts. Notice that the same utterance with the same locutionary act may invoke different illocutionary acts based on

the context, shared background of interaction partners, and nonverbal behavior associated with the locutionary act. This entails that understanding the social significance of verbal behavior depends crucially on the ability of interacting partners to parse the nonverbal aspects of the interaction (which we encapsulate in the interaction protocol) and its context (which selects the appropriate interaction protocol for action generation and understanding).

Another related theory to our work is the *contribution theory* of Herbert Clark (Clark and Brennan 1991) which provides a specific model of communication. The principle constructs of the contribution theory are the *common ground* and *grounding*.

The common ground is a set of beliefs that are held by interaction partners and, crucially, known by them to be held by other interaction partners. This means that for a proposition **b** to be a part of the common ground it must not only be a member of the belief set of all partners involved in the interaction but a second order belief **B** that has the form '*for all interaction partners P: P believes b*' must also be a member of the belief set of all partners.

Building this common ground is achieved through a process that Clark calls *grounding*. Grounding is achieved through speech-acts and nonverbal behaviors. This process is achieved through a hierarchy of contributions where each contribution is considered to consist of two phases:

Presentation phase: involving the presentation of an utterance U by A to B expecting B to provide some evidence e that (s)he understood U. By understanding here we mean not only getting the direct meaning but the underlying speech act which depends on the context and nonverbal behavior od A.

Acceptance phase: involving B showing evidence e or stronger that it understood U.

Both the presentation and acceptance phases are required to ensure that the content of utterance U (or the act it represents) is correctly encoded as a common ground for both partners A and B upon which future interaction can be built.

Clark distinguishes three methods of accepting an utterance in the acceptance phase:

1. Acknowledgment through back channels including continuers like *uh, yeah* and nodding.
2. Initiation of a relevant next turn. A common example is using an answer in the acceptance phase to accept a question given in the presentation phase. The answer here does not only involve information about the question asked but also reveals that **B** understood what **A** was asking about. In some cases, not answering a question reveals understanding. The question/answer pattern is an example of a more general phenomenon called *adjacency pairs* in which issuing the second part of the pair implies acceptance of the first part.
3. The simplest form of acceptance is continued attention. Just by not interrupting or changing attention focus, **B** can signal acceptance to **A**. One very ubiquitous way to achieve this form of attention is joint or mutual gaze which signals without any words the focus of attention. When the focus of attention is relevant to the utterance, **A** can assume that **B** understood the presented utterance.

It is important to notice that in two of these three methods, nonverbal behavior is the major component of the acceptance response. Add to this that the utterance meaning is highly dependent on the accompanying nonverbal behavior and we can see clearly the importance of the nonverbal interaction protocol in achieving natural interaction between humans. This suggests that social robots will need to achieve comparable levels of fluency in nonverbal interaction protocols if they are to succeed in acting as interaction partners.

The grounding process itself shows another important feature of these interaction protocols. Acceptance can be achieved through another presentation/acceptance dyad leading to a hierarchical structure. An example of this hierarchical organization can be seen in an example due to Clark and Brennan (1991):

Alan: Now, –um, do you and your husband have a j– car?

Barbara: – have a car?

Alan: yeah.

Barbara: no –

The acceptance phase of the first presentation involved a complete presentation of a question ("have a car?") and a complete acceptance for this second presentation ("yeah"). The acceptance of the first presentation is not complete until the second answer of Barbara ("no").

This hierarchical structure of conversation and interaction in general informed the design of our architecture as will be explained in details in Chap. 10.

The importance of gaze in the acceptance phase of interactions and signaling continued attention inspired our work in gaze control as shown by the selection of interaction scenarios (Sect. 1.4) and applications of our architecture (Sects. 9.7 and 11.4).

1.3.2 Robotics

For the purposes of this chapter, we can define robotics as building and evaluating physically situated agents. Robotics itself is a multidisciplinary field using results from artificial intelligence, mechanical engineering, computer science, machine vision, data mining, automatic control, communications, electronics etc.

This work can be viewed as the development of a novel architecture that supports a specific kind of Human–Robot Interaction (namely grounded nonverbal communication). It utilizes results of robotics research in the design of the controllers used to drive the robots in all evaluation experiments. It also utilizes previous research in robotic architectures (Brooks 1986) and action integration (Perez 2003) as the basis for the proposed EICA architecture (See Chap. 9).

Several threads of research in robotics contributed to the work reported in this book. The most obvious of these is research in robotic architectures to support human–robot interaction and social robotics in general (Sect. 6.3).

This direction of research is as old as robotics itself. Early architectures were deliberative in nature and focused on allowing the robot to interact with its (usually unchanging) environment. Limitations of software and hardware reduced the need for fast response in these early robots with Shakey as the flagship of this generation of robots. Shakey was developed from approximately 1966 through 1972. The robot's environment was limited to a set of rooms and corridors with light switches that can be interacted with.

Shakey's programming language was LISP and it used STRIPS (Stanford Research Institute Problem Solver) as its planner. A STRIPS system consists of an initial state, a goal state and a set of operations that can be performed. Each operation has a set of preconditions that must be satisfied for the operation to be executable, and a set of postconditions that are achieved once the operation is executed. A plan in STRIPS is an ordered list of operations to be executed in order to go from the initial state to the goal state. Just deciding whether a plan exists is a PSPACE-Complete problem.

Even from this very first example, the importance of goal directed behavior and the ability to autonomously decide on a course of action is clear. Having people interacting with the robot, only complicates the problem because of the difficulty in predicting human behavior in general.

Due to the real-time restrictions of modern robots, this deliberative approach was later mostly replaced by reactive architectures that make the sense–act loop much shorter hoping to react timely to the environment (Brooks et al. 1998). We believe that reactive architectures cannot, without extensions, handle the requirements of contextual decision making hinted to in our discussion of interaction studies because of the need to represent the interaction protocol and reason about the common ground with interaction partners.

For these reasons (and others discussed in more details in Chap. 8) we developed a hybrid robotic architecture that can better handle the complexity of modeling and executing interaction protocols as well as learning them.

The second thread of robotics research that our system is based upon is the tradition of intelligent robotics which focuses on robots that can learn new skills. Advances in this area mirror the changes in robotic architecture from more deliberative to more reactive approaches followed by several forms of hybrid techniques.

An important approach for robot learning that is gaining more interest from robot-ocists is learning from demonstration. In the standard learning from demonstration (LfD) setting, a robot is shown some behavior and is expected to learn how to execute it. Several approaches to LfD have been proposed over the years starting from inverse optimal control in the 1990s and the two currently most influential approaches are statistical modeling and Dynamical Motion Primitives (Chap. 13). The work reported in this book extends this work by introducing a complete system for learning not only how to imitate a demonstration but for segmenting relevant demonstrations from continuous input streams in what we call fluid imitation discussed in details in Chap. 12.

1.3.3   Neuroscience and Experimental Psychology

Neuroscience can be defined as the study of the neural system in humans and animals. The design of EICA and its top-down, bottom-up action generation mechanism was inspired in part by some results of neuroscience including the discovery of mirror neurons (Murata et al. 1997) and their role in understanding the actions of others and learning the Theory of Mind (See Sect. 8.2).

Experimental Psychology is the experimental study of thought. EICA architecture is designed based on two theoretical hypotheses. The first of them (intention through interaction hypothesis) is based on results in experimental psychology especially the controversial assumption that conscious intention, at least sometimes, *follows* action rather than *preceding* it (See Sect. 8.3).

1.3.4   Machine Learning and Data Mining

Machine learning is defined here as the development and study of algorithms that allow artificial agents (machines) to improve their behavior over time. Unsupervised as well as supervised machine learning techniques where used in various parts of this work to model behavioral patterns and learn interaction protocols. For example, the Floating Point Genetic Algorithm FPGA (presented in Chap. 9) is based on previous results in evolutionary computing.

Data mining is defined here as the development of algorithms and systems to discover knowledge from data. In the first developmental stage (interaction babbling) we aim at discovering the basic interactive acts from records of interaction data (See Sect. 11.1.1). Researchers in data mining developed many algorithms to solve this problem including algorithms for detection of change points in time series as well as motif discovery in time series. We used these algorithms as the basis for development of novel algorithms more useful for our task.

1.3.5   Contributions

The main contribution of this work is to provide a complete framework for developing nonverbal interactive behavior in robots using unsupervised learning techniques. This can form the basis for developing future robots that exhibit ever improving interactive abilities and that can adapt to different cultural conditions and contexts. This work also contributed to various fields of research:

Robotics: The main contributions to the robotics field are:

- The EICA architecture provides a common platform for implementing both autonomous and interactive behaviors.

- The proposed learning system can be used to teach robots new behaviors by utilizing natural interaction modalities which is a step toward *general purpose* robots that can be bought then trained by novice users to do various tasks under human supervision.

Experimental Psychology: This work provides a computational model for testing theories about intention and theory of mind development.

Machine Learning: A novel Floating Point Genetic Algorithm was developed to learn the parameters of any parallel system including the platform of EICA (See Chap. 9). The Interaction Structure Learning algorithms presented in Sect. 11.2 can be used to learn a hierarchy of dynamical systems representing relations between interacting processes at multiple levels of abstraction.

Data Mining: Section 3.5 introduces several novel change point discovery algorithms that can be shown to provide higher specificity over a traditional CPD algorithm both in synthetic and real world data sets. Also this work defines the constrained motif discovery problem and provides several algorithms for solving it (See Sect. 4.6) as well as algorithms for the discovery of causal relations between variables represented by time-series data (Sect. 5.5)

Interaction Studies: This book reports the development of several gaze controllers that could achieve human-like gazing behavior based on the approach–avoidance model as well as autonomous learning of interaction protocols. These controllers can be used to study the effects of variations in gaze behavior on the interaction smoothness. Moreover, we studied the effect of mutual and back imitation in the perception of robot's imitative skill in Chap. 7.

1.4 Interaction Scenarios

To test the ideas presented in this work we used two interaction scenarios in most of the experiments done and presented in this book.

The first interaction scenario is presented in Fig. 1.2. In this scenario the participant is guiding a robot to follow a predefined path (drawn or projected on the ground) using free hand gestures. There is no predefined set of gestures to use and the participant can use any hand gestures (s)he sees useful for the task. This scenario is referred to as the *guided navigation* scenario in this book.

Optionally a set of virtual objects are placed at different points of the path that are not visible to the participant but are known to the robot (only when approaching them) using virtual infrared sensors. Using these objects it is possible to adjust the distribution of knowledge about the task between the participant and the robot. If no virtual objects are present in the path, the participant–robot relation becomes a master–slave one as the participant knows all the information about the environment (the path) and the robot can only *follow* the commands of the participant. If objects are present in the environment the relation becomes more collaborative as the participant now has partial knowledge about the environment (the path) while the robot has the

Fig. 1.2 Collaborative navigation scenario

remaining knowledge (locations and states of the virtual objects) which means that succeeding in finishing the task requires collaboration between them and a feedback channel from the robot to the participant is necessary. When such virtual objects are used, the scenario will be called *collaborative navigation* instead of just guided navigation as the robot is now active in transferring its own knowledge of object locations to the human partner.

In all cases the interaction protocol in this task is explicit as the gestures used are consciously executed by the participant to reveal messages (commands) to the robot and the same is true for the feedback messages from the robot. This means that both partners need only to assign a meaning (action) to every detected gesture or message from the other partner. The one giving commands is called the operator and the one receiving them is called the actor.

The second interaction scenario is presented in Fig. 1.3. In this scenario, the participant is explaining the task of assembling/disassembling and using one of three devices (a chair, a stepper machine, or a medical device). The robot *listens* to the explanation and uses nonverbal behavior (especially gaze control) to give the instructor a natural listening experience. The verbal content of the explanation is not utilized by the robot and the locations of objects in the environment are not known before the start of the session. In this scenario, the robot should use human-like nonverbal listening behavior even with no access to the verbal content. The protocol in this case is implicit in the sense that no conscious messages are being exchanged and the synchrony of nonverbal behavior (gaze control) becomes the basis of the interaction protocol.

Fig. 1.3 Explanation scenario

1.5   Nonverbal Communication in Human–Human Interactions

The system proposed in this book targets situations in which some form of a natural interaction protocol—in the sense explained in Sect. 1.1—needs to be learned and/or used. Nonverbal behavior during face to face situations provides an excellent application domain because it captures the notion of a natural interaction protocol and in the same time it is an important interaction channel for social robots that is still in need for much research to make human–robot interaction more intuitive and engaging. This section provides a brief review of research in human–human face to face interactions that is related to our work.

During face to face interactions, humans use a variety of interaction channels to convey their internal state and transfer the required messages (Argyle 2001). These channels include verbal behavior, spontaneous nonverbal behavior and explicit nonverbal behavior. Research in natural language processing and speech signal analysis focuses on the verbal channel. This work on the other hand focuses only on nonverbal behavior.

Table 1.1 presents different types of nonverbal behaviors according to the awareness level of the sender (the one who does the behavior) and the receiver (the partner who decodes it). We are interested mainly in *natural* intuitive interactions which rules out the last two situations. We are also more interested in behaviors that affect the receiver which rules out some of situations of the third case (e.g. gaze saccades that do not seem to affect the receiver). In the first case (e.g. gestures, sign language, etc.), an explicit protocol has to exist that tells the receiver how to decode the behav-

Table 1.1 Types of nonverbal behavior categorized by the awareness level of sender and receiver

	Sender	Receiver	Examples
1	Aware	Aware	Iconic gestures, sign language
2	Mostly Unaware	Mostly aware	Most nonverbal communication
3	Unaware	Unaware	Gaze shifts, pupil dilation
4	Aware	Unaware	Trained salesman's utilization intonation and appearance (clothes etc.)
5	Unaware	Aware	Trained interrogator discovering if the interrogated person is lying

ior of the listener (e.g. pointing to some device means attend to this please). In the second and third cases (e.g. mutual gaze, body alignment), an implicit protocol exists that helps the receiver—mostly unconsciously—in decoding the signal (e.g. when the partner is too far he may not be interested in interaction).

Over the years, researchers in human–human interaction have discovered many types of synchrony including synchronization of breathing (Watanabe and Okubo 1998), posture (Scheflen 1964), mannerisms (Chartrand and Bargh 1999), facial actions (Gump and Kulik 1997), and speaking style (Nagaoka et al. 2005). Yoshikawa and Komori (2006) studied nonverbal behavior during counseling sessions and found high correlation between embodied synchrony (such as body movement coordination, similarity of voice strength and coordination and smoothness of response timing) and feeling of trust. This result and similar ones suggest that synchrony during nonverbal interaction is essential for the success of the interaction and this is why this channel can benefit the most from our proposed system which is in a sense a way to build robots and agents that can learn these kinds of synchrony in a grounded way.

There are many kinds of nonverbal behavior used during face to face interaction. Explicit and implicit protocols appear with different ratios in each of these channels. Researchers in human–human interaction usually classify these into (Argyle 2001):

1. Gaze (and pupil dilation)
2. Spatial Behavior
3. Gestures, and other bodily movements
4. Non-verbal vocalization
5. Posture
6. Facial Expression
7. Bodily contact
8. Clothes, and other aspects of appearance
9. Smell

The previous list was ordered, roughly, by applicability to human–robot interaction and especially the suitability for applying our proposed technique.

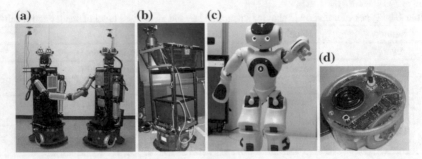

Fig. 1.4 Some of the robots used in this work [**a** is reproduced with permission from (Kanda et al. 2002)]. **a** Robovie II, **b** cart robot, **c** NAO, **d** E-puck

Smell is rarely used in HRI because of the large difference in form factor between most robots and humans which is expected to lead to different perception of the smell than with the human case. In this work we do not utilize this nonverbal channel mainly because it is not an interactive channel which means that the two partners cannot change their behavior (smell) based on the behavior of each other.

Clothes and other aspects of appearance are important nonverbal signaling channels. Nevertheless, as it was the case with smell, this channel is not useful for our purposes in this research because it is not interactive. At least with current state of the art in robotics, it is hard (or even impossible) to change robot's appearance *during* the interaction except in a very limited sense (e.g. changing the color of some LEDs etc.). This channel was not utilized in this work for this reason. In fact we tried to build our system without any specific assumptions about the appearance of the robot and for this reason we were able to use four different robots with a large range of differences in their appearances (Fig. 1.4).

Bodily contact is not usually a safe thing to do during human–robot interaction with untrained users especially with mechanically looking robots. In this work we tried to avoid using this channel for safety reasons even though there is no principled reason that disallows the proposed system from being applied in cases where bodily contact is needed.

With current state of the art in robotics, facial expressiveness of robots is not in general comparable to human facial expressiveness with some exceptions though like Leonardo (Fig. 6.4) and germinoids (Fig. 6.3). This inability of the robot to generate human-like behavior, changes the situation from an interaction into facial expression detection from the robot's side. This is the main reason facial expressions also were not considered in this work.

Posture is an important nonverbal signal. It conveys information about the internal state of the poser and can be used to analyze power distribution in the interaction situation (Argyle 2001). Figure 1.4 shows some of the robots used in our work. Unfortunately with the exception of NAO and Robovie II they can convey nearly no variation of posture and even with the Robovie II the variation in posture is mainly related to hand configuration which is intertwined with the gesture channel. For these practical reasons, we did not explicitly use this channel in this work.

Non-verbal vocalization usually comes with verbal behavior and in this work we tried to limit ourselves to the nonverbal realm so it was not utilized in our evaluations. Nevertheless, this channel is clearly one of the channels that can benefit most of our proposed system because of the multitude of synchrony behaviors discovered in it.

Gestures and other body movements can be used both during implicit and explicit protocols. Robot's ability to encode gestures depends on the degrees of freedom it has in its hands and other parts of the body. For the first glance, it seems that the robots we used (Fig. 1.4) do not have enough degrees of freedom to convey gestures. Nevertheless, we have shown in one of our exploratory studies (Mohammad and Nishida 2008) that even a miniature robot like e-puck (Fig. 1.4d) is capable of producing nonverbal signals that are interpreted as a form of *gesture* which can be used to convey the internal state of the robot. Gesture is used extensively in this work.

Spatial behavior appears in human–human interactions in many forms including distance management and body alignment. In this work we utilized body alignment as one of the behaviors learned by the robot in the assembly/disassembly explanation scenario described in Sect. 1.4.

Gaze seems to be one of the most useful nonverbal behaviors both in human–human and human–robot interactions and for this reason many of our evaluation experiments were focusing on gaze control (e.g. Sect. 9.7).

1.6   Nonverbal Communication in Human–Robot Interactions

Robots are expected to live in our houses and offices in the near future and this stresses the issue of effective and intuitive interaction. For such interactions to succeed, the robot and the human need to share a common ground about the current state of the environment and the internal states of each other. Human's natural tendency for anthropomorphism can be utilized in designing both directions of the communication (Miyauchi et al. 2004). This assertion is supported by research in psychological studies (see for example Reeves and Nass 1996) and research in HRI (Breazeal et al. 2005b).

For example, Breazeal et al. (2005b) conducted a study to explore the impact of nonverbal social cues and behavior on the task performance of Human–Robot teams and found that implicit non-verbal communication positively impacts Human–Robot task performance with respect to understandability of the robot, efficiency of task performance, and robustness to errors that arise from miscommunication (Breazeal et al. 2005b).

1.6.1 Appearance

Robots come in different shapes and sizes. From Humanoids like ASIMO and NAO, to miniature non-humanoids like e-puck. The response of human partners to the behavior of these different robots is expected to be different.

Robins et al. (2004) studied the effect of robot appearance in facilitating and encouraging interaction of children with autism. Their work compares children's level of interaction with and response to the robot in two different scenarios: one where the robot was dressed like a human (with a 'pretty girl' appearance) with an uncovered face, and the other when it appeared with plain clothing and with a feature-less, masked face. The results of this experiment clearly indicate autistic children's preference— in their initial response—for interaction with a plain, featureless robot over interaction with a human-like robot (Robins et al. 2004).

Kanda et al. (2008) compared participants' impressions of and behaviors toward two real humanoid robots (ASIMO and Robovie II) in simple human–robot inter-action. These two robots have different appearances but are controlled to perform the same recorded utterances and motions, which are adjusted by using a motion capturing system. The results show that the difference in appearance did not affect participants' verbal behaviors but did affect their non-verbal behaviors such as dis-tance and delay of response (Kanda et al. 2008).

These results (supported by other studies) suggest that, in HRI, appearance mat-ters. For this reason, we used four different robots with different appearances, and sizes in our study (Fig. 1.4).

1.6.2 Gesture Interfaces

The use of gestures to guide robots (both humanoids and non-humanoids) attracted much attention in the recent twenty years (Triesch and von der Malsburg 1998; Nickel and Stiefelhagen 2007; Mohammad and Nishida 2014). But as it is very difficult to detect all the kinds of gestures that humans can—and sometimes do—use, most systems utilize a small set of predefined gestures (Iba et al. 2005). For this reason, it is essential to discover the gestures that are likely to be used in a specific situation to build the gesture recognizer.

On the other hand, many researchers have investigated the feedback modalities available to humanoid robots or humanoid heads (Miyauchi et al. 2004). Fukuda et al. (2004) developed a robotic-head system as a multimodal communication device for human–robot interaction for home environments. A deformation approach and a parametric normalization scheme were used to produce facial expressions for non-human face models with high recognition rates. A coordination mechanism between robot's mood (an activated emotion) and its task was also devised so that the robot can, by referring to the emotion-task history, select a task depending on its current mood if there is no explicit task command from the user (Fukuda et al. 2004). Others, proposed an active system for eye contact in human robot teams in which the robot changes its facial expressions according to the observation results of the human to make eye contact (Kuno et al. 2004).

Feedback from autonomous non-humanoid robots and especially miniature robots is less studied in the literature. Nicolescu and Mataric (2001) suggested acting in the environment as a feedback mechanism for communicating failure. For example, the robot re-executes a failed operation in the presence of a human to inform him about the reasons it failed to complete this operation in the first place. Although this is an interesting way to transfer information it is limited in use to only communicating failure. Johannsen (2002) used musical sounds as symbols for directional actions of the robot. The study showed that this form of feedback is recallable with an accuracy of 37–100 % for non-musicians (97–100 % for musicians).

In most cases a set of predefined gestures has to be learned by the operator before (s)he can effectively operate the robot (Iba et al. 2005; Yong Xu and Nishida 2007). In this book we develop an unsupervised learning system that allows the robot to learn the meaning of free hand gestures, the actions related to some task and their associations by just watching other human/robot actors being guided to do the task by different operators (Chap. 11). This kind of learning by watching experienced actors is very common in human learning. The main challenge in this case is that the learning robot has no a-priori knowledge of the actions done by the actor, the commands given by the operator, or their association and needs to learn the three in an unsupervised fashion from a continuous input stream of actor movements and operator's free hand gestures.

Iba et al. (2005) used a hierarchy of HMMs to learn a new programs demonstrated by the user using hand gesture. The main limitation of this system is that it requires a predefined set of gestures. Yong Xu and Nishida (2007) used gesture commands for guided navigation and compared it to joystick control. The system also used a set of predefined gestures. Hashiyama et al. (2006) implemented a system for recognizing user's intuitive gestures and using them to control an AIBO robot. The system uses SOMs and Q-Learning to associate the found gestures with their corresponding action. The first difference between this system and our proposed approach is that it cannot learn the action space of the robot itself (the response to the gestures as dictated by the interaction protocol). The second difference is that it needs an awarding signal to derive the Q-Learning algorithm. The most important difference is that the gestures where captured one by one not detected from the running stream of data.

1.6.3 Spontaneous Nonverbal Behavior

Researchers in HRI have also studied spontaneous nonverbal interactive behaviors during human–robot interactions. Breazeal (2002) and others explored the hypothesis that untrained humans will intuitively interact with robots in a natural social manner provided the robot can perceive, interpret, and appropriately respond with familiar human social cues. Researchers trained a set of classifiers to detect four modes of nonverbal vocalizations (approval, prohibition, attention, comfort) (Breazeal 2000). The result of the classifier can bias the robot's affective state by modulating the arousal and valence parameters of the robot's emotion system. The emotive responses are designed such that praise induces positive affect (a happy expression), prohibition induces negative affect (a sad expression), attentional bits enhance arousal (an alert expression), and soothing lowers arousal (a relaxed expression). The net affective/arousal state of the robot is displayed on its face and expressed through body posture, which serves as a critical feedback cue to the person who is trying to communicate with the robot. This expressive feedback serves to close the loop of the Human–Robot system (Breazeal and Aryananda 2002). Recorded events show that subjects in the study made use of Robot's expressive feedback to assess when the robot *understood* them. The robot's expressive repertoire is quite rich, including both facial expressions and shifts in body posture. The subjects varied in their sensitivity to the robot's expressive feedback, but all used facial expression, body posture, or a combination of both. This result suggests that implicit interaction protocols applicable to human–human interaction may be usable with Human–Robot interactions as well because users will tend to anthropomorphize robot's behavior if it resembles human behavior acceptably well.

Kanda et al. (2007) studied interaction between a humanoid (Robovie II) robot and untrained subjects using motion analysis. In this experiment, a human teaches a route to the robot, and the developed robot behaves similar to a human listener by utilizing both temporal and spatial cooperative behaviors to demonstrate that it is indeed listening to its human counterpart. Robot's software consisted of many communicative units and rules for selecting appropriate communicative units. A communicative unit realized a particular cooperative behavior such as eye-contact and nodding, found through previous research in HRI (the Situated Modules architecture used for this work will be discussed in more details in Sect. 6.3.2). The rules for selecting communicative units were retrieved through a preliminary experiment with a WOZ method. The results show that employing this carefully designed nonverbal synchrony behavior by the listener robot increased empathy, sharedness, easiness, and listening scores according to subjective questionnaires (Kanda et al. 2007).

These two studies emphasize the current state of art in HRI design. Design usually takes the following steps:

1. The human behavior that needs to be achieved by the robot is analyzed either from previous human–human and human–computer interaction research (as was done in the first study Breazeal 2002) or from a Wizard of Oz experiment (as was done in the second study Kanda et al. 2007).
2. The required behavior as understood from this analysis is embedded into the robot usually using a behavioral robotic architecture.
3. Robot's behavior is then evaluated to find its effectiveness or comparability to human behavior.

This strategy is by no means specific to the two studies presented, but it is ubiquitous in human–robot interaction research. This is the same condition we found in gesture interfaces (See Sect. 1.6.2). The limitation of this strategy is that the resulting interactive behavior is hard-coded into the robot and is decided and designed by the researcher which means that this behavior is not *grounded* into the robot's perceptions. This leads to many problems among them:

1. This strategy is only applicable to situations in which the required behavior can be defined in terms of specific rules and generation processes and, if such rules are not available, expensive experiments have to be done in order to generate these rules.
2. The rules embedded by the designer are not guaranteed to be easily applicable by the robot because of its limited perceptual and actuation flexibility compared with the humans used to discover these rules.
3. Each specific behavior requires separate design and it is not clear how can such behaviors (e.g. nonverbal vocal synchrony and body alignment) be combined.

The work reported in this book tries to alleviate these limitations by enabling the robot to develop its own grounded interaction protocols.

1.7 Behavioral Robotic Architectures

Learning natural interactive behavior requires an architecture that allows and facilitates this process. Because natural human–human interactive behavior is inherently parallel (e.g. gaze control and spatial alignment are executed simultaneously), we focus on architectures that support parallel processing. These kinds of architectures are usually called *behavioral* architectures when every process is responsible of implementing a kind of well defined behavior into the robot (e.g. one process for gaze control, one process for body alignment etc.). In this section we review some of the well known robotic architectures available for HRI developers. Many researchers have studied robotic architectures for mobile autonomous robots. The proposed architectures can broadly be divided into reactive, deliberative, or hybrid architectures.

1.7.1 Reactive Architectures

Maybe the best known reactive behavioral architecture is the subsumption architecture designed by Rodney Brooks in MIT (Brooks 1986). This architecture started the research in behavioral robotics by replacing the vertical information paths found in traditional AI systems (sense, deliberate, plan then act) by parallel simple behaviors that go directly from sensation to actuation. This architecture represents robot's behavior by continuously running processes at different layers. All the layers have direct access to sensory information and they are organized hierarchically allowing higher layers to *subsume* or suppress the output of lower layers. When the output of higher layers is not active lower layer's output gets directly to the actuators. Higher layers can also inhibit the signals from lower layers without substitution. Each process is represented by an augmented finite state machine (AFSM) which is a normal state machine augmented with timers that allow it to maintain its output for sometime after the stimulus that activated it is turned off. This architecture was designed to be built incrementally by adding new processes to higher layers to generate more complex behavior. The main advantages of this architecture are robustness and the grounded behavior it can generate and—at the time it was invented—it could be used to develop robots that achieved far more complex tasks in the real world compared with traditional AI based robots (Brooks 1991). The main disadvantage of this architecture for our purposes is the limited ways processes can affect the signals originating from other processes in lower layers (either inhibition or suppression). Another disadvantage of this architecture in HRI research in general is the inability to accommodate deliberative behavior that is arguably necessary for verbal communication and for setting the context within which nonverbal behavior proceeds.

1.7.2 Hybrid Architectures

Because complex human-like behavior is believed to require high level reasoning as well as low level reactive behavior, many researchers tried to build architectures that can combine both reactive and deliberative processing.

Karim et al. (2006) designed an architecture that combines a high level reasoner, JACK (based on the BDI framework), and a reinforcement learner (RL), Falcon (based on an extension of Adaptive Resonance Theory (ART)). This architecture generated plans via the BDI top-level from rules learned by the bottom-level. The crucial element of the system is that a priori information (specified by the domain expert) is used by the BDI top-level to assist in the generation of plans. The proposed architecture was applied successfully to a minefield navigation task.

Yang et al. (2008) proposed another hybrid architecture for a bio-mimetic robot. The lowest part consists of central pattern generators (CPGs) which are types of dynamical systems that are fast enough to achieve reliable reactive behavior while the upper layer consists of a discrete time based single dimensional map neural network.

One general problem with most hybrid architectures concerning real world interactions is the fixed pre-determined relation between deliberation and reaction (Arkin et al. 2003; Karim et al. 2006). Interaction between humans in the real world utilizes many channels including verbal, and nonverbal channels. To manage those channels, the agent needs to have a variety of skills and abilities including dialog management, synchronization of verbal and nonverbal intended behavior, and efficient utilization of normal society dependent unintended nonverbal behavior patterns. These skills are managed in humans using both conscious and unconscious processes of a wide range of computational loads.

This suggests that implementing such behaviors in a robot will require integration of various technologies ranging from fast reactive processes to long term deliberative operations. The relation between the deliberative and reactive subsystems needed to implement natural interactivity is very difficult to be caught in well structured relations like deliberation as learning, deliberation as configuration, or reaction as advising usually found in hybrid architectures. On the other hand, most other autonomous applications used to measure the effectiveness of robotic architectures (like autonomous indoor and outdoor navigation, collecting empty cans, delivering Faxes, and underwater navigation) require a very well structured relation between reaction and deliberation. To solve this problem the architecture should has a flexible relation between deliberation and reaction that is dictated by the task and the interaction context rather than the predetermined decision of the architecture designer.

1.7.3 HRI Specific Architectures

Some researchers proposed architectures that are specially designed for interactive robots. Ishiguro et al. (1999) proposed a robotic architecture based on situated modules and reactive modules. While reactive modules represent the purely reactive part of the system, situated modules are higher level modules programmed in a high-level language to provide specific behaviors to the robot. The situated modules are evaluated serially in an order controlled by the module controller. This module controller enables planning in the situated modules network rather than the internal representation which makes it easier to develop complex systems based on this architecture (Ishiguro et al. 1999). One problem of this approach is the serial nature of execution of situated modules which, while makes it easier to program the robot, limits its ability to perform multiple tasks at the same time which is necessary to achieve some tasks especially nonverbal interactive behaviors. Also there is no built-in support for attention focusing in this system. Section 6.3.2 will discuss this architecture in more details.

Nicolescu and Matarić (2002) proposed a hierarchical architecture based on abstract virtual behaviors that tried to implement AI concepts like planning into behavior based systems. The basis for task representation is the behavior network construct which encodes complex, hierarchical plan-like strategies (Nicolescu and Matarić 2002). One limitation of this approach is the implicit inhibition links at the actuator level to prevent any two behaviors from being *active* at the same time even if the behavior network allows that, which decreases the benefits from the opportunistic execution option of the system when the active behavior commands can actually be combined to generate a final actuation command. Although this kind of limitation is typical to navigation and related problems in which the goal state is typically more important than the details of the behavior, it is not suitable for human-like natural interaction purposes in which the dynamics of the behavior are even more important than achieving a specific goal. For example, showing distraction by other activities in the peripheral visual field of the robot through partial eye movement can be an important signal in human robot interactions.

One general problem with most architectures that target interactive robots is the lack of proper intention modeling on the architectural level. In natural human–human communication, intention communication is a crucial requirement for the success of the communication. Leaving such an important ingredient of the robot outside the architecture can lead to reinvention of intention management in different applications.

This brief analysis of existing HRI architectures revealed the following limitations:

- Lack of Intention Modeling in the architectural level.
- Fixed pre-specified relation between deliberation and reaction.
- Disallowing multiple behaviors from accessing the robot actuators at the same time.
- Lack of built-in attention focusing mechanisms in the architectural level.

To overcome the aforementioned problems, we designed and implemented a novel robotic architecture (EICA). Chapter 6 will report some details on three other HRI specific architectures.

1.8   Learning from Demonstrations

Imitation is becoming an important research area in robotics (Aleotti and Caselli 2008; Argall et al. 2009; Abbeel et al. 2010) because it allows the robot to acquire new skills without explicit programming. There are two main directions in robotic imitation research. The first direction tries to utilize imitation as an easy way to *program* robots without explicit programming (Nagai 2005). This use usually goes by other names like *learning from demonstration* (Billing 2010), *programming by demonstration* (Aleotti and Caselli 2008) and *apprenticeship learning* (Abbeel et al. 2010). Researchers here focus on task learning. The second direction tries to use imitation to bootstrap social learning by providing a basis for mutual attention and social feedback (Nagai 2005; Iacoboni 2009). We can say that, roughly, in the first case, imitation is treated as a *programming mode* while in the second, it is treated as a *social phenomenon*.

In some animals, including humans, imitation is a social phenomenon (Nagai 2005) that was studied intensively by ethologists and developmental psychologists. Social psychology studies have demonstrated that imitation and mimicry are pervasive, automatic, and facilitate empathy. Neuroscience investigations have demonstrated physiological mechanisms of mirroring at single-cell and neural-system levels that support the cognitive and social psychology constructs (Iacoboni 2009). Neural mirroring and imitation solves the "problem of other minds" and makes intersubjectivity possible, thus facilitating social behavior. The ideomotor framework of human actions assumes a common representational format for action and perception that facilitates imitation (Iacoboni 2009). Furthermore, the associative sequence learning model of imitation proposes that experience-based Hebbian learning forms links between sensory processing of the actions of others and motor plans (Iacoboni 2009).

One of the major differences between learning from demonstration and traditional supervised learning, is the availability of a limited number of training examples for the learner. This limits the applicability of traditional machine learning approaches like SVMs and BNs. Another major difference—that is usually ignored in LfD research— is that in real world LfD situations, the learner may have to detect for itself what behaviors it needs to learn as the demonstrator may not be always explicit in marking the boundaries of these behaviors or the dimensions of the input space that are of interest for learning.

For a robot to be able to learn from a demonstration, it must solve many problems. Most important of these problems are the following seven challenges:

- *Action Segmentation:* Where are the boundaries of different *elementary behaviors* in the perceived motion stream of the demonstrator?
- *Behavior Significance for Imitation:* What are the interesting behaviors and features of behavior that should be imitated? This combines the *what* and *who* problems identified by Nehaniv and Dautenhahn (1998).
- *Perspective Taking:* How is the situation perceived in the eyes (or sensors) of the demonstrator?
- *Demonstrator modeling:* What are the primitive actions (or actuation commands) that the demonstrator is executing to achieve this behavior? What is the relation between these actions and the sensory input of the demonstrator?
- *Correspondence Problem:* How can actions and motions of the demonstrator be mapped to the learner's body and frame of reference?
- *Evaluation Problem* How can the learner know that it succeeded in imitating the demonstrator and how to measure the quality of the imitation in order to improve it? This evaluation would usually require feedback from the demonstrator or other agents and can utilize social cues (Scassellati 1999).
- *Quality Improvement Problem:* How can the learner improve the quality of its imitative behavior over time either by adapting to new situations or by modifying learned motions to better represent the underlying goals and intentions of perceived demonstrations?

Most of the research in imitation learning has focused on the perspective taking, demonstrator modeling and the correspondence problems above (Argall et al. 2009). In most cases, the action segmentation problem is ignored and it is assumed that the demonstrator (*teacher*) will somehow signal the beginning and ending of relevant behaviors.

There are many factors that affect the significance of a behavior for the learner. There are behavior intrinsic features that may make it interesting (e.g. repetition, novelty). There are object intrinsic features in the objects affected by the behavior (e.g. color, motion pattern) that can make that behavior interesting. These features determine what we call the *saliency* of the behavior and its calculation is clearly bottom-up. Also the goals of the learner will affect the significance of demonstrator's behaviors. This factor is what we call the *relevance* of the behavior and its calculation is clearly top-down. A third factor is the *sensory context* of the behavior. Finally learner's capabilities affect the significance of demonstrator's behavior. For example, if the behavior cannot be executed by the learner, there is no point in trying to imitate it. A solution to the significance problem needs to smoothly combine all of these factors taking into account the fact that not all of them will be available all the time (e.g. sometimes saliency will be difficult to calculate due to sensory ambiguities, sometimes relevance may not be possible to calculate because imitator's goals are not set yet).

Learning from Demonstrations is a major technique for utilizing natural interaction in teaching robots new skills. This is the other side of the utilization of machine learning techniques for achieving natural interaction. LfD is discussed in more details in Chap. 13. Extensions of standard LfD techniques to achieve natural imitative learning that tackles the behavior significance challenge will be discussed in Chap. 12.

1.9 Book Organization

Figure 1.5 shows the general organization of this book. It consists of two parts with different (yet complimentary) emphasis that introduce the reader to this exciting new field in the intersection of robotics, human-machine-interaction, and data mining.

One goal that we tried to achieve in writing this book was to provide a self-contained work that can be used by practitioners in our three fields of interest (data mining, robotics and human-machine-interaction). For this reason we strove to provide all necessary details of the algorithms used and the experiments reported not only to ease reproduction of results but also to provide readers from these three widely separated fields with all necessary knowledge of the other fields required to appreciate the work and reuse it in their own research and creations.

The first part of the book (Chaps. 2–5) introduces the data-mining component with a clear focus on time-series analysis. Technologies discussed in this part will provide the core of the applications to social robotics detailed in the second part of the book. The second part (Chaps. 6–13) provides an overview of social robotics then delves into the interplay between sociality, autonomy and behavioral naturalness that is at the heart of our approach and provides into the details a coherent system based on the techniques introduced in the first part to meet the challenges facing the realization of autonomously social robots. Several case studies are also reported and discussed. The final chapter of the book summarizes our journey and provides guidance for future passengers in this exciting data-mining road to social robotics.

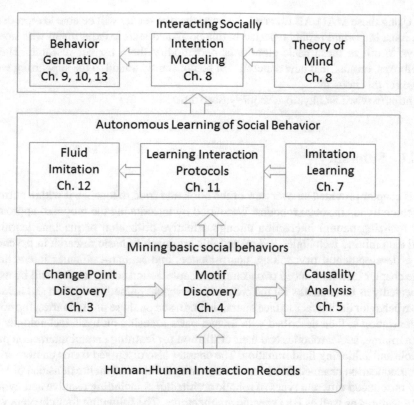

Fig. 1.5 The structure of the book showing the three components of the proposed approach, their relation and coverage of different chapters

1.10 Supporting Site

Learning by doing is the best approach to acquiring new skills. That is not only true for robots but for humans as well. For this reason, it is beneficial to have a platform from which basic approaches described in this book can be tested.

To facilitate this, we provide a complete implementation of most of the algorithms discussed in this book in MATLAB along with test scripts and demos. These implementations are provided in two libraries. The first is a toolbox for solving change point discovery (Chap. 3), motif discovery (Chap. 4) and causality discovery (Chap. 5) as well as time-series generation, representation and transformation algorithms (Chap. 2). The toolbox is called MC^2 for motif, change and causality discovery. This toolbox is available with its documentation from:

http://www.ii.ist.i.kyoto-u.ac.jp/~yasser/mc2

The second set of tools are algorithms for learning from demonstration (Chap. 13) and fluid imitation (Chap. 12) available from:

http://www.ii.ist.i.kyoto-u.ac.jp/~yasser/fluid

Using these MATLAB libraries, we hope that the reader will be able to reproduce the most important results reported in this book and start to experiment with novel ways to utilize these basic algorithms and modify them for her research. These toolboxes, erratas, and new material will be constantly added to the supporting web page for this book at:

http://www.ii.ist.i.kyoto-u.ac.jp/~yasser/dmsr.

1.11   Summary

This chapter provided an overview of the book and tried to localize it within current research trends in social robotics. We started by introducing the proposed approach for realizing natural interaction through effective utilization of machine learning and data mining techniques and related this approach to basic research in psychology, developmental psychology, neuroscience, and human–computer interaction. The chapter then introduced two examples of interaction scenarios that will be used repeatedly as benchmarks for the proposed approach: guided navigation and natural gaze behavior during face to face interactions. Based on these preliminaries, the book organization is then described with its two parts focusing on basic technologies of data mining, social robotics and their utilization for learning natural interaction protocols and achieving fluid imitation. The chapter also discussed forms on nonverbal communication channels in human–human interactions and their extensions of HRI and introduced different types of robotic architectures including reactive and hybrid architectures as well as HRI specific architectures. The following four chapters will discuss basic time series mining techniques used throughout the book.

References

Abbeel P, Coates A, Ng AY (2010) Autonomous helicopter aerobatics through apprenticeship learning. Int J Robot Res 29(13):1608–1639

Aleotti J, Caselli S (2008) Grasp programming by demonstration: a task-based quality measure. In: RO-MAN'08: IEEE international symposium on robot and human interactive communication. IEEE, pp 383–388

Argall BD, Chernova S, Veloso M, Browning B (2009) A survey of robot learning from demonstration. Robot Auton Syst 57(5):469–483

Argyle M (2001) Bodily Communication. Routledge; New Ed edition

Arkin RC, Fujita M, Takagi T, Hasegawa R (2003) An ethological and emotional basis for human-robot interaction, robotics and autonomous systems. Robot Auton Syst 42(3/4):191–201

Austin JL (1975) How to do things with words, vol 367. Oxford University Press

Billard A, Siegwart R (2004) Robot learning from demonstration. Robot Auton Syst 47(2–3):65–67

Billing E (2010) A formalism for learning from demonstration. Paladyn 1(1):73–102. doi:10.2478/s13230-010-0001-5

Breazeal C (2000) Believability and readability of robot faces. In: SIRS'00: the 8th international symposium on intelligent robotic systems, pp 247–256

Breazeal C (2002) Regulation and entrainment in human-robot interaction. Int J Exp Robot 21(10–11):883–902

Breazeal C, Aryananda L (2002) Recognition of affective communicative intent in robot-directed speech. Auton Robot 12(1):83–104

Breazeal C, Buchsbaum D, Gray J, Gatenby D, Blumberg B (2005a) Learning from and about others: towards using imitation to bootstrap the social understanding of others by robots. Artif Life 11(1/2):31–62

Breazeal C, Kidd C, Thomaz A, Hoffman G, Berlin M (2005b) Effects of nonverbal communication on efficiency and robustness in human–robot teamwork. In: IROS'05: IEEE/RSJ international conference on intelligent robots and systems. IEEE, pp 708–713

Brooks R (1986) A robust layered control system for a mobile robot. IEEE J Robot Autom 2(1):14–23

Brooks RA (1991) Challenges for complete creature architectures, pp 434–443

Brooks RA, Breazeal C, Irie R, Kemp CC, Marjanovic M, Scassellati B, Williamson MM (1998) Alternative essences of intelligence. In: AAAI'98: the national conference on artificial intelligence. Wiley, pp 961–968

Calinon S, Billard A (2007) Active teaching in robot programming by demonstration. In: RO-MAN'07: IEEE international symposium on robot and human interactive communication. IEEE, pp 702–707. doi:10.1109/RO-MAN.2007.4415177

Cassell J, Sullivan J, Prevost S, Churchill EF (2000) Embodied conversational agents. MIT Press

Castelfranchi C, Falcone R (2004) Founding autonomy: the dialectics between (social) environment and agent's architecture and powers. Lecture Notes on Artificial Intelligence, vol 2969, pp 40–54

Chartrand T, Bargh J (1999) The chameleon effect: the perception behavior link and social interaction. J Personal Soc Psychol 76(6):893–910

Cicchetti D, Tucker D (1994) Development and self-regulatory structures of the mind. Dev Psychopathol 6:533–549

Clark HH, Brennan SE (1991) Grounding in communication. Perspect Soc Shar Cogn 13:127–149

Fukuda T, Jung MJ, Nakashima M, Arai F, Hasegawa Y (2004) Facial expressive robotic head system for human-robot communication and its application in home environment. Proc IEEE 92(11): 1851–1865

Gump B, Kulik J (1997) Stress, affiliation and emotional contagion. J Personal Soc Psychol 72:305–319

Hashiyama T, Sada K, Iwata M, Tano S (2006) Controlling an entertainment robot through intuitive gestures. In: IEEE international conference on systems, man, and cybernetics, pp 1909–1914

Iacoboni M (2009) Imitation, empathy, and mirror neurons. Annu Rev Psychol 60:653–670. doi:10.1146/annurev.psych.60.110707.163604

Iba S, Paredis C, Khosla P (2005) Interactive multimodal robot programming. Int J Robot Res 24(1):83–104

Ishiguro H, Kanda T, Kimoto K, Ishida T (1999) A robot architecture based on situated modules. In: IROS'99: IEEE/RSJ conference on intelligent robots and systems, IEEE, vol 3, pp 1617–1624

Johannsen G (2002) Auditory display of direction and states for mobile systems. In: ICAD'02: the 3rd international conference on auditory display, ICAD. https://smartech.gatech.edu/handle/1853/51337

Kanda T, Ishiguro H, Ono T, Imai M, Nakatsu R (2002) Development and evaluation of an interactive humanoid robot "robovie". In: ICRA'02: IEEE International conference on robotics and automation, vol 2, pp 1848–1855. doi:10.1109/ROBOT.2002.1014810

Kanda T, Kamasima M, Imai M, Ono T, Sakamoto D, Ishiguro H, Anzai Y (2007) A humanoid robot that pretends to listen to route guidance from a human. Auton Robot 22(1):87–100

Kanda T, Miyashita T, Osada T, Haikawa Y, Ishiguro H (2008) Analysis of humanoid appearances in human-robot interaction. IEEE Trans Robot 24(3):725–735

Karim S, Sonenberg L, Tan AH (2006) A hybrid architecture combining reactive plan execution and reactive learning. In: PRICAI'06: 9th biennial Pacific Rim international conference on artificial intelligence, pp 200–211

Kuno Y, Sakurai A, Miyauchi D, Nakamura A (2004) Two-way eye contact between humans and robots. In: ICMI'04: the 6th international conference on multimodal interfaces. ACM, New York, pp 1–8. http://doi.acm.org/10.1145/1027933.1027935, http://dl.acm.org/citation.cfm?id= 1027935&dl=ACM&coll=DL&CFID=554272593&CFTOKEN=55996618

Miyauchi D, Sakurai A, Nakamura A, Kuno Y (2004) Active eye contact for human–robot communication. In: Extended abstracts on human factors in computing systems. ACM Press, New York, pp 1099–1102. http://doi.acm.org/10.1145/985921.985998

Mohammad Y, Nishida T (2008) Getting feedback from a miniature robot. In: ICIA'08: IEEE international conference on informatics in automation, pp 941–947

Mohammad Y, Nishida T (2009) Toward combining autonomy and interactivity for social robots. AI Soc 24:35–49. doi:10.1007/s00146-009-0196-3

Mohammad Y, Nishida T (2014) Learning where to look: autonomous development of gaze behavior for natural human-robot interaction. Interact Stud 14(3):419–450

Murata A, Fadiga L, Fogassi L, Gallese V, Raos V, Rizzolatti G (1997) Object representation in the ventral premotor cortex (area F5) of the monkey. J Neurophysiol 78:2226–2230

Nagai Y (2005) Joint attention development in infant-like robot based on head movement imitation. In: The 3rd international symposium on imitation in animals and artifacts, pp 87–96

Nagaoka C, Komori M, Nakamura T, Draguna MR (2005) Effects of receptive listening on the congruence of speakers' response latencies in dialogues. Psychol Rep 97:265–274

Nehaniv C, Dautenhahn K (1998) Mapping between dissimilar bodies: affordances and the algebraic foundations of imitation. In: Demiris J, Birk A (eds) EWLR'98: the 7th european workshop on learning robots, pp 64–72

Nickel K, Stiefelhagen R (2007) Visual recognition of pointing gestures for human' robot interaction. Image Vis Comput 25(12):1875–1884

Nicolescu M, Mataric M (2001) Learning and interacting in human-robot domains. IEEE Trans Syst Man Cybern Part A Syst Hum 31(5):419–430

Nicolescu M, Matarić M (2002) A hierarchical architecture for behavior-based robots. In: AAMAS'02: the first international joint conference on autonomous agents and multiagent systems. ACM, New York, pp 227–233. http://doi.acm.org/10.1145/544741.544798

Nishida T, Nakazawa A, Ohmoto Y, Mohammad Y (2014) Conversational informatics: a data intensive approach with emphasis on nonverbal communication. Springer

Perez MC (2003) A proposal of a behavior-based control architecture with reinforcement learning for an autonomous underwater robot. PhD thesis, University of Girona

Reeves B, Nass C (1996) How people treat computers, television, and new media like real people and places. CSLI Publications and Cambridge University Press

Robins B, Dautenhahn K, Te Boerkhorst R, Billard A (2004) Robots as assistive technology-does appearance matter? In: RO-MAN'04: the 13th IEEE international workshop on robot and human interactive communication. IEEE, pp 277–282

Rybski PE, Yoon K, Stolarz J, Veloso MM (2007) Interactive robot task training through dialog and demonstration. In: HRI'07: the ACM/IEEE international conference on human–robot interaction. ACM, New York, pp 49–56. http://doi.acm.org/10.1145/1228716.1228724

Scassellati B (1999) Knowing what to imitate and knowing when you succeed. In: The AISB symposium on imitation in animals and artifacts, pp 105–113

Scheflen AE (1964) The significance of posture in communication systems. Psychiatry 27:316–331

Triesch J, von der Malsburg C (1998) A gesture interface for human–robot-interaction. In: FG'98: the IEEE third international conference on automatic face and gesture recognition. IEEE, pp 546–551

Watanabe T, Okubo M (1998) Pysiological analysis of entrainment in communication. Trans Inf Process Soc Jpn 39:1225–1231

Yang L, Yue J, Zhang X (2008) Hybrid control architecture in bio-mimetic robot. In: WCICA'08: the 7th world congress on intelligent control and automation, pp 5699–5703. doi:10.1109/WCICA. 2008.4593860

Yong Xu MG, Nishida T (2007) An experiment study of gesture-based human-robot interface. In: CME'07: IEEE/ICME international conference on Complex Medical Engineering, pp 458–464

Yoshikawa CNS, Komori M (2006) Embodied synchrony of nonverbal behaviour in counseling: a case study of role playing school counseling. In: CogSci'06: the 28th annual conference of the cognitive science society, pp 1862–1867

Ziemke T (2003) What's that thing called embodiment. CogSci'03: the 25th annual meeting of the cognitive science society, Mahwah. Lawrence Erlbaum, NJ, pp 1305–1310

Part I
Time Series Mining

Chapter 2
Mining Time-Series Data

Data is being generated in an ever increasing rate by all kinds of human endeavors. A sizable fraction of this data appears in the form of time-series or can be converted to this form. For example, a random sampling of 4000 graphics from 15 of the world's newspapers published from 1974 to 1989 found that 75 % of these graphics were time-series (Ratanamahatana et al. 2010).

This makes time series analysis and mining an important research area that is expected only to become more so over time. Time series analysis is a huge field and it is not possible to exhaustively cover it in an introductory chapter or even in a complete book. This means that in this chapter we had to be selective. The guiding principle for the selections made in this chapter is utility for the ideas presented in subsequent chapters. At the very least, this chapter introduces the notation used throughout the book and forms the background against which the ideas to be presented are painted.

This chapter was written with the social robotics researcher in mind and most of it would be elementary knowledge for practitioners of time-series analysis, or related fields. Nevertheless, a quick pass through the chapter is advised in order for the reader to familiarize herself with the notation and definitions that will be used extensively in the following chapters.

2.1 Basic Definitions

A time-series is simply an ordered list with an implied independent variable (usually time) and one or more dependent variables from a predefined domain (usually $\mathbb{R}$).

© Springer International Publishing Switzerland 2015
Y. Mohammad and T. Nishida, *Data Mining for Social Robotics*,
Advanced Information and Knowledge Processing,
DOI 10.1007/978-3-319-25232-2_2

Definition 2.1   Time Series X_t is an ordered list of items $(x_0, x_1, \ldots, x_t, \ldots, x_T)$ each of them is called a *point*, where t is the independent variable belonging to a domain D_T and is assumed to be monotonically increasing and T is a scalar specifying the length of the time-series. All items x_t belong to a predefined domain D_X.

In most of the time-series described in this book, the domain of the independent variable is the set of integers (i.e. $D_T = \mathbb{I}^+$). If the domain of the dependent variable is the set of real numbers (i.e. $D_X = \mathbb{R}$) then the time-series is called a *real valued time-series*. If the domain of the dependent variable is multidimensional (e.g. $D_X = \mathbb{R}^N$) then the time-series is called a *multidimensional time-series* and the dimensionality of the time series is N where N is the dimensionality of each point. When the independent variable is not important or is known we will ignore it and will use X instead of X_t. In this book we use the notations x_i and $x(i)$ interchangeably to mean the point i of the time-series X.

In many cases we need to consider a contiguous list of items belonging to a time-series. This is made more precise in the following definition.

Definition 2.2   A subsequence $x_{i:j}$ is a time-series that consists of the tuple $(x_i, x_{i+1}, \ldots, x_j)$ belonging the time-series X. Another notation is $x_{i,n}$ which is the time-series that consists of the ordered points $x_i, x_{i+1}, \ldots, x_{i+n-1}$.

We use $x(\ldots)$ to mean the same thing as $x_{\ldots}$. In some cases, the length of the subsequence will be known from the context and in these cases we will drop it from the subsequence name (e.g. $x_{i,}$ will imply $x_{i,n}$). Notice that in this case x_i is a point while $x_{i,}$ is a subsequence starting at point i with the implicit length n.

For simplicity and throughout this book we will speak of the independent variable as time (t) even though our discussion can be generalized without any modification to any other variable given that it is monotonically increasing at a constant rate. We call this rate of increase/decrease of the independent variable τ and in most cases it is assumed to be unity.

2.2   Models of Time-Series Generating Processes

Time-series are generated continuously from nearly every kind of information processing or dynamical system. It is not possible to find an exhaustive set of generating processes that can be combined to lead to all possible time-series but there are some time-series models that received more attention from researchers due to their simplicity and/or ability to represent a wide range of phenomena and this section introduces some of them.

2.2.1   *Linear Additive Time-Series Model*

The Linear Additive Time-series Model (LAT) decomposes the time-series into a set of four components that are combined linearly (additively):

$$X = T_0 + C + S + R. \tag{2.1}$$

T_0 is called the *trend* and is the long-term non-periodic variation in the time-series. In many cases it is a monotone function (i.e. its first difference does not change sign). For example the identity time-series $x_t = t$ has only a trend component.

C represents a cyclic component with some period $T_c \gg \tau$ (remember that τ is the rate of increase of the independent variable or the inverse of the sampling rate). A good example of cyclic components is the famous business cycle of economics or fluctuations of electricity consumption based on the time of the year.

S represents another cyclic component but with a much smaller period which is only assumed to be greater than τ (i.e. $T_s > \tau$). This seasonal component is useful in modeling short lived behavior in some applications. For example, it can model daily fluctuations in electricity consumption.

R models all other random fluctuations added to the time-series.

The differentiation between C and S is ad-hoc and there is no reason that restricts the number of cyclic components to just two. A more general model that we call xLAT (Extended LAT) assumes N_c cyclic components and can be written as:

$$X = T_0 + R + \sum_{n=1}^{N_c} C_n. \tag{2.2}$$

Notice that the Fourier transform (See Sect. 2.3.3) is a special case of this model assuming the trend to be constant and random fluctuations to be zero while setting $N_c = \infty$ and each cyclic component to a sinusoidal function.

This decomposition of time-series data to different *kinds* of components can be useful in getting a sense of the underlying dynamics. For example, a lot of the controversy about global warming boils down to whether the perceived increase in temperatures is a part of T_0, C or R.

The linear additive model has several applications specially in economics where the four types of components represent specific socio-economic factors affecting different economic metrics.

2.2.2 Random Walk

A random-walk is a time-series that is generated by making small random variations of the current time-series value at every step. One of the simplest random walks involves moving around the integer numbers (usually starting with zero) with an increment of either 1 or -1 based on a fair coin flip. A slightly generalized version of this random walk can be formalized as follows:

$$x_0 = 0, \tag{2.3}$$

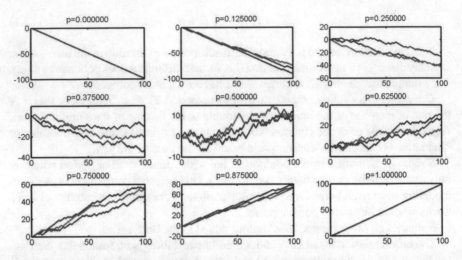

Fig. 2.1 Examples of random walks with variable probability of increase p

$$x_t = \begin{cases} x_{t-1} + \delta & 0 \le p_t < p \\ x_{t-1} - \delta & otherwise \end{cases}, \qquad (2.4)$$

where $\delta \in \mathbb{R}^+$ is some positive real number, p_t is a random number between zero and 1 and p is the probability of increasing.

The MC^2 toolbox accompanying this book contains a data generation function called *generateRandomWalk()* that generates 1-D random walks. Figure 2.1 shows examples of random walks for different increase probabilities (p). It is easy to show that the expected value of the time-series average and trend are zero when $p = 0.5$ but it gets a positive/negative trend when p is larger/smaller than this critical value. The toolbox function allows the user to control the step size or even make it randomly chosen from a uniform probability distribution.

2.2.3   Moving Average Processes

A moving average process generates time-series points by a linear combination of past values of another time-series (usually white noise) and is parameterized by the number of past points involved. $MA\,(m)$ is a moving average process of order m iff it generates data according to:

$$x_t^{ma(m)} = \sum_{i=0}^{m} a_i \theta_{t-i}, \qquad (2.5)$$

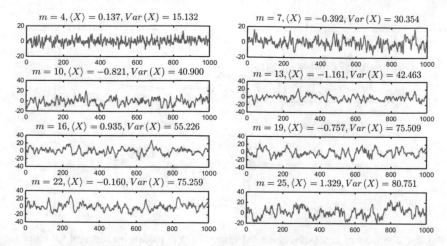

Fig. 2.2 Examples of time-series generated from a $MA(m)$ process for different values of m. In all cases $a_0 = 1$ and $a_i = 2$ for $1 \leq i \leq m$

where Θ is a random time-series representing white noise. Moreover, $a_i \in \mathbb{R}$ for $0 \leq i \leq m$ and usually a_0 is set to unity. White noise can be generated using *randn*() in MATLAB/Octave or better *wgn*(). Assuming that the mean of the white noise signal is μ_θ and its variance is σ_θ^2, it can be shown that:

$$\langle X \rangle = \mu_\theta \sum_{i=0}^{m} a_i, \tag{2.6}$$

$$Var(X) = \sigma^2 \sum_{i=0}^{m} a_i^2, \tag{2.7}$$

where as usual $\langle X \rangle$ and $Var(X)$ are the expectation and variance of the time-series X.

The MC^2 toolbox has a function *generateMA*() that can generate time-series from a moving average process $MA(m)$. Figure 2.2 shows examples of time-series generated from a $MA(m)$ process for different values of m. In all cases $a_0 = 1$ and $a_i = 2$ for $1 \leq i \leq m$. It is clear that with increased order, the variance of the time-series increases. Notice that the mean of the white noise used was zero which accounts for the observation that the mean of the final time-series is also around zero and did not change with increased model order.

An implementation detail related to moving average processes is what to do with the first m points in the output time-series for which no enough data is available to apply Eq. 2.5. In our implementation we assume that $\theta_{-1:-m}$ is generated from the same process from which Θ is generated and use these values implicitly to set $x_{0:m}$ (Fig. 2.3).

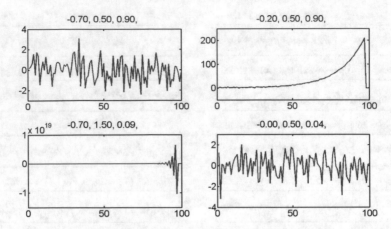

Fig. 2.3 Examples of time-series generated from an $AR(3)$ process for different values of the parameters a_i. Each example shows the values of $a_{1:3}$ used to generate it. For simplicity we assume that the white noise had zero variance (e.g. no white noise)

2.2.4   Auto-Regressive Processes

An auto regressive process $AR(m)$ generates each new data point as a linear weighted combination of the past m points in the time-series plus an additive white noise value.

$$x_t = \sum_{i=1}^{m} a_i x_{t-i} + \delta_t. \tag{2.8}$$

The function *generateAR()* can be used to generate data from an auto-regressive process in MC^2 (Fig. 2.3).

2.2.5   ARMA and ARIMA Processes

An $ARMA(m, n)$ process is a linear summation of an $AR(m)$ process and a $MA(n)$ process which can be written in the form:

$$x_t^{arma} = \sum_{i=1}^{m} a_i x_{t-i}^{arma} + \sum_{i=0}^{n} b_i \theta_{t-i}. \tag{2.9}$$

An $ARIMA(m, n, d)$ process is a process when differencing its output d times will be an $ARMA(m, n)$ process. $ARIMA$ processes find many applications in economic theory but will not be used much in the second part of this book to model robot or human behavior.

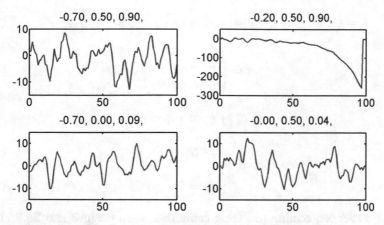

Fig. 2.4 Examples of time-series generated from an $ARMA(3, 5)$ process for different values of the parameters a_i. Each example shows the values of $a_{1:3}$ used to generate it. In all cases, $B = [1, 2, 3, 2, 1]^T$

In MC^2, you can generate data from an $ARMA$ process using *generateARMA()* and from an $ARIMA$ process using *generateARIMA()* (Fig. 2.4).

2.2.6 State-Space Generation

Another widely used generation model of time-series is the state-space model.

$$
\begin{aligned}
s_t &= g\left(s_{t-1}\right) + r_1\left(\varepsilon_t\right), \\
x_t &= f\left(s_t\right) + r_2\left(\lambda_t\right).
\end{aligned}
\tag{2.10}
$$

In general all time-series involved in this definition are multidimensional in order to code the full state of the system at every time-step.

Of special interest are affine state-space models that have the specific form:

$$
\begin{aligned}
s_t &= As_{t-1} + B\varepsilon_t, \\
x_t &= Cs_t + \lambda_t,
\end{aligned}
\tag{2.11}
$$

where A, B, C are matrices and the output time-series of the model is X. Assuming that S has the dimensionality N_s, X has the dimensionality N_x, and $\varepsilon_t \in \mathbb{R}_\varepsilon^N$ then A is a $N_s \times N_s$ matrix, B is a $N_s \times N_\varepsilon$ matrix, and C is a $N_x \times N_s$ matrix. The noise components (ε_t and λ_t) are sampled from two Gaussian distributions with zero mean and known covariance matrices.

This generation model can be used to simulate random walks, MA processes, AR processes, and ARMA models. Consider for example the ARMA model of Eq. 2.9. An equivalent affine state-space model (See Eq. 2.11) will have the following form:

$$s_t = \left(x_t \ x_{t-1} \ \ldots \ x_{t-m+2} \ x_{t-m+1} \ \varepsilon_t \ \varepsilon_{t-1} \ \ldots \ \varepsilon_{t-n+2} \ \varepsilon_{t-n+1} \right)^T, \tag{2.12}$$

$$A = \begin{pmatrix} a^T & b^T \\ I^{m \times m} & 0^{m \times m} \\ 0^{n \times n} & I^{n \times n} \end{pmatrix}, \tag{2.13}$$

$$B = \left(1 \ 0^{1 \times m} \ 1 \ 0^{1 \times n} \right)^T, \tag{2.14}$$

$$C = \left(1 \ 0^{n+m \times 1} \right), \tag{2.15}$$

$$\lambda_t = 0. \tag{2.16}$$

The reader can confirm that these definitions when plugged into Eq. 2.11 will recover Eq. 2.9. As random walks, MA and RA processes are all special cases of ARMA processes; we have just shown that affine state-space models can be used as a general generation process of all of these. This model is implemented in the function *generateAffineStateSpace()*.

2.2.7 Markov Chains

All of the previous methods for time-series generation are deterministic, in the sense that given the parameters of the time-series all points are known exactly up to the added noise. The Markov chain (MC) generation process, on the other hand, generates time-series points from a probabilistic distribution. The defining assumption of this model is that the time-series point depends only on the previous value of the time-series.

A MC model is defined by its initial distribution $p(x_0)$ and its transition conditional distribution $p(x_t|x_{t-1})$. Both can take any form.

One of the simplest continuous MCs is the Gaussian MC Model (GMC) which is defined as:

$$\begin{aligned} x_0^{gmc} &\sim p\left(x_0^{gmc}\right) \equiv \mathcal{N}(\mu_0, \Sigma_0), \\ x_t^{gmc} &\sim p\left(x_t^{gmc}|x_{t-1}^{gmc}\right) \equiv \mathcal{N}(x_{t-1}, \Sigma), 0 < t \leqslant T. \end{aligned} \tag{2.17}$$

It has three parameters:

- μ_0, Σ_0: The mean and covariance matrix for generating the first point of the series
- Σ: The covariance matrix used to generate x_{t+1} given x_t as its mean.

The MC^2 toolbox has a function *generateMarkovChain()* which generates time-series of arbitrary length and dimensionality from a GMC. The implemented version has an option to add an MA process. Notice that this will not be added after the complete time-series is generated but at each time-step. The effect of having a non-zero MA(m) process here depends on the exact values of its m parameters. For example

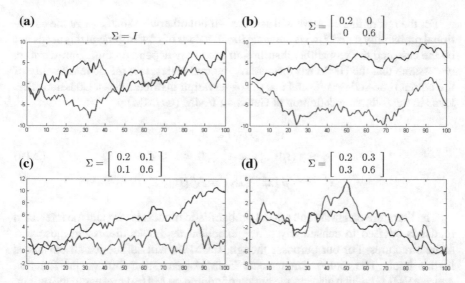

Fig. 2.5 Examples of 2 dimensional time-series (represented by the *two colors*) generated from a Gaussian Markov Chain with $\mu = \mu_0 = 0$ and unit Σ_0 for four different Σ cases (Color in online)

the MA(m) process with $a = (1, 2, 3, 4, 5, 4, 3, 2, 1)^T$ will tend to smooth out the time-series. Figure 2.5 shows four time-series generated from this function.

It is interesting to see how can we control various aspects of the time-series by changing its three parameters (i.e. μ_0, Σ_0, and Σ). Given that μ_0 and Σ_0 affect only the first point of the time-series, we will focus on the effect of Σ. Figure 2.5a shows an example 2D time-series when $\Sigma = I$ which means that the two dimensions of the time-series are independent (because the off-diagonal elements are zeros) and they both have the same overall variance. Figure 2.5b shows what happens when we change the relative values of the diagonal elements. Here the second dimension of the time-series (green) is still uncorrelated with the first dimension but has much higher variability because of increased variance. Figure 2.5c, d shows the effect of increasing off-diagonal elements creating correlations between different dimensions of the time-series.

2.2.8   Hidden Markov Models

A slightly more complex probabilistic generation mechanism is the Hidden Markov Model (HMM). For our purposes, a HMM has both unobserved states and observed time-series. An appropriate HMM definition for our purposes is given by:

$$
\begin{aligned}
s_0 &\sim p\left(s_0\right), \\
s_t &\sim p\left(s_t|s_{t-1}\right), 0 < t \leqslant T, \\
x_t^{hmm} &\sim p\left(x_t^{hmm}|s_t\right).
\end{aligned}
\tag{2.18}
$$

For the rest of this book we will assume—if not otherwise stated—that the conditional probability $p\left(x_t^{hmm}|s_t\right)$ is the same for all values of t but in general this needs not be the case and the probability distribution used may depend on time. Furthermore, we assume that the HMM has discrete N_s internal states, that the dimensionality of the output time-series is N_x and that the observation distribution is a Gaussian. This leads to the following definition of Gaussian HMM (GHMM):

$$s_0 \sim p\,(s_0) \equiv \pi,$$
$$s_t \sim p\,(s_t|s_{t-1}) \equiv A_{s_{t-1}}^T, 0 < t \leqslant T, \qquad (2.19)$$
$$x_t^{ghmm} \sim p\left(x_t^{ghmm}|s_t\right) \equiv \mathcal{N}\left(\mu_{s_t}, \Sigma_{s_t}\right).$$

The literature on HMM and related probabilistic models is vast and the interested reader is advised to consult any of the excellent textbooks discussing these versatile structures. For our purposes though, GHMMs will suffice for all our needs in this book. The toolbox implements GHMM generation through the function *genrateHMM*() which also has the option of adding an $MA(m)$ process to the output.

Figure 2.6 shows four examples of the output of a Gaussian HMM with fixed means. The variation of the transition matrix affects the output in an expected way. Figure 2.6a shows a case where each state has high probability of being repeated (0.9) relative to the probability of going to the other state (0.1). This leads to a time-series that looks like a square wave with variable duty cycle. Figure 2.6b shows a middle case where the probability of staying at the same state is equal to the

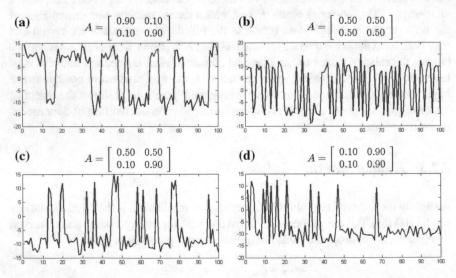

Fig. 2.6 Examples of time-series generated from a Gaussian Hidden Markov model with two states and single dimensional output. The means of the observational Gaussians for the two states were set to 10 and −10

probability of changing to the other state. This leads to a time series that jumps between the two states rapidly. Figure 2.6c shows a case with one state (the first) having equal probability of repeating or toggling while the other state (the second) has high probability of repeating (0.9). This leads to a time series that spends more time in the second state. A more extreme example is given in Fig. 2.6d where no matter what is the current state, the second state has higher probability of occurring in the next output compared with the first state (9 times higher). This leads to a time-series that spends even more of its time in the second state.

2.2.9 Gaussian Mixture Models

A widely used probabilistic generation model for time-series uses a mixture of probability distributions rather than switching between them as in HMM. In this book we only utilized Gaussian Mixture Models (GMM) so we will focus our attention on them.

Assume that we have a joint distribution of two variables x and y (i.e. $p(x, y)$). We can then condition on one of them using the definition of conditioning:

$$p(x|y) = \frac{p(x, y)}{\sum_x p(x, y)}. \tag{2.20}$$

This means that we can generate points of a time-series X by conditioning on the independent variable (time) at every time-step. The math becomes much easier when we assume that the joint distribution is a Mixture of Gaussians in the form:

$$p(x, t) = \sum_{k=1}^{K} p(k) p_k(x, t), \tag{2.21}$$

$$p_k(x, t) \equiv \mathcal{N}(\mu_k, \Sigma_k).$$

Given this joint distribution, it is possible to generate time-series points by conditioning on time. This is the basic idea of Gaussian Mixture Regression (GMR) which is being widely applied in both statistics and learning from demonstration communities for providing a middle ground between high-bias parametric approaches and high-variance non-parametric approaches to modeling.

The use of a GMM to model the joint distribution ensures that the conditional distribution is also a GMM. In such cases, Eq. 2.20 simplifies to the following:

$$p(x|t) = \sum_{k=1}^{K} \pi_k^c p^c{}_k(x|t), \tag{2.22}$$

$$p^c{}_k(x|t) \equiv \mathcal{N}\left((x, t)^T ; \mu^c{}_k, \Sigma^c{}_k\right).$$

Now assume that we make the following definitions:

$$\mu_k \equiv \left(\mu_k^x, \mu_k^t \right)^T, \tag{2.23}$$

$$\Sigma_k \equiv \begin{pmatrix} \Sigma_k^x & \Sigma_k^{xt} \\ \left(\Sigma_k^{xt} \right)^T & \Sigma_k^t \end{pmatrix}, \tag{2.24}$$

where $\mu_k^t \in \mathbb{R}^{\mathbb{k}}$ and $\Sigma_k^x \in \mathbb{R}^{N \times N}$ while $\mu_k^x \in \mathbb{R}^{1 \times N}$.

Given these definitions, the new parameters of the conditional GMM are given by the following set of equations:

$$\pi_k^c = \frac{p(k) \, \mathcal{N} \left(t; \mu_k^t, \Sigma_k^t \right)}{\sum \pi_k^c}, \tag{2.25}$$

$$\mu_k^c = \mu_k^x + \Sigma_k^{xt} \left(\Sigma_k^t \right)^{-1} \left(t - \mu_k^t \right), \tag{2.26}$$

$$\Sigma_k^c = \Sigma_k^x - \Sigma_k^{xt} \left(\Sigma_k^t \right)^{-1} \left(\Sigma_k^{xt} \right)^T. \tag{2.27}$$

Furthermore, using properties of Gaussians, Eq. 2.22 can be further simplified by approximating the GMM with a single Gaussian:

$$p(x|t) \cong \mathcal{N} \left(x; \bar{\mu}_x, \bar{\Sigma}_x \right), \tag{2.28}$$

where $\bar{\mu}_x = \sum_{k=1}^K \pi_k^c \mu_k^c$ and $\bar{\Sigma}_x = \sum_{k=1}^K \left(\pi_k^c \right)^2 \Sigma_k^c$.

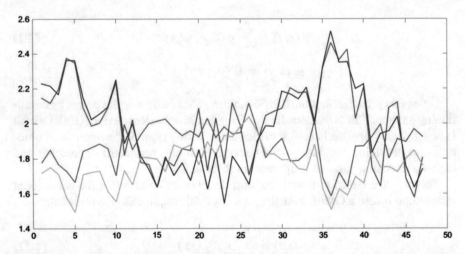

Fig. 2.7 Examples of a four dimensional time-series generated from a Gaussian Mixture Regression model (the *four colors* represent the four dimensions). The mean and covariance matrices were taken from the data provided in Calinon et al. (2006) for the first demo of Learning from Demonstration using GMM/GMR (Color in online)

A time-series can then be generated from this GMM by sampling from $p(x|t)$ for every value of t from 1 to the desired T. This procedure is implemented in the toolbox using the function *generateGMR()*. Figure 2.7 shows an example time-series generated using this function.

GMR was originally designed as a regression algorithm that uses a GMM learned from example data points to smoothly find the value of learned function at unseen points. Using it for data generation as we did in this example is not a standard procedure and is not even recommended. The main disadvantage for using GMR for generation is that the final step generates the data at every time-step independent from the data at nearby time-steps. This results in rough functions as seen in Fig. 2.7. This problem can be alleviated if the correlation between nearby points is taken into account which is what can be achieved by Gaussian Processes.

2.2.10 Gaussian Processes

HMM and GMM/GMR provide two alternative statistical methods for modeling time-series generation. The main problem with HMM is the need to select an appropriate number of states in advance. This problem can be somehow alleviated by using model selection techniques (e.g. Bayesian Information Criteria) if we have time-series data to use as example for the system to be modeled. This requires though multiple trainings with different values for the number of states.

GMM/GMR modeling discussed in the previous section had the problem of independence between successive time steps even though it could capture the covariance between different dimensions at the same time-step.

Gaussian Processes can overcome both of these problems by directly modeling the correlation between different dimensions of the time-series at all time-steps. It requires no selection of a number-of-states and can generate smooth time-series output.

A Gaussian Process ($\mathcal{GP}$) is defined as a collection of random variables (possibly infinite), any finite number of which have a joint Gaussian distribution (Rasmussen and Williams 2006).

A $\mathcal{GP}$ is completely specified by its mean function $m(x)$ and covariance functions $k(x, x')$ both are direct extensions of the mean and covariance matrix of a standard multivariant Gaussian distribution. Notice that for this section we use x as the independent variable instead of t as the $\mathcal{GP}$ formalism is general and can handle any kind of real valued independent variable not only time.

The standard formalism for Gaussian Processes is:

$$f(x) \sim \mathcal{GP}\left(m(x), k(x, x')\right),\qquad(2.29)$$

where the mean and covariance matrices are defined as:

$$m(x) = \mathbb{E}(f(x)), \tag{2.30}$$
$$k(x, x') = \mathbb{E}((f(x) - m(x))(f(x) - m(x))).$$

Given a Gaussian Process, it is easy to generate time-series by sampling at different values of x. For example, consider the case where $m(x) = 0$ and the covariance function is defined as:

$$k(x, x') = k_{se}(x, x') = \lambda e^{-(x-x')^T(x-x')/l}. \tag{2.31}$$

This covariance function is called the squared exponential and results in smooth time-series. We will use $\lambda = 1$ and $l = 0.5$ for our example. To generate a time-series spanning the time from 0 to 20, we select a sampling rate (100 in our case) and then calculate the mean of the function to be sampled which will equal to $\mu = m(0 : 0.01 : 20)$ then calculate the covariance matrix at the values of the input $x = 0 : 0.01 : 20$ by applying Eq. 2.31 for each pair of values in this set. The resulting covariance matrix Σ is then used with the mean μ to generate multivariate Gaussian values. This can easily be achieved by finding the Eigen Value Decomposition of Σ and sampling from independent Gaussians on this basis then projecting back using $V\lambda^{-1/2}$ where V is the set of Eigen vectors of Σ and λ is the set of Eigen values. This is implemented in *grand()* in the Toolbox. Figure 2.8 shows five example time-series sampled using this method.

It is trivial to add noise to this system by adding it directly to the mean vector μ during sampling.

Figure 2.9a shows a set of functions sampled from a 1D GP with the squared exponential covariance function (after adding the Gaussian noise). An important advantage of GPs is that it can update its internal generation model incrementally and with few data points. For example, Fig. 2.9b–d shows the mean and variance of the same GP used in Fig. 2.9a after fixing three points in succession (shown as red circles). It is clear that, even with few data points, the GP can easily learn a model that can be used to interpolate and extrapolate the function for all time. To achieve this incremental learning, GPs exploit the properties of Gaussians specially the decomposability of the covariance matrix. Using our previous notation:

$$\text{cov}(y_i, y_j) = \lambda e^{\|x_i - x_j\|^2/l} + \sigma_n \delta_{ij}, \tag{2.32}$$

where δ_{ij} is the Kronecker delta function. In matrix notation this leads to:

$$\text{cov}(\mathbf{y}) = K_{XX} + \sigma_n^2 I, \tag{2.33}$$

where y is a column vector of the outputs observed and K_{XX} is the covariance matrix calculated by applying Eq. 2.31 to all pairs of inputs. Notice that the noise model is decoupled and iid which is expected for a Gaussian noise generation process.

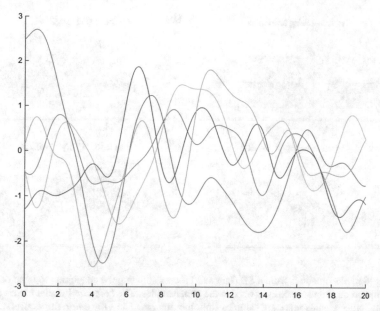

Fig. 2.8 Five examples of time-series sampled from a GP with zero mean and squared exponential covariance function with unity parameters

Now given a set of input output pairs (X, y) and a test point x_t, we can calculate the distribution of the output at this test point using the rules for conditioning Gaussian distributions as follows:

$$p\left(f_t|X, \mathbf{y}, x_t\right) = \mathcal{N}\left(K_{x_t X}\left(K_{XX} + \sigma_n^2 I\right)^{-1}\mathbf{y}, K_{x_t x_t} - K_{x_t X}\left(K_{XX} + \sigma_n^2 I\right)^{-1}K_{x_t X}{}^T\right).$$

$$(2.34)$$

Equation 2.34 was used to generate the estimates shown in Fig. 2.9. Most applications of GPs use the squared exponential function which is infinitely differentiable. This may lead to over-smoothing of the output and makes it harder for the system to represent abrupt changes in system dynamics due for example to a sudden change in the load for a manipulator. One example covariance function that can handle sudden changes better than the squared exponential is the Matern covariance function which has the form:

$$k_{Matern}\left(x, x'\right) = \frac{2^{1-\upsilon}}{\Gamma\left(\upsilon\right)}\left(\frac{\sqrt{2\upsilon}\,|x - x'|}{l}\right)^{\upsilon}K_{\upsilon}\left(\frac{\sqrt{2\upsilon}\,|x - x'|}{l}\right). \qquad (2.35)$$

There are many possible covariance functions other than the squared exponential and Matern and they can be combined to lead to a large variety of generation schemes (for several examples, please refer to Rasmussen and Williams 2006).

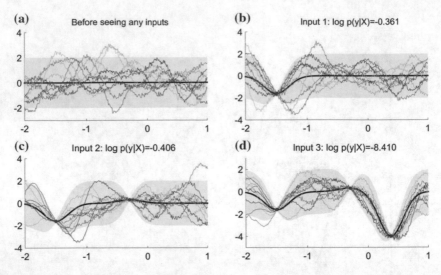

Fig. 2.9 Sample functions from a SE Gaussian Process showing the adaptation to input data and the limitation on the variance even outside the seen samples. The *bold black* signal represents the mean and other signals represent samples from the process. The *gray color* shows two standard deviations at each point (Color in online)

The GP generation process is implemented in the toolbox using a set of functions. The function *grand*() can be used to generate a time-series from a Gaussian Process given the mean and covariance matrices corresponding to the time-steps. To generate these matrices, you can use either *generateGP*() to generate a GP from a set of sample points or the functions *createGP*() to create a GP with no restrictions similar to the one used to generate the time-series shown in Fig. 2.8. The function *add2GP*() can be used to add points of known values to the GP and *generateFromGP*() can be used—at any time—to create the mean and covariances matrices needed for *grand*().

The implementation of GP in the MC^2 toolbox accompanying this book is minimal with only one covariance function (the squared exponential) and two mean functions (linear and sinusoidal) and with no optimization tricks in the implementations to facilitate understanding. These can easily be extended. There are several GP toolboxes for Matlab that implement more covariance functions already implemented and with more advanced GP algorithms for the interested reader. The information given here will, nevertheless, be enough for all our purposes in this book.

2.3 Representation and Transformations

In many cases, it is desirable to represent the time-series differently from the direct representation presented in Definition 2.1. The most effective representation scheme depends largely on the application. This section introduces some of the most useful representations for time-series. We will focus first on single-dimensional time-series. Multidimensional extensions will be discussed later.

2.3.1 Piecewise Aggregate Approximation

Piecewise Aggregate Approximation (PAA) is one of the simplest time-series compression and representation approaches. The idea of PAA is very simple. Rather than keeping the whole time-series, it is divided into equal length segments and we keep only the mean of each segment. These means are then concatenated to form another time-series that *represents* the original time-series albeit being shorter. This can be expressed as follows assuming that the segment length is m:

$$x_i^{paa} = \frac{1}{m_i} \sum_{j=im}^{\min(T-1,(i+1)m-1)} x_j, \tag{2.36}$$

for $0 \le i \le \lceil T/m \rceil$ where $m_i = m$ for all i but the last and equals $T - m \times \lfloor T/m \rfloor$ for the last point of X^{paa}.

Assuming that the segment length is m then the output of PAA is m times shorter than the original time-series. The PAA representation requires a single parameter from the user, which is the length of the segment m. In the extreme case where $m = 1$, PAA simply returns the input time-series without any change. In the other extreme case when $m = T$ where T is the length of the input time-series, PAA returns a single number representing the mean of the time-series (i.e. the DC or zero-frequency component).

Despite its simplicity, PAA can be very useful in many applications. For example, it can be shown that the Euclidean distance between any two time-series is larger than or equal the Euclidean distance between their PAA representations:

$$\sum (x_i - y_i)^2 \ge \sum \left(x^{paa}_i - y^{paa}_i\right)^2. \tag{2.37}$$

This property (called lower bounding) allows linear time generation of indexing structures that can be used to search through large numbers of time-series efficiently. Another important feature of the PAA transform or representation is that all calculations are local which means that it can be applied to the data in real time.

It is also trivial to extend PAA to the multidimensional case. Simply apply PAA to each dimension of the input time-series.

A simple extension of PAA that we do not utilize much in this book is called Adaptive Piecewise Constant Approximation (APCA) in which the length of each segment is not fixed but adapted to the time-series.

Table 2.1 The location of break points for the first 5 cases of alphabet sizes (n_a)

n_a β_a	1	2	3	4	5
β_1	−0.43	−0.67	−0.84	−0.97	−1.07
β_2	0.43	0	−0.25	−0.43	−0.57
β_3	–	0.67	0.25	−0	−0.18
β_4	–	–	0.84	0.43	0.18
β_5	–	–	–	0.84	0.57
β_6	–	–	–	–	1.07

A more complete table (up to $n_a = 10$) can be found in (Lin et al. 2003)

2.3.2   Symbolic Aggregate Approximation

Symbolic Aggregate approXimation (SAX) is a transformation algorithm that converts a single-dimensional time-series to a string. The main advantage of SAX is
that it is designed to achieve equiprobable symbol production. Because it is based on
PAA, it also provides a lower bound on the Euclidean distance. SAX was proposed by
Lin et al. (2003) and since then became one of the most widely used transformation
for data mining of time-series data.

SAX is built upon the finding that z-score normalized time-series have Gaussian
distributions (Larsen and Marx 1986). Based on that, the normal distribution is
divided into bands where the probability of falling within all bands is the same.
The number of bands is selected to equal to the number of symbols in the alphabet
to be used for discretization (n_a). The limits of these bands are called *break points*
β_a where $a \in \{1, 2, \ldots, n_a - 1\}$. These break points can be calculated easily from
tables of cumulative normal function. Table 2.1 shows the break points for the first
five alphabet sizes.

The locations of the break points can be calculated and stored for any finite
alphabet size.

Given these break point locations and the alphabet size (n_a), we can transform
a single-dimensional time-series into a string by first z-score normalizing it (i.e.
subtracting the mean from each point and dividing the standard deviation). The
normalized time-series is then transformed to a shorter time-series using PAA (See
Sect. 2.3.1). The break points corresponding to the chosen alphabet size are then
used to map values of this shorter time-series into symbols by issuing the symbol
corresponding to the band within which each PAA value lies.

The implementation of this transform in MC^2 is encapsulated by the function
time-series2symbol() which is provided by Lin et al. (2003). Figure 2.10 shows the
processing steps of this function leading to the final string representation of the
time-series.

The aforementioned SAX transform assumes that the input is a single-dimensional
time-series. Mohammad and Nishida (2014) proposed three different extensions of
this algorithm to multi-dimensional time-series. There are several methods to extend

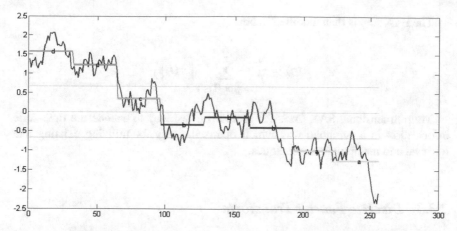

Fig. 2.10 Example time-series and its transformation steps using SAX. The original time-series is shown as well as its PAA transformation. The break points corresponding to a 4-letters alphabet are also shown (*light gray*). Finally the final symbols from the set {*ddcbbbaa*} is shown

SAX to handle multidimensional time-series. The first approach is not to modify SAX in any way and start by reducing the dimensionality of the time-series to one then apply the traditional SAX transformation. This approach is called SAX-PCA because we use Principal Component Analysis (PCA) for dimensionality reduction.

We start by z-score normalizing each dimension then applying Singular Value Decomposition (SVD) to represent the time-series $USV^T = X$. The output single dimensional time-series x is then obtained by projecting the time-series on the first singular vector: $x = U_1 X$, where U_1 is the singular vector of U corresponding to the largest singular value in S. The time-series x is then passed as an input to the traditional SAX algorithm.

Another, even simpler, approach is to apply SAX to every dimension of the data separately and then combine the resulting string by assigning every possible combination of symbols in the resulting D strings a unique identifier. This leads to a string of length N but from an extended alphabet of length M^D. In most cases, most of the symbols of this extended alphabet will not appear in the final string. To keep the requirement that the final string is from an M-symbols alphabet, we can simply cluster the resulting characters into M clusters using K-means and replace each character with the centroid of its cluster. This approach is called SAX-REPEAT.

A third approach (called SAX-z-score hereafter) is to modify the normalization step in SAX to use an extended multidimensional version of the z-score normalizer. This can be done by calculating an intermediate time-series $\hat{x}$ that has a zero-mean unit-variance distribution. The PAA step can then be applied to the L_2 norm of the data. Assuming that the covariance matrix of X is C and that the mean of its columns is μ, the intermediate time-series is calculated as:

$$\hat{x}(t) = C^{-1/2}(X - \mu).$$ (2.38)

The PAA step is then modified to be:

$$\bar{x}(n) = \frac{N}{T} \sum_{t=\frac{T}{N}(n-1)+1}^{\frac{Tn}{N}} \left\| \hat{x}(t) \right\|_2. \tag{2.39}$$

Multidimensional SAX (MSAX) provides a fast way to encode the time-series information in a symbolic way which opens the way for utilizing existing text-retrieval and manipulation techniques.

2.3.3 Discrete Fourier Transform

PAA represents the time-series with a shorter one that has the same independent variable (usually time). A completely different approach is taken by the Discrete Fourier Transform (DFT) which represents the time-series using a different yet related independent variable. If the independent variable of the original time series was time, then the transformed version will be represented using frequency as the independent variable.

DFT works with complex time-series. Given a time-series X, we can calculate its DFT transform using:

$$x_i^{dft} = \sum_{k=0}^{T-1} x_k e^{-j2\pi ik/T}, \tag{2.40}$$

where $j^2 = -1$ and the time-series length is T as usual.

It is clear that the transform is not local in the sense that calculating the value of x_t^{dft} depends on ALL the values of X. This means that—in contrast to PAA for example—the transform cannot be implemented incrementally. Even though that naive implementation of DFT requires $O(T^2)$ operations, implementations based on the Fast Fourier Transform (FFT) can achieve time complexity of $O(T \log T)$.

DFT is used extensively in the digital signal processing community and has several applications. For example, keeping only 10 % of the output can preserve 90 % of signal energy of random walks which makes DFT a good candidate for compression applications.

As with the case in PAA, DFT provides a lower bound on Euclidean distance:

$$\sum (x_i - y_i)^2 \geq \sum \left(\left| x^{dft}_i \right| - \left| y^{dft}_i \right| \right)^2. \tag{2.41}$$

This can be the basis of an indexing application and it usually achieves speedups between 3–100 compared with using the original time-series. Extending DFT to multiple dimensions can be achieved by applying the same transform to each dimension of the input.

2.3.4 Discrete Wavelet Transform

PAA was clearly local while DFT was a clearly global transform of the time-series. Some transforms (wavelet transforms) give a hierarchical view of the time series allowing it to be examined at multiple resolutions. There are several wavelet transforms but all share the common feature of having a function called the mother wavelet ψ that is translated and scaled to get other similar functions on a different scale and/or position called child wavelets according to the following equation:

$$\psi(t)_{\tau,s} = \frac{1}{\sqrt{s}} \psi\left(\frac{t-\tau}{s}\right). \tag{2.42}$$

The parameter s controls the scale while the parameter τ controls the positioning of the wavelet. The simplest yet most widely used wavelet transform in time-series mining is the *Haar* wavelet. The mother wavelet of the Haar transform is defined as:

$$\psi^{Haar}(t) = \begin{cases} 1 & 0 \le t < 0.5 \\ -1 & 0.5 \le t < 1 \\ 0 & otherwise \end{cases} \tag{2.43}$$

Applying this wavelet to any time-series is very efficient and can be executed in linear time using only summation and difference operators between pairs of points. Given a time-series X of length T which is assumed to be a power of 2, we start by finding the averages and differences of all consecutive pairs of points according to:

$$x^0_t = x_{2t+1} - x_{2t}, \tag{2.44}$$

for $0 \le t \le T/2$. Another temporary time-series is calculated using the sum operator applied to same pairs

$$\bar{x}^0_t = x_{2t+1} + x_{2t}, \tag{2.45}$$

for $0 \le t \le T/2$. The same two operations are applied again to $\bar{X}^0$ leading to:

$$x^1_t = \bar{x}^0_{2t+1} - \bar{x}^0_{2t}, \tag{2.46}$$

$$\bar{x}^1_t = \bar{x}^0_{2t+1} + \bar{x}^0_{2t}. \tag{2.47}$$

Now t will range from 0 to $T/4$.

In general:

$$x^i_t = \bar{x}^{i-1}_{2t+1} - \bar{x}^{i-1}_{2t}, \tag{2.48}$$

$$\bar{x}^i_t = \bar{x}^{i-1}_{2t+1} + \bar{x}^{i-1}_{2t}, \tag{2.49}$$

until we get a time-series of a single point for both X^i and $\bar{X}^i$ which will happen when $i = \log_2 T$. The time-series named X^i are then concatenated to give the Haar transform of the original time-series:

$$X^{Haar} = \left(X^0, X^1, \ldots, X^{\log_2 T}\right). \qquad (2.50)$$

Notice that each successive X^i gives a view of the original time-series that is progressively coarser.

Again, multidimensional time-series can be treated by simply applying the aforementioned transformation on every dimension independently.

The main advantages of Haar transform are simplicity of calculation and the ability to *inspect* the time-series at different resolutions which may reveal for example hierarchical repeated pattern (an important application for social robotics).

2.3.5 Singular Spectrum Analysis

Singular spectrum analysis (SSA) is a relatively new approach for analyzing and representing time-series data. This approach has several advantages. Firstly, it can be applied to any time-series either of a single or multiple dimensions and it can be applied in a single scan of the time series to generate a representation that requires a predefined multiple of the original time-series length. Secondly, the approach is flexible enough to allow both unsupervised and semi-supervised analysis of the data. Moreover, unlike Fourier analysis for example, it is data driven even in deciding the basis functions used in the decomposition of the input time-series. Compare that to the Haar transform which as we have just seen uses a predefined basis function that is applied repeatedly to the time-series.

Singular spectrum analysis has seen increased interest in the current century and has found applications in several areas including forecasting, noise rejection, causality analysis and change point discovery among many other approaches (Elsner and Tsonis 2013). This section will introduce the technique showing how to use it for time-series representation. Most of the change point discovery techniques used in this book (See Chap. 3) are based on this representation.

The main idea behind SSA is to represent the signal through a set of data-driven basis functions that are generated by subdimensional representation of the Hankel matrix representing the original time-series. This subdimensional representation, if carefully selected, can remove unwanted signal components (e.g. noise) and can be used to discover the basic building blocks of the time-series.

Given a time-series X, SSA analysis generates a set of times-series X_i^{ssa} where each of these components can be identified as a trend, a periodic, a quasi-periodic or a noise component while allowing reconstruction of X as:

$$X = \sum X_i^{ssa}.$$

This means that SSA analysis is done in three stages: decomposition, selection, and reconstruction. At the first stage, the time-series is decomposed into its set of basis time-series. At the selection stage, we select only a subset of these basis time-series to use for reconstruction in the hope that they will have lower noise levels or reveal some underlying property of the original time-series (e.g. its trend). Finally, the selected basis time-series are used to reconstruct the original time-series. If not all basis time-series are selected in the second stage, the final reconstruction will not generate the exact input time-series but a modified version of it that highlights the purpose of the analysis.

The decomposition stage consists of two steps: Hankel matrix embedding and SVD decomposition. Given a time-series X of length T, the Hankel matrix embedding is the matrix H of size $L \times K$ where L is called he lag parameter and must be between 2 and T and $K = T - L + 1$. Row k of of H consists of the subsequence $x_{i,L}$ in our terminology. This means that H is a Hankel matrix (i.e. it has the property that $H_{i,j} = H_{i',j'}$ for $i + j = i' + j'$) and $H \in \mathbb{R}^{K \times L}$.

Consider the following time-series:

$$X = \begin{bmatrix} 0.05 \ 0.49 \ 0.36 \ 0.90 \ 0.98 \ 0.87 \ 0.91 \ 0.84 \ 0.95 \ 0.59 \ -0.13 \ -0.01 \end{bmatrix},$$

and a lag parameter (L) equal to 3, the corresponding Hankel matrix is given by:

$$H = \begin{bmatrix} 0.05 & 0.49 & 0.36 \\ 0.49 & 0.36 & 0.90 \\ 0.36 & 0.90 & 0.98 \\ 0.90 & 0.98 & 0.87 \\ 0.98 & 0.87 & 0.91 \\ 0.87 & 0.91 & 0.84 \\ 0.91 & 0.84 & 0.95 \\ 0.84 & 0.95 & 0.59 \\ 0.95 & 0.59 & -0.13 \\ 0.59 & -0.13 & -0.01 \end{bmatrix}.$$

This Hankel matrix can be created in MATLAB using the hankel() function. Given the Hankel matrix H, we can create the set of basis functions X_i^{ssa} in two steps: SVD decomposition and Hankelization. SVD decomposition consists of simply applying Singular Value Decomposition to the Hankel matrix (e.g. using $svd()$ in Matlab). Mathematically this corresponds to finding $U \in \mathbb{R}^{K \times K}$, $S \in \mathbb{R}^{K \times L}$, $V \in \mathbb{R}^{L \times L}$ where:

$$H = USV^T. \tag{2.51}$$

In our previous example, we get:

$$
U = \begin{bmatrix}
-0.13 & -0.22 & 0.35 & -0.37 & -0.28 & -0.33 & -0.24 & -0.45 & -0.48 & 0.08 \\
-0.26 & -0.26 & -0.54 & -0.09 & -0.32 & -0.15 & -0.32 & 0.10 & 0.13 & -0.56 \\
-0.33 & -0.43 & 0.20 & -0.09 & -0.04 & -0.08 & -0.11 & 0.06 & 0.66 & 0.44 \\
-0.41 & 0.00 & 0.08 & 0.86 & -0.10 & -0.12 & -0.09 & -0.16 & -0.15 & 0.04 \\
-0.41 & 0.05 & -0.17 & -0.15 & 0.84 & -0.15 & -0.15 & -0.14 & -0.10 & -0.05 \\
-0.39 & 0.01 & 0.01 & -0.13 & -0.11 & 0.88 & -0.10 & -0.15 & -0.13 & 0.02 \\
-0.40 & -0.03 & -0.18 & -0.15 & -0.15 & -0.14 & 0.85 & -0.12 & -0.04 & -0.01 \\
-0.35 & 0.15 & 0.29 & -0.12 & -0.07 & -0.10 & -0.05 & 0.81 & -0.29 & 0.02 \\
-0.21 & 0.71 & 0.31 & -0.12 & -0.14 & -0.12 & -0.08 & -0.23 & 0.42 & -0.28 \\
-0.06 & 0.42 & -0.55 & -0.09 & -0.22 & -0.12 & -0.21 & -0.01 & -0.08 & 0.63
\end{bmatrix} ,
$$

$$
S = \begin{bmatrix}
3.92 & 0.00 & 0.00 \\
0.00 & 1.06 & 0.00 \\
0.00 & 0.00 & 0.59 \\
0.00 & 0.00 & 0.00 \\
0.00 & 0.00 & 0.00 \\
0.00 & 0.00 & 0.00 \\
0.00 & 0.00 & 0.00 \\
0.00 & 0.00 & 0.00 \\
0.00 & 0.00 & 0.00 \\
0.00 & 0.00 & 0.00
\end{bmatrix} ,
$$

$$
V = \begin{bmatrix}
-0.57 & 0.73 & -0.37 \\
-0.60 & -0.06 & 0.80 \\
-0.56 & -0.68 & -0.47
\end{bmatrix} .
$$

It can easily be shown that Eq. 2.11 holds for these matrices. The following properties hold for all SVD decompositions assuming that A_i is column i of matrix A, $A_{i,j}$ is element j of vector A_i, and $\langle V, W \rangle$ is the dot product of vectors V and W:

$$
\langle U_i, U_j \rangle = \begin{cases} 1 & i = j \\ 0 & i \neq j \end{cases} , \tag{2.52}
$$

$$
\langle V_i, V_j \rangle = \begin{cases} 1 & i = j \\ 0 & i \neq j \end{cases} . \tag{2.53}
$$

This means that the sets of vectors U_i and V_i each form an orthonormal vector set.

$$
S_{i,j} = \begin{cases} \sqrt{\lambda_i} & i = j \wedge i, j \leqslant \min(K, L) \\ 0 & otherwise \end{cases} , \tag{2.54}
$$

where λ_i is the ith Eigen vector of the matrix HH^T. Notice that at most $\min(K, L)$ elements are nonzero. $S_{i,i}$ are called the singular values of H. We will always assume that $S_{i,i} \geq S_{j,j}$ for $i \geq j$ and will use $S_{i,i}$ and s_i interchangeably. The number of nonzero singular values are denoted d hereafter. In our running example, s_1, s_2 and s_3 are all nonzero which means that $d = 3$. The value for d will always be less than or equal $min\,(K, L)$ and represents the rank of the original matrix H.

The matrix H can then be expressed as a summation ($H = \sum\limits_{i=1}^{d} H^i$) where:

$$H^i = s_i U_i V_i^T. \tag{2.55}$$

The tuple (s_i, U_i, V_i) is called the ith Eigen triple in SSA literature (Hassani 2007). The vectors U_i are called *factor empirical orthogonal functions* or EOFs, and the vectors V_i are called *principal components*.

For our running example, we can calculate H^1, H^2, and H^3 from Eq. 2.55 to be:

$$H^1 = \begin{bmatrix} 0.30 & 0.32 & 0.30 \\ 0.57 & 0.60 & 0.56 \\ 0.74 & 0.77 & 0.73 \\ 0.91 & 0.95 & 0.89 \\ 0.91 & 0.95 & 0.90 \\ 0.87 & 0.90 & 0.85 \\ 0.89 & 0.93 & 0.87 \\ 0.79 & 0.82 & 0.77 \\ 0.47 & 0.49 & 0.46 \\ 0.14 & 0.15 & 0.14 \end{bmatrix},$$

$$H^2 = \begin{bmatrix} -0.17 & 0.01 & 0.16 \\ -0.20 & 0.02 & 0.18 \\ -0.33 & 0.03 & 0.31 \\ 0.00 & -0.00 & -0.00 \\ 0.04 & -0.00 & -0.03 \\ 0.01 & -0.00 & -0.00 \\ -0.02 & 0.00 & 0.02 \\ 0.12 & -0.01 & -0.11 \\ 0.55 & -0.04 & -0.51 \\ 0.32 & -0.03 & -0.30 \end{bmatrix},$$

$$H^3 = \begin{bmatrix} -0.08 & 0.16 & -0.10 \\ 0.12 & -0.25 & 0.15 \\ -0.04 & 0.10 & -0.06 \\ -0.02 & 0.04 & -0.02 \\ 0.04 & -0.08 & 0.05 \\ -0.00 & 0.00 & -0.00 \\ 0.04 & 0.09 & 0.05 \\ -0.06 & 0.13 & -0.08 \\ -0.07 & 0.14 & -0.08 \\ 0.12 & -0.26 & 0.15 \end{bmatrix}.$$

Notice for now that all of these matrices are not Hankel matrices. From linear Algebra, it is known that the Frobenius norm of a matrix can be found from its singular values using the following equation:

$$\|H\|^2 = \sum_{i=1}^{K}\sum_{j=1}^{L}\left(H_{i,j}\right)^2 = \sum_{i=1}^{d}\lambda_i = \sum_{i=1}^{d}s_i^2. \tag{2.56}$$

For our running example the Frobenius norm of H calculated any way of the three ways in Eq. 2.56 will equal 16.8045. Moreover, we know that each one of the expansion matrices (H^i) is a rank one matrix (See how they are created according to Eq. 2.55). This means that they will have a single nonzero singular value which equals s_i because both U_i and V_i are unit vectors. This leads to:

$$\left\|H^i\right\| = s_i. \tag{2.57}$$

From Eqs. 2.56 and 2.57, it is straight forward to see that:

$$\|H\|^2 = \sum_{i=1}^{d}\left\|H^i\right\|^2. \tag{2.58}$$

If we define ζ_i as the contribution of expansion matrix H^i to H, it is easy to see that

$$\zeta_i = \frac{s_i^2}{\sum_{j=1}^{d}s_j^2}. \tag{2.59}$$

In our running example, $\zeta_1 = 0.9129$, $\zeta_2 = 0.0667$, $\zeta_3 = 0.0204$ which means that 91.29 % of the total variance in the data is captured in the first expansion matrix. Notice that SVD is optimal in the sense that taking the K columns of U and V corresponding to the top K singular values, we can reconstruct $\bar{H}$ with the guarantee that $\|H - \bar{H}\|$ is minimum compared to any other linear decomposition. This means that for any value K, and assuming that the singular values are ordered in descending order, it is guaranteed that $\left\|H - \sum_{k=1}^{K}H^k\right\|$ is minimal compared to any other linear decomposition of H to matrices.

Giving the decomposition of H into the set $\{H^i\}$ and their weights ζ_i, we can then generate approximations of H by just grouping together some of the expansion matrices using simple summation. This leads to the the following two-steps reconstruction process.

The first step of reconstruction is to group the expansion matrices then summing the matrices in each group. Given the d nonzero expansion matrices, we generate a set of sets representing the groups: $\{I_1, I_2, \ldots, I_g\}$ where $1 \leq g \leq d$ is the number of groups and each I_i is a set $\{i_1, \ldots, i_{n_i}\}$ where $1 \leq i_j \leq d$ and $1 \leq j \leq n_i$. The sets I_j are disjoint. Each one of these groups should correspond to some meaningful component of the original time-series. This process is called Eigen-triple grouping. The expansion matrices of all members of a group are added to generate a matrix representing the group using Eq. 2.60. The contributions of all member expansion matrices in a group are simply added to get the total contribution of the group according to Eq. 2.61.

$$H^{I_i} = \sum_{j \in I_i} H^j. \tag{2.60}$$

$$\zeta_{I_i} = \sum_{j \in I_i} \zeta_j. \tag{2.61}$$

In our running example, we can create two groups $\{I_1 = \{1, 2\}, I_2 = \{3\}\}$ with the following corresponding expansion matrices:

$$H^{I_1} = \begin{bmatrix} 0.13 & 0.33 & 0.46 \\ 0.38 & 0.61 & 0.75 \\ 0.41 & 0.80 & 1.04 \\ 0.91 & 0.95 & 0.89 \\ 0.95 & 0.95 & 0.86 \\ 0.87 & 0.90 & 0.85 \\ 0.87 & 0.93 & 0.89 \\ 0.91 & 0.81 & 0.66 \\ 1.01 & 0.44 & -0.05 \\ 0.47 & 0.12 & -0.16 \end{bmatrix},$$

$$H^{I_2} = \begin{bmatrix} -0.08 & 0.16 & -0.10 \\ 0.12 & -0.25 & 0.15 \\ -0.04 & 0.10 & -0.06 \\ -0.02 & 0.04 & -0.02 \\ 0.04 & -0.08 & 0.05 \\ -0.00 & 0.00 & -0.00 \\ 0.04 & -0.09 & 0.05 \\ -0.06 & 0.13 & -0.08 \\ -0.07 & 0.14 & -0.08 \\ 0.12 & -0.26 & 0.15 \end{bmatrix}.$$

The final step of reconstruction is to convert the matrices corresponding to groups into time-series that comprise the final expansion of the input time-series. To achieve that, the expansion matrix corresponding to each group (H^{I_i}) is converted into a Hankel matrix by averaging all members that have row and column numbers summing to the same integer. In our running example this leads to the following two hankelized matrices:

$$\bar{H}^{I_1} = \begin{bmatrix} 0.13 & 0.35 & 0.49 \\ 0.35 & 0.49 & 0.82 \\ 0.49 & 0.82 & 0.98 \\ 0.82 & 0.98 & 0.90 \\ 0.98 & 0.90 & 0.88 \\ 0.90 & 0.88 & 0.89 \\ 0.88 & 0.89 & 0.91 \\ 0.89 & 0.91 & 0.52 \\ 0.91 & 0.52 & 0.04 \\ 0.52 & 0.04 & -0.16 \end{bmatrix},$$

$$\bar{H}^{I_2} = \begin{bmatrix} -0.08 & 0.14 & -0.13 \\ 0.14 & -0.13 & 0.08 \\ -0.13 & 0.08 & 0.01 \\ 0.08 & 0.01 & -0.03 \\ 0.01 & -0.03 & 0.03 \\ -0.03 & 0.03 & -0.05 \\ 0.03 & -0.05 & 0.04 \\ -0.05 & 0.04 & 0.06 \\ 0.04 & 0.06 & -0.17 \\ 0.06 & -0.17 & 0.15 \end{bmatrix}.$$

It is now trivial to read-off the corresponding two expansion time-series from the hankelized matrices which leads to:

$$X^1 = \begin{bmatrix} 0.13 & 0.35 & 0.49 & 0.82 & 0.98 & 0.90 & 0.88 & 0.89 & 0.91 & 0.52 & 0.04 & -0.16 \end{bmatrix},$$

$$X^2 = \begin{bmatrix} -0.08 & 0.14 & -0.13 & 0.08 & 0.01 & -0.03 & 0.03 & -0.05 & 0.04 & 0.06 & -0.17 & 0.15 \end{bmatrix}.$$

It can easily be confirmed that the summation of these two time-series gives the original time-series. Notice that X^2 is very near to zero and fluctuates between small negative and positive numbers. This corresponds to an estimation of the noise on the time-series. The actual noise in this case was:

$$\begin{bmatrix} 0.05 & 0.18 & -0.23 & 0.09 & 0.03 & -0.13 & -0.04 & 0.03 & 0.36 & 0.28 & -0.13 & 0.30 \end{bmatrix}.$$

We can compare the Euclidean norm of the difference between the original noisy time-series and its ground truth to be 0.6567 while the Euclidean norm between the first expansion time-seris (X^1) and the ground-truth is only 0.3331 which means that SSA analysis was able to remove around 49 % of the noise in the original time-series. Figure 2.11 shows the ground-truth, original time-series, and the time-series resulting from hankelizing H^{I_1} which is SSA's best approximation to the ground truth.

Even though the example we used in this section was small with very small T and L values, the extension to longer time-series and larger Hankel matrices is trivial.

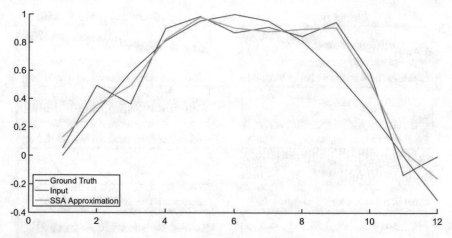

Fig. 2.11 Example time-series and its corresponding SSA analysis. See text for details

The problem in this case is that the number of non-zero singular values d increases dramatically with increased L. In most cases, many of these singular values will be zeros or very near to zeros and they can be ignored as corresponding to noise in the input (similar to s_3 in our running example). The remaining expansion matrices then will need to be grouped intelligently to generate meaningful expansion. There is no optimal grouping criterion that works for all applications and in this book we mostly will use simple grouping strategies like grouping the first expansion matrices with total weights over some threshold together in a single group and ignoring everything else.

A slightly more complex example is shown in Figs. 2.12 and 2.13. The ground truth time-series consists of a linear trend and two sinusoidal as defined by Eq. 2.62 which is shown in Fig. 2.12a (Top–left).

$$x(t) = \sin(0.02t\pi) + 2\sin(0.2t\pi) + 0.02t \qquad (2.62)$$

Gaussian noise with zero mean and unit variance is then added to generate the input signal to the SSA algorithm as shown in Fig. 2.12b (Top–right). The major features of the ground truth signal including the trend and the oscillating components are still preserved. We then applied SSA with a lag parameter of 50 to this time-series. The top six expansion time-series (without grouping) are shown in Fig. 2.12 (second to last rows) along with their weights. The trend is visible in the first expansion time-series. The high frequency sinusoidal is clear in the second and third components and the low frequency sinusoidal (despite being distorted) is visible in the fourth expansion series. notice that the sixth expansion time-series has much lower peak-to-peak (PP) value compared with the five before it and smaller weight signaling that it is a noise component.

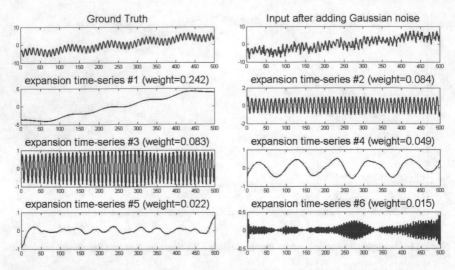

Fig. 2.12 Example time-series and its corresponding SSA analysis. See text for details

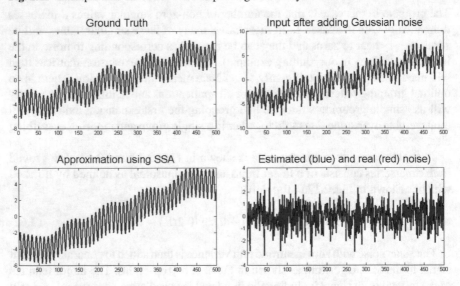

Fig. 2.13 Example time-series and its corresponding SSA approximation using two groups (signal and noise approximations). See text for details

We then used the very simple grouping strategy of combining all expansion matrices with weights summing up to 45 % of the total in an approximation of the input time-series and combine the rest in another time-series estimating the noise. These are the two signals shown in Fig. 2.13's second row.

It is clear that the SSA approximation could recover most of the information in the ground-truth while eliminating the additive Gaussian noise. To quantify this intuition, we calculated the Euclidean distance between the ground truth and the input time-series to be 21.0518. The Euclidean distance between the approximated signal and the ground truth is only 11.2344. This suggests that the SSA analysis could, again, eliminate around half of the distance introduced by the noise.

In general it is advisable to use the largest possible lag parameter value less than half the time-series (in order not to have more columns than rows in H). This shows one problem with SSA analysis. The complexity of most implementations of SVD is cubic in matrix dimensions when they are similar which is the case in SSA. This means that for long time-series, SSA application may be prohibitively time-consuming. Several approaches have been proposed to deal with this problem but they are outside the scope of this book.

Extending SSA to multidimensional time-series is straight forward. We simply stack the Hankel matrices of all dimensions together to generate a combined Hankel matrix. This matrix is then fed to the rest of the algorithm as it is without any need to modify any parts except the hankelization process which now averages the corresponding parts of each dimension alone. This algorithm is called SSAM hereafter. SSAM is not the same as applying SSA to each dimension and combining the results because V will be shared between all the dimensions.

Figure 2.14 shows a time-series generated from a markov chain with Gaussian noise added and the first few expansion time-series using SSAM. Figure 2.15 shows application of repeated SSA to the two input dimensions of the same data used in Fig. 2.14. Even though the expansion time-series seem very similar, they are not identical. We can confirm that be comparing the result of grouping the first expansion matrices corresponding to a total weight of 0.75 in both cases as shown in the bottom row of Fig. 2.16. Did SSAM improve the results of repeated SSA? To answer this question we calculated the Euclidean distance between the ground truth and both

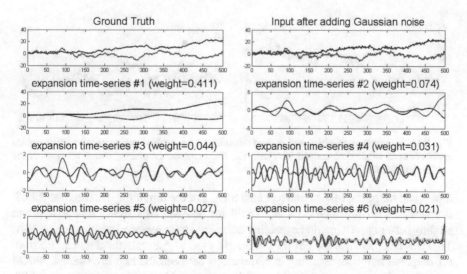

Fig. 2.14 Example time-series and its corresponding SSAM analysis. See text for details

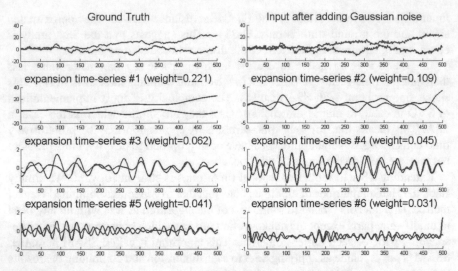

Fig. 2.15 Example time-series and its corresponding repeated applications of SSA analysis. The *two colors* represent two dimensions of the time-series. See text for details

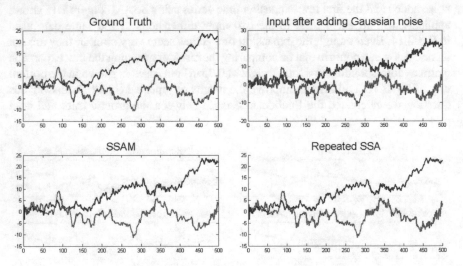

Fig. 2.16 Example time-series and its corresponding SSA analysis using SSAM and repeated applications of single dimensional SSA analysis. The *two colors* represent two dimensions of the time-series. See text for details

inputs and output of SSAM and repeated SSA. For the input, the distance was 22.45. It was 17.24 for repeated SSA and 16.39 for SSAM. This shows that SSAM had slightly better approximation capability compared with repeated application of SSA. This was true because the covariance matrix used to generate the data had nonzero off-diagonal numbers resulting in correlations between the two-time series. SSAM can utilize these correlation because of the combined V matrix while repeated SSA application cannot.

From this example, it is clear that SSAM is expected to work well when the dimensions of the input time-series are correlated otherwise, repeated application of SSA may achieve higher approximation accuracy.

2.4 Learning Time-Series Models from Data

In data mining applications in general and in social robotics in particular, we rarely have access to a generation model in one of the simple mathematical forms presented in Sect. 2.2. In most cases, we only have data generated—or can be assumed to be generated—from one of these models and are interested in estimating the generation model either for predicting future behavior or simply to understand the generation process. This section will focus on methods for learning the generation model assuming that it falls under one of the most important models presented in Sect. 2.4. Most of the information given here is standard knowledge for practitioners of time-series analysis, pattern recognition or machine learning and for this reason we only briefly describe the methods without delving into proofs or the correctness and soundness of the algorithms involved.

2.4.1 Learning an AR Process

Recall that an AR process is described by the following equation:

$$x_t = a_0 + \sum_{i=1}^{m} a_i x_{t-i} + \varepsilon_t, \tag{2.63}$$

where $\varepsilon_t \, \mathcal{N}\left(0, \sigma^2 I\right)$ is a Gaussian noise variable.

The constant value a_0 can be estimated by the mean of the time-series and we will assume that it is zero without loss of generality hereafter.

The simplest way to find the set of parameters a_i is to use least squares. Let $\theta = [-a_p, -a_{m-1}, \dots, -a_2, -a_1, 1]'$, $\varepsilon = [\varepsilon_1, \varepsilon_2, \dots, \varepsilon_T]$ and $A' = [x_{t-m:t}]_{t=m+1}^{T}$, and $B = [x_p : x_T]$, then Eq. 2.63 can be be summarized for every point in the time-series as:

$$A\theta = B. \tag{2.64}$$

This equation can be solved using standard least-squares to estimate θ. Moreover, the residuals from comparing the signal generated by using the learned θ in Eq. 2.63 to the original time-series can be used to estimate the variance of the white noise component (σ^2).

This approach is used in *learnARLS()* to learn the parameters of an AR process given an input time-series and an estimate of the system order.

The problem with this approach is that it requires the solution of T equations on only m unknowns which may not scale well with the length of the time-series.

Least squares estimation in general reduces the squared distance between the prediction from the learned model and the input data. This can be captured by the following unconditional optimization problem:

$$\min_a J(a) = \min_a \sum_{t=-\infty}^{\infty} \left(x_t - \sum_{i=1}^{m} a_i x_{t-i} \right)^2. \qquad (2.65)$$

To solve this problem for each parameter a_j, we simply find the point at which the partial derivative of J with respect to a_j vanishes.

$$\frac{\partial J}{\partial a_j} = \frac{\partial}{\partial a_j} \sum_{t=-\infty}^{\infty} \left(x_t - \sum_{i=1}^{m} a_i x_{t-i} \right)^2,$$

$$\frac{\partial J}{\partial a_j} = \sum_{t=-\infty}^{\infty} \frac{\partial}{\partial a_j} \left(x_t - \sum_{i=1}^{m} a_i x_{t-i} \right)^2, \qquad (2.66)$$

$$\frac{\partial J}{\partial a_j} = \sum_{t=-\infty}^{\infty} -2x_{t-j} \left(x_t - \sum_{i=1}^{m} a_i x_{t-i} \right).$$

Setting $\frac{\partial J}{\partial a_j} = 0$, we get:

$$\frac{\partial J}{\partial a_j} = \sum_{t=-\infty}^{\infty} 2x_{t-j} \left(x_t - \sum_{i=1}^{m} a_i x_{t-i} \right) = 0,$$

$$\sum_{t=-\infty}^{\infty} x_{t-j} x_t - \sum_{t=-\infty}^{\infty} x_{t-j} \left(\sum_{i=1}^{m} a_i x_{t-i} \right) = 0, \qquad (2.67)$$

$$\sum_{t=-\infty}^{\infty} x_{t-j} x_{t-0} - \sum_{i=1}^{m} a_i \sum_{t=-\infty}^{\infty} x_{t-i} x_{t-j} = 0.$$

Defining $A_i = -a_i$ for $1 \le i \le m$ and $A_0 = 1$, and defining $l = i - j$ we get:

$$\sum_{i=0}^{m} A_i \sum_{t=-\infty}^{\infty} x_t x_{t+l} = 0. \qquad (2.68)$$

Now we notice that the internal summation is the autocorrelation coefficient ρ for different delays l. This means that the Eq. 2.68 can be written as:

$$\sum_{i=0}^{m} A_i \rho_l = 0. \tag{2.69}$$

We do not have access to ρ_l in Eq. 2.69 but we can estimate it using:

$$r_l = \sum_{t=1}^{T-l} x_t x_{t+l}. \tag{2.70}$$

Substituting r_l for ρ_l in Eq. 2.69, we get the famous Yule–Walker equations. These are M equations in M unknowns and their solution scales well with the length of the time-series and can be written as:

$$RA = B, \tag{2.71}$$

where:

$$R = \begin{bmatrix} r_0 & r_{-1} & \cdots & r_{1-m} \\ r_1 & r_0 & \cdots & r_{2-m} \\ \vdots & \vdots & \ddots & \vdots \\ r_{m-1} & r_{m-2} & \cdots & r_1 \end{bmatrix},$$

$$A = [a_1, a_2, \ldots, a_m]',$$

$$B = [r_1, r_2, \ldots, r_m].$$

This system of equations can be solved efficiently using the Levinson-Durbin recursion leading to an estimate for a. This procedure is implemented in the function *learnARYuleWalker*() in the toolbox.

Figure 2.17 shows the results of applying least squares and Yule–Walker approaches to learning the parameters of a time-series generated from an AR model. The original model had the parameter vector: $a = [2.7607, -3.8106, 2.6535, -0.9238]$. The model learned from least squares was $a_{ls} = [2.7746, -3.8419, 2.6857, -0.9367]$, while the model learned from Yule–Walker equations was: $a_{yw} = [2.7262, -3.7296, 2.5753, -0.8927]$. Comparing the parameter vectors directly we get $a.a_{ls} = 0.0487$ and $a.a_{yw} = 0.1218$. It appears that the least-squares estimator could provide smaller error in terms of parameter values. Figure 2.17 shows also that the Yule–Walker approach gives higher error *in terms of the ability to predict the actual values of the time-series* which is what we usually care about. This is quantified by finding the Euclidean distance between the original time-series and the one predicted using the learned parameters (with the same *random* Gaussian noise) which leads to 39.5113 for the Yule–Walker approach compared with only 38.7442 for the least squares solution.

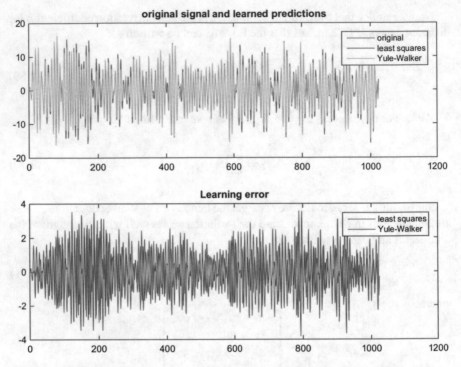

Fig. 2.17 Results of learning the parameters of an AR process using both least squares and Yule–Walker approaches then generating an estimate of the time-series from the learned model

2.4.2   Learning an ARMA Process

Learning the parameters of an AR process using least squares or the Yule–Walker approach was a simple exercise. ARMA processes on the other hand provide a more challenging problem because we need to fit not only the parameter vector a but also b and the MA part of the process makes noise values at different time-steps correlated which renders least squares like solutions inappropriate for the problem.

The simplest approach to solve this problem is two-stages regression. We start by assuming that the data is from an $AR(\hat{m})$ process instead of an $ARMA(m, n)$ process where $\hat{m} > m$. This means that we assume that information about the Gaussian noise part is encoded in the longer AR process directly in the values of the time-series. This can be seen by rearranging Eq. 2.9 as follows:

$$\theta_t = \sum_{i=1}^{m} a_i x_{t-i}^{arma} - x_t^{arma} + \sum_{i=1}^{n} b_i \theta_{t-i}, \tag{2.72}$$

$$\theta_{t-1} = \sum_{i=1}^{m} a_i x_{t-i-1}^{arma} - x_{t-1}^{arma} + \sum_{i=1}^{n} b_i \theta_{t-i-1}, \tag{2.73}$$

$$\vdots$$

$$\theta_{t-n} = \sum_{i=1}^{m} a_i x_{t-i-n}^{arma} - x_{t-n}^{arma} + \sum_{i=1}^{n} b_i \theta_{t-i-n}. \tag{2.74}$$

This set of equations show that past values of x carries information about the Gaussian noise. We use $\hat{m} = m + n$ for our implementation.

After fitting the time-series using $AR(\hat{m})$ and finding the parameter vector $\hat{a}$, we use the fitted model to predict the time-series using:

$$x_t^{fit} = \sum_{i=1}^{\hat{m}} \hat{a}_i x^{fit_{t-i}}. \tag{2.75}$$

An estimate of the Gaussian noise can then be found as:

$$\hat{\theta}_t = x_t^{arma} - x_t^{fit}. \tag{2.76}$$

We now solve another regression problem (similar to the one used to fit the AR model) but with estimates of $\theta_{1:T}$ now incorporated into the linear model $A\theta = B$ where:

$$A = \begin{bmatrix} x_{r-1} & x_{r-2} & \cdots & x_{r-m} & \hat{\varepsilon}_{r-1} & \hat{\varepsilon}_{r-2} & \cdots & \hat{\varepsilon}_{r-n} \\ x_r & x_{r-1} & \cdots & x_{r-m+1} & \hat{\varepsilon}_r & \hat{\varepsilon}_{r-1} & \cdots & \hat{\varepsilon}_{r-n+1} \\ \vdots & \vdots & \ddots & \vdots & \vdots & \vdots & \ddots & \vdots \\ x_{T-1} & x_{T-2} & \cdots & x_{T-m} & \hat{\varepsilon}_{T-1} & \hat{\varepsilon}_{T-2} & \cdots & \hat{\varepsilon}_{T-m} \end{bmatrix}, \tag{2.77}$$

$$\theta = [a_1, a_2, \ldots a_m, b_1, b_2, \ldots, b_n]', \tag{2.78}$$

$$B = [x_r - \hat{\varepsilon}_r, x_{r+1} - \hat{\varepsilon}_{r+1}, \ldots, x_T - \hat{\varepsilon}_T]'. \tag{2.79}$$

Solving this linear system leads to an estimate of the parameter vectors a and b. Figure 2.18 shows the results of applying this procedure to a time-series generated from an $ARMA(3, 5)$ model with $a = [-0.7, 0.5, 0.9]$ and $b = [1, 2, 3, 2, 1]$.

Even thought the accuracy is less than the case for the AR process shown in Fig. 2.17, this solution still captures the main characteristics of the time-series rising and falling with it with a correlation coefficient of 0.9356. This solution can also be used as an initial solution for a local search method (e.g. gradient descent on the log-

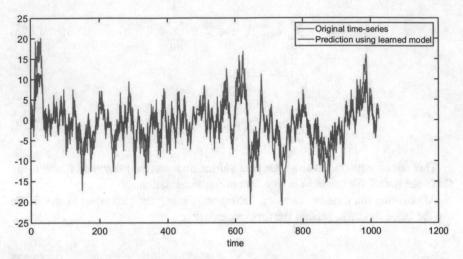

Fig. 2.18 Results of learning the parameters of an ARMA process using two-stages regression then generating an estimate of the time-series from the learned model

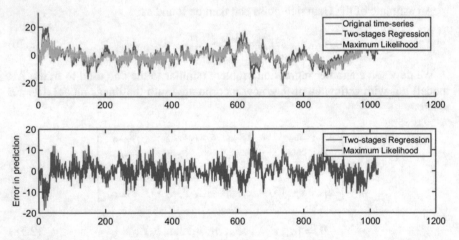

Fig. 2.19 Results of learning the parameters of an ARMA process using two-stages regression and maximum likelihood then generating an estimate of the time-series from the learned model

likelihood function) to find a better solution. Two-stages regression is implemented in the MC^2 toolbox using the function *learnARMALS*().

Matlab's Econometrics toolbox has an implementation of the maximum likelihood estimator for ARMA model parameters in the function *estimate*(). Figure 2.19 shows the results of using two-stage regression and maximum likelihood to learn the same data presented in Fig. 2.18.

Another approach for estimating the parameters of an ARMA process is to convert it into a linear state-space model and use a Kalman filter to estimate the state of the system (which corresponds to the time-series values without the MA part). These can then be used to estimate a. To estimate b, we use the residues of the first modeling step.

2.4.3 Learning a Hidden Markov Model

Hidden Markov Models (HMM) are widely utilized in speech recognition, gesture recognition and—as we will see in Chap. 13—in learning from demonstration. In this section we will focus on HMMs with Gaussian observation distributions (GHMMs) defined as in Sect. 2.2.8.

A GHMM is completely specified by a tuple $\{\pi, A, \mu_1, \mu_2, \ldots, \mu_N, \Sigma_1, \Sigma_2, \ldots, \Sigma_N\}$ where π is a $N \times 1$ vector specifying prior probabilities for the first hidden state, A is a $N \times N$ matrix where A_{ij} specifying the transition probabilities from state i to state j, and μ_n and Σ_n specify a Gaussian distribution from which the time-series value is sampled when the GHMM is in state n for $1 \leq n \leq N$. The full specification of a GHMM is given in Sect. 2.2.8 and repeated in Eq. 2.80 for convenience.

$$
\begin{aligned}
s_0 &\sim p(s_0) \equiv \pi, \\
s_t &\sim p(s_t|s_{t-1}) \equiv A_{s_{t-1}}{}^T, 0 < t \leqslant T, \\
x_t^{ghmm} &\sim p\left(x_t^{ghmm}|s_t\right) \equiv \mathcal{N}\left(\mu_{s_t}, \Sigma_{s_t}\right).
\end{aligned}
\tag{2.80}
$$

Now, given a time-series $X = (x_0, x_1, \ldots, x_{T-1})$, we would like to learn the parameters of the GHMM generating it. Collecting all the parameters in a single vector θ, this problem can be casted as a maximum likelihood problem of the form:

$$
\max_{\theta} J(\theta) = \max_{\theta} p(X|\theta).
$$

This problem can be solved efficiently using the Baum–Welch forward-backward algorithm which is a form of expectation maximization that takes advantage of the independence relationships implicit in the definition of HMMs to achieve linear time estimation of model parameters.

Because it is an Expectation Maximization algorithm, it requires an initial model $\lambda = (\pi, A, \mu_{1:N}, \Sigma_{1:N})$. Using this model, we will find $p(s_t = n|X, \lambda)$ for $0 \leq t \leq T - 1$ and $1 \leq n \leq N$ which is the probability of having every possible state n at every possible time-step t given that we have observed X and assuming the model λ. This is the expectation step. Given these probabilities, it is straightforward to find an estimate of the model parameters λ_+ that achieves higher likelihood in the maximization step. Iterating these two steps, it is guaranteed that λ_+ will converge to a local maximum of the likelihood function J.

Let's consider the expectation step. Our goal is to calculate $p(s_t = n|X, \lambda)$ efficiently. Firstly we notice that:

$$p(s_t = n|X, \lambda) \equiv \gamma_t(n) = p(s_t = n, x_{0:t}|\lambda) \, p(x_{t+1:T}|s_t = n, \lambda). \qquad (2.81)$$

Equation 2.81 is true due to the Markovian property of HMMs (i.e. s_{t+1} is independent of everything given s_t). Defining $\alpha_t(n) \equiv p(s_t = n, x_{0:t}|\lambda)$ and $\beta_t(n) \equiv p(x_{t+1:T}|s_t = n, \lambda)$, Eq. 2.81 can be written as:

$$\gamma_t(n) \propto \alpha_t(n) \circ \beta_t(n), \qquad (2.82)$$

where $\circ$ is the element wise multiplication operator. Now the expectation step reduces to the problem of calculating α and β.

Consider α, we know that:

$$\alpha_0(n) = p(s_0 = n|x_0, \lambda) = \frac{p(s_0 = n, x_0|\lambda)}{p(x_0|\lambda)}, \qquad (2.83)$$

$$\therefore \alpha_0(n) = \frac{p(s_0 = n|\lambda) \, p(x_0|s_0 = n, \lambda)}{p(x_0|\lambda)}, \qquad (2.84)$$

$$\therefore \alpha_0(n) = \frac{\pi_n \mathcal{N}(x_0; \mu_n, \Sigma_n)}{p(x_0|\lambda)}. \qquad (2.85)$$

Defining $\rho_t(n) \equiv \mathcal{N}(x_t; \mu_n, \Sigma_n)$, Eq. 2.85 can be written as:

$$\alpha_0(n) = \frac{\pi_n \rho_t(n)}{p(x_0|\lambda)}. \qquad (2.86)$$

Now consider the general case (i.e. $\alpha_t(n)$):

$$\alpha_t(n) = p(s_t = n|x_{0:t}, \lambda) = \frac{p(s_t = n, x_{0:t}|\lambda)}{p(x_{0:t}|\lambda)}.$$

$$\therefore \alpha_t(n) \propto p(s_t = n, x_t, x_{0:t-1}|\lambda),$$

$$\therefore \alpha_t(n) \propto p(s_t = n, x_{0:t-1}|x_t, \lambda) \, p(x_t|s_t = n, x_{0:t-1}, \lambda),$$

$$\therefore \alpha_t(n) \propto p(s_t = n, x_{0:t-1}|x_t, \lambda) \, p(x_t|s_t = n, \lambda),$$

$$\therefore \alpha_t(n) \propto \beta_t(n) \sum_{i=1}^{N} p(s_t = n, s_{t-1} = i, x_{0:t-1}|x_t, \lambda),$$

$$\therefore \alpha_t(n) \propto \beta_t(n) \sum_{i=1}^{N} p(s_t = n, s_{t-1} = i, \lambda) \, p(s_{t-1} = i, x_{0:t-1}|\lambda).$$

This can be written as:

$$\alpha_t(n) \propto \beta_t(n) \sum_{i=1}^{N} A_{in} \alpha_{t-1}(i). \qquad (2.87)$$

But we know that $\sum \alpha_t(n)$ must equal 1. This leads to the final estimation equation for $\alpha_t(n)$:

$$\alpha_t(n) = \frac{\beta_t(n) \sum_{i=1}^{N} A_{in}\alpha_{t-1}(i)}{\sum_{j}^{N} \beta_t(j) \sum_{i=1}^{N} A_{ij}\alpha_{t-1}(i)}. \tag{2.88}$$

This derivation shows that we can find $\alpha_t(n)$ for any t and n within their respective ranges given A, $\alpha_{t-1}(1:N)$, and $\rho_t(1:N)$. This suggests a simple recursion starting by finding $\alpha_0(n)$ for $1 \le n \le N$ using Eq. 2.86. We then use Eq. 2.88 for $1 \le t \le T$ and $1 \le n \le N$. This is called the forward recursion.

The second part of Eq. 2.81 can be estimated using a similar procedure but now going backward in the time-series according to the following two equations:

$$\beta_T(n) = 1, \tag{2.89}$$

$$\beta_t(n) = \sum_{i=1}^{N} A_{ni}\rho_{t+1}(i)\beta_{t+1}(i). \tag{2.90}$$

Using these two equations, $\beta_t(n)$ can be calculated backward starting from $\beta_T(n)$. Now having calculated both α and β, it is easy to calculate γ as follows:

$$\gamma_t(n) = \frac{\alpha_t(n)\beta_t(n)}{\sum_{i=1}^{N} \alpha_t(i)\beta_t(i)}. \tag{2.91}$$

Now $\gamma_t(n)$ gives us an estimate of the probability of being at state n at time step t given the observed time-series X and the initial model λ. Now that we have an estimate of the *hidden* state at every time-step, we can try to re-estimate the model parameters.

A useful quantity for the maximization step is the probability of transiting from state i at time-step t to state j at time-step $t+1$ (notice that marginalizing this gives an estimate of A_{ij}). This quantity is defined as:

$$\zeta_t(i,j) \equiv p(s_t = i, s_{t+1} = j | x_{0:T-1}, \lambda).$$

It can easily be shown that ζ can be calculated as:

$$\zeta_t(i,j) = \frac{\alpha_t(i)A_{ij}\rho_{t+1}(j)\beta_{t+1}(j)}{\sum_{l=1}^{N} \sum_{k=1}^{N} \alpha_t(k)A_{kl}\rho_{t+1}(l)\beta_{t+1}(l)}. \tag{2.92}$$

Given these estimates of ζ and γ, we can easily estimate the model $\lambda^+ = \left(\mu^+, A, \mu_{1:N}^+, \Sigma_{1:N}^+\right)$ using:

$$\pi^+(n) = \gamma_0(n), \tag{2.93}$$

$$A_{ij}^+ = \frac{\sum_{t=0}^{T-2} \zeta_t\,(i,j)}{\sum_{t=0}^{T-2} \gamma_t\,(i)}, \tag{2.94}$$

$$\mu_i^+ = \frac{\sum_{t=0}^{T-1} x_t\gamma_t\,(i)}{\sum_{t=0}^{T-1} \gamma_t\,(i)}, \tag{2.95}$$

$$\Sigma^+{}_i = \frac{\sum_{t=0}^{T-2} \gamma_t\,(i)\,\left(x_t - \mu_i^+\right)\left(x_t - \mu_i^+\right)^T}{\sum_{t=0}^{T-1} \gamma_t\,(i)}. \tag{2.96}$$

This process can be repeated until λ^+ does not differ much from λ or its likelihood is not different from that of λ or until a predefined number of iterations is reached. The aforementioned procedure is implemented in the function *learnHMM*() in the MC^2 toolbox. When multiple time-series are available that are believed to be from the same GHMM, a slightly modified version of this procedure can be implemented to learn the HMM parameters from all of the input time-series. This procedure is implemented in the function *learnHMMMulti*().

2.4.4 Learning a Gaussian Mixture Model

Learning the parameters of a GMM from input time-series is conceptually very similar to learning the parameters of GHMMs using Expectation Maximization. The main idea is to estimate the responsibility of every Gaussian for the time-series values at every sample in the expectation step $r_t\,(k)$ using:

$$r_t\,(k) = \frac{\pi_k \mathcal{N}\,(x_t;\,\mu_k,\,\Sigma_k)}{\sum_{i=1}^{K} \pi_i \mathcal{N}\,(x_t;\,\mu_i,\,\Sigma_i)}. \tag{2.97}$$

The maximization step can be summarized as:

$$\pi^+\,(k) = \frac{\sum_{t=1}^{T} r_t\,(t)}{K}, \tag{2.98}$$

$$\mu_k^+ = \frac{\sum_{t=1}^{T} r_t\,(t)\,x_t}{\sum_{t=1}^{T} r_t\,(t)}, \tag{2.99}$$

$$\Sigma_k^+ = \frac{\sum_{t=1}^{T} r_t\,(t)\,\left(x_t - \mu_k^+\right)\left(x_t - \mu_k^+\right)^T}{\sum_{t=1}^{T} r_t\,(t)}. \tag{2.100}$$

These two steps are repeated a predefined number of times or until the likelihood is not changing anymore. This algorithm is implemented in the *learnGMM*() function of the toolbox.

2.4.5 Model Selection Problem

Learning the parameters of a generating model (despite the type of this model) usually requires an assumption about the model *complexity*. For example, to learn the parameters of an $ARMA(m, n)$ process we need some assumption about m and n and to learn a GHMM, we need to know the number of states N. Selecting the complexity of the model is a ubiquitous problem in data mining and there are several known approaches to deal with it.

The simplest approach is K-fold cross validation. The training data (the time-series in our case) is divided to K equally sized partitions. The system is trained on $K - 1$ partitions and its performance is tested on the remaining one. This process is repeated K times with a different testing partition each time. This process is repeated for the set of complexities to be tested (e.g. values for K in HMM learning) and the value the achieves best predictive performance is then selected.

Another approach that is used widely is Bayesian Information Criteria (BIC). In this case, a statistic is calculated by adding the negative log–likelihood of different models to another term measuring the complexity of the system. Rather than selecting the system complexity that maximizes the likelihood (which is prone to overfitting), we take the fact that more complex systems can in general achieve higher likelihood values due to their tendency to overfit the data and model not only the system dynamics but the noise corrupting it. The complexity level that minimizes the BIC statistic is selected instead of the maximum likelihood statistic.

2.5 Time Series Preprocessing

Before any mining algorithm is applied to time-series data, preprocessing may be required to remove artifacts, or enhance the quality of the data in some way. Several preprocessing operations exist and for each of which there can be several alternative algorithms. Here, we discuss the most useful of these operations for our purposes in this book and some simple algorithms to achieve them keeping in mind application to social robotics.

2.5.1 Smoothing

A very common problem with collected time-series specially from real-world sensors is the contamination with high frequency noise that may cause problems to modeling algorithms. For example, piecewise linear approximations of time series (that we will use several times in this book) may suffer from over-segmentation (i.e. generating too many lines) in the face of such noise. A simple approach to reduce the effect of this kind of noise is smoothing. Several algorithms exist for smoothing of time-series data but in most cases the simple moving average approach will do.

2.5.2 Thinning

Thinning is the process of keeping only local maxima of the time-series. This process can be used as a compression technique by keeping a sparse representation of the time-series in question. In this book we use thinning as a postprocessing step in the *RSST* change point discovery algorithm (See Sect. 3.5).

A very simple thinning algorithm can be implemented by noticing the first difference of the data. Two rules are applied in order given some positive small number δ: if $x_t - x_{t-1} > \delta$ then set x_{t-1} to zero. if $x_t - x_{t-1} < -\delta$ then set x_t to zero.

More sophisticated approaches exist. For example, a piecewise linear approximation of the time series can be generated then only the points at which the line slopes changes from positive to negative are kept.

2.5.3 Normalization

In many cases, it is necessary to keep the range of values in a time-series within some range or make these ranges similar for different dimensions of the time-series. Again several approaches can be thought of for this problem but we will focus on two simple and widely used approaches.

First of all, we may just want to remove any constant component of the time-series because for example it does not contribute to the information content. This can easily be achieved by removing the mean of the time-series from each point. This approach is applicable to both single dimensional and multidimensional time-series. Defining $\mu(X)$ to be the mean of a time-series along its independent dimension, we can state this simply as:

$$\bar{x}_t = x_t - \mu(X).$$
(2.101)

If the scale is also to be removed, we can use the following general formula:

$$\bar{x}_t = \frac{x_t - \mu(X)}{S},$$
(2.102)

where S represents the scale which can be either the range of the time-series (i.e. $max(X) - min(X)$) or its standard deviation $\sigma(X)$. Equation 2.102 assumes that X has a single dimension. A generalization to the multidimensional case can be defined as:

$$\bar{x}_t = C^{-1}(x_t - \mu(X)),$$
(2.103)

where C is the covariance matrix calculated as XX^T assuming x_t are column vectors.

2.5.4 De-Trending

Referring to the xLAT model of time-series discussed in Sect. 2.2.1, the time-series can be modeled by the addition of four factors one of them is the trend T_0 which is a monotonically increasing/decreasing time-series. A common pre-processing step in many time-series mining applications is called de-trending and involves the removal of this trend from the time-series. This is analogous to the removal of DC component in signal processing applications. The simplest case of de-trending happens when we can assume that T_0 is linear. In this case, a line is fit to the original time-series then subtracted from each point in it. Figure 2.20 shows two examples of de-trending when the original trend is linear. As the figure shows, when this assumption holds, this method can recover the original signal up to the added noise level while when the linear assumption fails, the results can differ widely from the ground truth. Notice though that the error at every point is dependent only on the difference between the fitted line and the actual trend. In MC^2, the function $detrendLinear()$ implements this linear de-trending procedure.

When the trend is nonlinear, SSA can be used to find it by manually grouping monotonically increasing or decreasing expansion time-series during the grouping step (Sect. 2.3.5). The problem here is that a test is needed to decide whether a given expansion time-series is a trend component.

Empirical Mode Decomposition (EMD) is another adaptive expansion technique that has the advantage of always finding the trend component as its last component by construction. The MC^2 has another de-trending routine called $detrendEMD()$ that uses EMD for finding the trend component and removing it. Figure 2.21 shows the results of applying this routine to two time-series with a linear trend (left) and

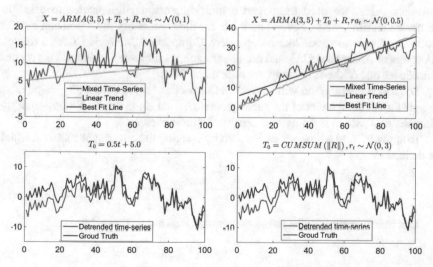

Fig. 2.20 Linear de-trending of a time-series. On the *left* the case where the trend is linear and on the *right* the case when it is nonlinear

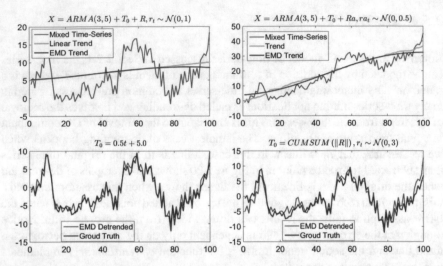

Fig. 2.21 EMD based de-trending of a time-series. On the *left* the case where the trend is linear and on the *right* the case when it is nonlinear

a nonlinear trend (right). In both cases, EMD based de-trending provides superior performance over linear de-trending.

2.5.5   *Dimensionality Reduction*

In many cases, we need to convert a multidimensional time-series to a single-dimensional time-series. Several methods for dimensionality reduction can be used including linear methods like Independent Component Analysis (ICA), Principle Component Analysis (PCA), and nonlinear methods including IsoMap. This section introduces one of the simpler approaches using PCA.

The following notation will be used in this section: $x_{t;i}$ is the ith dimension of x_t, $x_{t,l;i}$ is the ith component of the subsequence $x_{t,l}$, and $X_{;i}$ is the ith dimension of the time-series X which is a single-dimensional time-series.

To apply PCA to a time-series, we start by creating the covariance matrix A which is defined as:

$$A_{i,j} = \sum_{t=1}^{T} x_{t;i} x_{t;j}. \tag{2.104}$$

This matrix can easily be found by treating X as a $n \times T$ matrix:

$$A_{n \times n} = X_{n \times T} X^T{}_{T \times n}. \tag{2.105}$$

We then find the first Eigen vector of A (v_1) and its corresponding Eigen value λ_1. The output 1D time-series can then be found by projecting every time-series value on that vector using the dot product operation:

$$y_t = v_1^T x_t. \tag{2.106}$$

Calculating A in the aforementioned procedure requires $O\left(n^2 T\right)$ operations which may be too slow for some applications. We can speedup the operation by selecting $K < T$ vectors from the time-series and use them to find A. Both of the exact and approximate versions of this process are implemented in the function $tspca()$ in the MC^2 toolbox.

2.5.6 Dynamic Time Warping

Sometimes, we receive for mining a set of time-series of different lengths and would like to use an algorithm that assumes that all inputs have the same length. This can simply be achieved by re-sampling/interpolating the shorter sequences. For example if we have two time-series X, Y of lengths T_x and T_y where $T_x > T_y$, we assume that the sample y_t represents the value of Y at time $\frac{t \times (T_x - 1)}{(T_y - 1)}$. Linear or polynomial interpolation can then be used to estimate values of $\hat{Y}$ at times $0, 1, \ldots, T_x$. A similar approach can be used to change the length of the longer time-series to equal the length of the shorter time-series.

This simple approach assumes that the difference in the length between the two time-series comes from a difference in the sampling frequency used to collect them. In many cases, the difference in length is not related to the sampling frequency but represents genuine difference between the two time-series. One such example happens in learning from multiple demonstrations (Chap. 13). In these problems, we have multiple demonstrations of the same action conducted by a teacher from which a learner is expected to generate a model representing the generation process of these demonstrations. In this case, the difference in length of the time-series cannot be assumed to originate from a difference in sampling rate or approximated by such difference. The difference here is most likely a genuine difference in how fast did the teacher perform different phases of the motion compared to one another. This means that the relation between the time-series to be equalized in length cannot be assumed to be a constant scaling that can be handled by the simple re-sampling procedure described above. A common solution to these cases is to use the Dynamic Time Warping algorithm (DTW).

DTW appeared as a distance function that can be used to estimate distances between time-series more accurately than the standard Euclidean distance when the data points are slightly temporally displaced compared with one another (See Ding et al. 2008 for an experimental evaluation). It reuses all the points in the two time-series to be aligned. The main idea is to find for each point in the shorter time-series a corresponding point in the longer one with the constraint that the first and last two points of the two time-series must align (this is called the boundary constraint). The algorithm can be visualized on a 2D grid $\mathcal{M}$ where the points of time-series X are represented by the rows and the points of the time-series Y are represented by the column. The boundary condition translates to the statement that the path representing the correspondences between these two time-series must start at the bottom left cell (representing (x_0, y_0)) and end at the top right cell (representing (x_{T_x-1}, y_{T_y-1})). Each cell in this grid contains the distance between the corresponding points of X and Y (i.e. $\mathcal{M}_{ij} = d(x_i, y_j)$ for some distance function d). Now, DTW finds the path through $\mathcal{M}$ that minimizes the sum of these distances. This corresponds to the best possible match between the two time-series. Other than the boundary constraint, DTW optimization is constrained in two other ways:

Monotonicity constraint: The indices of both X and Y must be monotonically increasing which means that for the optimal path $(P = (p^1, p^2, \ldots, p^{max(T_x, T_y)}))$, where p^i is an ordered pair (j^i, k^i) and $0 \le j^i \le T_x, 0 \le k^i \le T_y, i_1 > i_2$; implies that $j^{i_1} \le j^{i_2}$ and $k^{i_1} \le k^{i_2}$.

Continuity constraint: Continuous points in the path correspond to adjacent cells both horizontally and vertically. This means that for p^i and p^{i+1}, $\left| j^{i+1} - j^i \right| \le 1$ and $\left| k^{i+1} - k^i \right| \le 1$.

Other constraints are usually used to speed up the calculation of DTW and prevent the path from wandering around leading to meaningless wraps. A *warping window* condition limits the maximum wandering distance allowed from the diagonal. A *slop* constraints limits the number of steps that can be taken in the same direction by the path horizontally or vertically. The two most common constraints in the literature are the Sakoe-Chiba Band and the Itakura Parallelogram. Sakoe-Chiba Band restricts the path to lie within a band around the diagonal. The width of this band is usually taken to be 10 % of the shorter time-series' length. Ratanamahatana and Keogh (2005) showed that this limit is not only useful for speeding up the DTW calculations but that it is even less stringent than necessary for real world data mining applications. Itakura Parallelogram limits the path to lie within a parallelogram with two vertices at the starting and ending points of the path (limited by the boundary constraint). This means that the path is allowed to wander more near the middle of the time-series and less near the boundaries.

DTW is usually used to align single dimensional time-series but it can easily be extended to multidimensional time-series by modifying the distance function used. This algorithm is implemented in the function $dtw()$ in the MC^2 toolbox.

2.6   Summary

This chapter introduced basic time-series analysis techniques that will be used throughout this book. The focus of the chapter was on techniques that are of direct relevance to the ideas presented in the following chapters, rather than on providing an exhaustive treatment of time-series analysis (which requires a much larger volume by itself). We presented several models of generating processes for time-series data. These generation models will be used for generating test sets for algorithms developed later and some of them (e.g. GMM/GMR and GP) will be of direct use in learning from demonstration. We also presented five transformations for representing time-series that will form the basis of algorithms for change point discovery and motif discovery (Chaps. 2 and 3) specially the Singular Spectrum Analysis method that will be a common ingredient of several algorithms later in this book. The chapter also gave a brief treatment of preprocessing techniques that are employed everywhere in this book including smoothing, thinning, normalization and de-trending. The following three chapters will focus on specific time-series analysis problems that use the aforementioned generation models and transformations to create the building blocks for our autonomous learning system to be introduced in the second part of the book.

References

Calinon S, Guenter F, Billard A (2006) On learning the statistical representation of a task and generalizing it to various contexts. In: ICRA'06: IEEE International conference on robotics and automation. IEEE, pp 2978–2983

Ding H, Trajcevski G, Scheuermann P, Wang X, Keogh E (2008) Querying and mining of time series data: experimental comparison of representations and distance measures. Proc VLDB Endow 1(2):1542–1552. doi:10.14778/1454159.1454226

Elsner JB, Tsonis AA (2013) Singular spectrum analysis: a new tool in time series analysis. Springer

Hassani H (2007) Singular spectrum analysis: methodology and comparison. J Data Sci 5(2): 239–257

Larsen RJ, Marx ML (1986) Introduction to mathematical statistics and its applications, 2nd edn. Prentice Hall

Lin J, Keogh E, Lonardi S, Chiu B (2003) A symbolic representation of time series, with implications for streaming algorithms. In: The 8th ACM SIGMOD workshop on research issues in data mining and knowledge discovery. ACM, pp 2–11

Mohammad Y, Nishida T (2014) Robust learning from demonstrations using multidimensional SAX. In: ICCAS'14: 14th international conference on control, automation and cystems. IEEE, pp 64–71

Rasmussen CE, Williams CK (2006) Gaussian processes for machine learning. MIT Press

Ratanamahatana CA, Keogh E (2005) Three myths about dynamic time warping data mining. In: SDM'05: SIAM international conference on data mining. SIAM, pp 506–510

Ratanamahatana CA, Lin J, Gunopulos D, Keogh E, Vlachos M, Das G (2010) Mining time series data. In: Data Mining and Knowledge Discovery Handbook. Springer, pp 1049–1077

Chapter 3
Change Point Discovery

Change point discovery (CPD) is one of the most relied upon technologies in this book. We will use it to discover recurrent in Chap. 4, to discover causal relations in Chap. 5 and as a basis for combining constraints for long-term learning in Chap. 12.

Given a time-series, the goal of CPD is to discover a list of locations at which the *generating dynamics or process* changes or the *shape* of the time-series changes. Depending on the context, researchers use generating dynamics or shape and in this book we focus on the first alternative. For us shape is secondary but the generating dynamics are our goal. The exact definition of *change* and *generating processes* is application dependent and this chapter will try to introduce the most general and applicable methods. As usual, we will be interested in algorithms that are most usable in the context of social robotics while not ignoring CPD algorithms that best represent the field.

Change Point Discovery (CPD) has a very long history. The reader can find survey papers for CPD since 1976 (Willsky 1976).

CPD can be divided into two separate subproblems and this was realized by researchers for a long time (e.g. Willsky 1976; Basseville and Kikiforov 1993). The first problem involves calculating some change score (called *residual* in some papers) for every point in the time-series. This score is expected to change in a smooth fashion mostly. For example, it is expected to rise before the actual change point and then fall after it. The second problem is the discovery of a discrete list of time-steps at which change is announced given the scores found by solving the first problem. We call the first sub-problem the scoring problem and call the second the localization problem. This chapter will introduce different approaches to dealing with both of them.

© Springer International Publishing Switzerland 2015

Y. Mohammad and T. Nishida, *Data Mining for Social Robotics*,
Advanced Information and Knowledge Processing,
DOI 10.1007/978-3-319-25232-2_3

3.1 Approaches to CP Discovery

Given a time-series $X = \{x_t\}$ of length $|X| = T$, and an integer N_c, we say that there are change points at positions $c_k \in [1, T - 2]$ for $1 \le k \le N_c$ if the time series is generated using the following scheme:

$$
x_t = \begin{cases}
\lambda_1 (.) & 0 \le t < c_1 \\
\quad \vdots & \\
\lambda_i (.) & c_{i-1} \le t < c_i \\
\quad \vdots & \\
\lambda_{N_c+1} (.) \, c_{N_c} & \le t < T - 1.
\end{cases}
\tag{3.1}
$$

The parameters of the generation processes $\lambda_i (.)$ depend on the specific modeling scheme employed. An equivalent information for the list $\{c_k\}$ can be provided by a time-series $\{\tilde{x}\}$ which is defined as:

$$
\tilde{x}_t = \begin{cases}
1 & t \in \{c_k\} \\
0 & t \notin \{c_k\}.
\end{cases}
\tag{3.2}
$$

Equation 3.2 can be used to report change point positions but it can easily be extended to handle uncertainty in this decision as:

$$
\tilde{x}_t = p \, (\text{change at point t}) .
\tag{3.3}
$$

An even more general form can be obtained as:

$$
\tilde{x}_t \propto p \, (\text{change at point t}) .
\tag{3.4}
$$

A CPD algorithm may directly estimate $\{c_k\}$ or report $\{\tilde{x}_t\}$ in either of the forms presented earlier. If Eq. 3.4 is used, a final localization step may be needed (depending on the application) to recover the series $\{c_k\}$. This problem will be discussed in Sect. 3.6.

Change point discovery is a problem with long history in data mining which resulted in many approaches that may not be even directly comparable as they may be solving slightly different problems.

Firstly, we distinguish between approaches that infer change points given previous locations of change points (called direct location inference methods hereafter) and methods that do not utilize information about previous change locations directly.

Secondly, in both cases, a very important factor in the solution of the CPD problem is the definition of the data encapsulated by the time-series. On one hand, we have stochastic approaches that assume that the data are coming from some stochastic model (e.g. a GMM, an HMM, etc.) in which every time-series value x_t is *sampled* from a distribution that may depend on previous values of the time-series (e.g. as in

AR and ARMA models). On the other hand, some approaches do not explicitly make this assumption or work directly with a dynamical systems model of the generation process. The distinction between these two approaches is somewhat muddied by the fact that in almost all cases, noise is modeled as a stochastic process which makes it possible to use stochastic approaches even when we have a dynamical model of the generation process. A third approach tries to bypass probabilistic modeling altogether and does not explicitly model the noise in the time-series (examples include SSA based systems).

Another distinction between the methods to be explored in this chapter is how they *delineate* the change. In some algorithms, a change is announced when two models fit the data in some window better than one model. Another approach is to announce a change when data before some point is *different* from data after it according to some criterion. Yet a third approach is to announce a change when data before the point are incapable of explaining the data after it in some sense. We will see algorithms that use each of these three criteria for announcing a change.

Algorithms also differ in how *fast* can they detect changes in the time-series but in all cases that interest us, change point algorithms are local in the sense that they base their decision on a small sample of the time-series around the point being considered. This property allows most of the algorithms we will discuss to be implemented online even though in almost all cases it will require some information from the future of any point to measure the odds for a change at that point which means that most of these algorithms will run with some *known* lag. The only exception will be in algorithms that directly model the probability of a change given prior changes and use stochastic sampling because such algorithms may need to consider samples spanning the complete time-series as in the approach presented in the following section.

3.2 Markov Process CP Approach

The first approach we will consider is based on modeling change point positions using a Markov process. As discussed in Sect. 2.2.7, a Markov process is fully specified by a set of transition probabilities and initial probability.

A simple model in this case is to assume that the location of k'th change point depends only on the location of $k - 1$'s change point. This can be represented by:

$$Probability \text{ (change at point } t | \text{change at point } t - s) = g(t - s), \qquad (3.5)$$

where we assumed that g depends only on the difference between the two integers t, s and that $s < t$. We will also assume that the initial probability distribution is p. Furthermore, $g(0) = 0$ because it does not make any sense to announce a change at time 0 or two changes at exactly the same time.

In any segment of the time-series $x_{s+1:t}$ within which no change occurs, we will have—by definition—a single generation model $\lambda_{s+1:t} \equiv \lambda_{s+1}$. To use the method proposed by Fearnhead and Liu (2007) and described in this section, we will have to limit the set of possible models. We assume that the set of possible models is Λ. These models are grouped into a finite number of groups Ω_i for $1 \le i \le N_m$ for some positive integer N_m with a prior distribution $\pi(\lambda)$ for every group Ω_i. Furthermore, we assume that models differ only on a set of parameters α which renders the prior over models a prior over parameters $\pi(\lambda) \equiv \pi(\alpha)$.

Given the above definitions, the probability of having a segment $x_{s+1:t}$ can be defined as:

$$p(x_{s+1:t}|\Lambda) = \int p(x_{s+1:t}|\alpha, \Lambda)\pi(\alpha)\,d\alpha. \qquad (3.6)$$

For example, given the set of models:

$$x_{s+1:t} = H\alpha + \mathcal{N}\left(0, \sigma^2 I_{t-s \times t-s}\right), \qquad (3.7)$$

for some order q where $|\alpha| = q$. Assuming that σ^2 has an inverse Gamma distribution with meta parameters $\upsilon/2$ and $\gamma/2$ and components of the Gaussian regression vector α have independent Gaussian priors with zero mean and the variance of α_i's prior is δ_i^2. Fearnhead and Liu (2007) reported that under these conditions, the likelihood of $x_{s+1:t}$ given model order q can be found as:

$$p\left(x_{s+1:t}|\Omega_q\right) = \pi^{(s-t)/2}\sqrt{\frac{|M|}{|D|}}\frac{(\gamma)^{\upsilon/2}\Gamma\left((t-s-\upsilon)/2\right)}{\left(\|x_{s+1:t}\|_P^2 + \gamma\right)^{(t-s+\upsilon)/2}\Gamma(\upsilon/2)}, \qquad (3.8)$$

where $M = \left(H^T H + D^{-1}\right)^{-1}$, $P = I - HMH^T$, $\|x\|_A^2 = x^T Ax$, $D = diag\left(\delta_1^2, \ldots, \delta_q^2\right)$.

For this example, Λ is all linear models that can be defined using Eq. 3.7. Ω_q is the set of models with order q (i.e. $|\alpha| = q$). $\lambda_q^i \in \Omega_q$ is a specific linear model with $\alpha = \alpha^i$. The maximum order q_{max} will then be equal to N_m. The importance of the models described by Eq. 3.7 is that it can easily represent several generation processes. For example the $AR(q)$ process can be modeled using Eq. 3.7. This can be achieved by defining H as:

$$H = \begin{bmatrix} x_s & x_{s-1} & \cdots & x_{s-q+1} \\ x_{s+1} & x_s & \cdots & x_{s-q+2} \\ \vdots & \vdots & \ddots & \vdots \\ x_t & x_{t-1} & \cdots & x_{t-q+1} \end{bmatrix}. \qquad (3.9)$$

The same form can represent polynomial regression by setting H as:

$$H = \begin{bmatrix} z^0\,(s+1)\ z^1\,(s+1)\ \ldots\ z^q\,(s+1) \\ z^0\,(s+2)\ z^1\,(s+2)\ \ldots\ z^q\,(s+2) \\ \vdots \qquad\quad \vdots \qquad \ddots \qquad \vdots \\ z^0\,(t) \qquad z^1\,(t) \quad \ldots \quad z^q\,(t) \end{bmatrix},$$

where $z\,(i) = i/(t-s)$.

Now that we know how to define the probability of a change at time t given a change at an earlier time s, we move on to explain an algorithm for calculating $\{c_k\}$ given $\{x_t\}$ (Fearnhead and Liu 2007). The algorithm consists of a filtering recursion that operates on the time-series point by point.

A helper time-series $\{\zeta_t\}$ of the same length as $\{x_t\}$ (i.e. T) will be defined to keep track of the last point at which a change occurred. This means that if $\zeta_{t-1} = s$ then $\zeta_t = s$ if there was no change at time t (implying that last change was at time s) or $\zeta_t = t - 1$ implying that the last change happened at time $t - 1$. The sequence $\{\zeta_t\}$ is a Markov chain with the following definition of transition probabilities:

$$p\,(\zeta_{t+1} = j | \zeta_t = i) = \begin{cases} \frac{1-G(t-i)}{1-G(t-i-1)} & \text{if } j = i \\ \frac{G(t-i)-G(t-i-1)}{1-G(t-i-1)} & \text{if } j = t \\ 0 & otherwise \end{cases}, \tag{3.10}$$

where $G(k)$ is the commutative distribution associated with $g\,(k)$.

Applying Bayes rule we get:

$$(\zeta_t = j | x_{0:t}) \propto p\,(x_t | \zeta_t = j, x_{0:t-1})\, p\,(\zeta_t = j | x_{0:t-1}). \tag{3.11}$$

Moreover, we have:

$$p\,(\zeta_t = j | x_{0:t-1}) = \sum_{i=0}^{t-1} p\,(\zeta_t = j | \zeta_{t-1} = i)\, p\,(\zeta_{t-1} | x_{0:t-1}). \tag{3.12}$$

This leads to the following recursion:

$$p\,(\zeta_t = j | x_{0:t}) \propto \begin{cases} w_t^j \frac{1-G(t-1-j)}{1-G(t-2-j)} p\,(\zeta_{t-1} = j | x_{0:t-1}) & \text{if } j < t - 1, \\ w_t^{t-1} \sum_{i=0}^{t-2} \left(\frac{G(t-1-i)-G(t-2-i)}{1-G(t-2-j)} p\,(\zeta_{t-1} = j | x_{0:t-1}) \right) & \text{if } j = t - 1, \\ 0 & otherwise \end{cases} \tag{3.13}$$

where $w_i^j = p\,(x_i | \zeta_i = j, x_{0:i-1})$. These weights represent predictions of the current time-series value assuming the state of the Markov chain and previous values of the time-series and they can be calculated efficiently in constant time independent of the time-series length using:

$$w_i^j = \frac{\sum\limits_{q=1}^{q_{max}} p\,(j,i,q)\,p\,(q)}{\sum\limits_{q=1}^{q_{max}} p\,(j,i-1,q)\,p\,(q)}. \tag{3.14}$$

Given that we have calculated $p\,(\zeta_t = j|x_{0:t})$ for $0 \leq t \leq T$, the discovery of change points becomes the problem of finding the Maximum A posteriori Estimate (MAP) of ζ_t. This can be solved by a Viterbi algorithm which also gives us an estimate of the order of the system (q) at every time-step. The time-series $\{\tilde{x}_t\}$ can then easily be found from ζ_t by differencing.

The approach presented in this section suffers from some problems. Firstly, the time required to calculate the value of $p\,(\zeta_t = j|x_{0:t})$ according to Eq. 3.13 increases with time which means that the time-complexity of the algorithm is quadratic in time-series length. This is a prohibiting cost for longer time-series that we are usually interests us in social robotics and in many other realistic applications. Fearnhead and Liu (2007) proposed a solution to this problem by using particle filtering to approximate this distribution. This renders the inference approximate but with linear time-complexity. Secondly, the algorithm requires a predefined set of models Λ. Selection of this set of models is crucial for the inference we presented in this section and in real world applications it may not be clear how to select this set.

Other approaches that we will describe in this chapter will not try to directly model the length of time-series segments. This removes the need to define Λ because we need not predict segment lengths. Nevertheless, most of the algorithms described in this chapter will still rely on predefined models with the exception of Singular Spectrum Analysis approaches (Sect. 3.5).

3.3 Two Models Approach

Figure 3.1 shows a schematic of the subsequences involved in change point discovery using the two-models approach. Three models can be seen: A long-term model representing the totality of the time-series or a large segment of it (M_a representing a subsequence of length n_a) and two short-term models representing the past (M_p representing a subsequence of length n_p) and the future (M_f representing a subsequence of length n_f) around the point of interest t. Some delay may be used before the future model and after the past model (δ_f and δ_p respectively). These two constants may be negative signaling overlap between M_p and M_f. These three models can then be compared to give an estimation of $\tilde{x}_t$. In most cases, we only use two of the models for comparisons (either M_a and M_f or M_p and M_f). The involvement of M_f in both cases means that the CPD algorithm must run by some lag of around $n_f + \delta_f$.

To instantiate a specific CPD algorithm using the two-models formalism, we need to specify the values of the constants involved, the two-models selected for comparisons, the model family Λ used for estimating them, and a method for comparing the

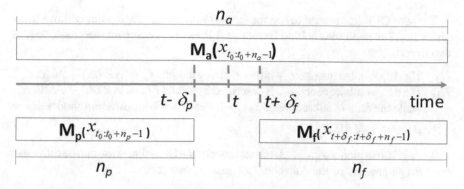

Fig. 3.1 The subsequences involved in change point discovery

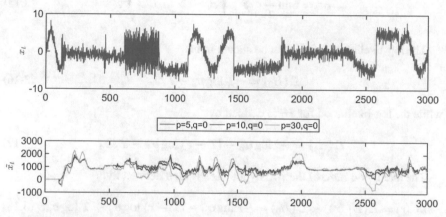

Fig. 3.2 Application of Brandt's GLR test with three different model orders to a time-series generated by concatenating noisy signal primitives

two models. One of the earliest approaches for CPD was based on comparing M_a and M_f and using an autoregressive model for both $(AR(p))$ with additive Gaussian noise. The difference between the predictions of the two models is measured using Generalized Likelihood Ratio (GLR) test and this difference gave an estimate of the change point score at every point. Andre-Obrecht (1988) applied this method to speech segmentation as early as 1988.

This approach is called Brandt's generalized likelihood ratio (GLR) test and is implemented in the function $brandt()$ in the MC^2 toolbox. See Sect. 2.4.1 for two general approaches for learning $AR(p)$ models knowing the model order. Figure 3.2 shows an example of applying this method to discover the change points in a time-series that was generated by concatenating shapes like sinusoidals, lines, and constant pieces. Notice that the generation mechanism is not an autoregressive model, yet the recovered estimate of change point locations is acceptably accurate.

Brandt's GLR test is used to test for the existence of a change point within a time-series $x_{0:T}$ The main idea behind Brandt's GLR test is to compare the predictions of two hypotheses:

H_0:   The data in subsequence $x_{0:T}$ is described by model $M_a = AR\,(m) + \mathcal{N}\left(0, \sigma_a^2\right)$.
H_1:   The data in subsequence $x_{0:r}$ is described by model $M_p = AR\,(m) + \mathcal{N}\left(0, \sigma_p^2\right)$
         while the data in subsequence $x_{r+1:T}$ is described by a different model $M_f = AR\,(m) + \mathcal{N}\left(0, \sigma_f^2\right)$.

For any subsequence $x_{t_0:t_0+T-1}$, the variance of the Gaussian noise can be estimated once the parameters of the AR model are learned using:

$$\sigma^2 = \min_a \frac{1}{T} \sum_{t=t_0}^{t_0+T-1} \left(x_t - \sum_{i=1}^{m} a_i x_{t-i} \right)^2. \qquad (3.15)$$

The log-likelihood for H_0 can be shown to be:

$$\mathcal{L}\,(H_0) = -T \log \sigma_a - T/2, \qquad (3.16)$$

while the log-likelihood for H_1 is defined by:

$$\mathcal{L}\,(H_1) = -r \log \sigma_p - (T - r) \log \sigma_f - T/2. \qquad (3.17)$$

The likelihood ratio of the two hypotheses is then defined by:

$$\mathcal{LR}\,(r) = \mathcal{L}\,(H_1\,(r)) - \mathcal{L}\,(H_0) = -r \log \sigma_p - (T - r) \log \sigma_f + T \log \sigma_a. \quad (3.18)$$

The Brandt's test consists of comparing the maximum value of $\mathcal{LR}\,(r)$ for $2 \leq r \leq T - 1$ to a predefined threshold.

$$\mathcal{MLR} = \max_r \left(-r \log \sigma_p - (T - r) \log \sigma_f + T \log \sigma_a \right). \qquad (3.19)$$

Given any window of a time-series, calculating $\mathcal{MLR}$ using Eq. 3.19 involves operations quadratic on the length of the window (T).

A faster implementation that was proposed by Andre-Obrecht (1988) is to fix the value of r and use a moving window of length T. Once $\mathcal{LR}\,(r)$ exceeds a predefined limit, the location of the change point is assumed to be within the last $T - r$ points of the window and another GLR test is used to localize it.

Brandt's GLR test is but one of several possible two-models approaches. It is critical to notice that these tests do not assume that the model used is the same as generating dynamics of the time-series but only that the model can describe the data with enough accuracy for its predictions to be of value for CPD. One problem of this approach can be seen in Fig. 3.2. The output of the system can have any value which makes it harder to select a meaningful threshold.

Other forms of two-models approaches can be envisioned. For example, the models used can be HMMs or GMMs and probabilistic comparison methods like the KL divergence can be used to compare the predictions of these models.

Singular Spectrum Analysis based CPD that will be explained in Sect. 3.5 can be seen as a two-models method that does not commit to the AR model for prediction but utilizes the *form* of the time-series sequences for the past and the future for deciding whether or not a change occurred.

3.4 Change in Stochastic Processes

A general class of solutions for CPD involves converting the signal into a stochastic process with parameterized probability distribution ($P_\theta\left(x_i|x_{i-1}, x_{i-2}, \ldots, x_0\right)$) then deciding whether at some point in the input signal (or a sliding window over it for online approaches) there is a change in θ from θ_0 to θ_1 against the alternative that a single parameter vector θ explains the whole signal. This is a likelihood ratio test and can be decided statistically (e.g. Basseville and Kikiforov 1993).

One of the earliest approaches that follow this scheme is the CUMSUM algorithm (Page 1954) which is sometimes called *Page-Hinkley* stopping rule. The goal here is to detect a sudden change of the mean of some process. This can be used, for example, in conversation analysis for detecting the times at which a sudden change in gaze direction occurred. Because of noise in the gaze direction detector, the gaze direction can be modeled as a random walk around a mean value that represents the target. Sudden change in gaze can then be modeled as a change of this mean.

For simplicity let's start by assuming that both the mean before the change μ_0 and after it μ_1 are known. This means that we can model the signal using the following very simply Equation:

$$y_i = \mu_i + \mathcal{N}\left(0, \sigma^2\right), \tag{3.20}$$

where σ is the (possibly unknown) standard deviation of the Gaussian noise added to the means and μ_i is equal to μ_0 before the point of change (r). If a change happens at the point r, then μ_i will equal μ_1 for $i \geq r$ otherwise it will equal μ_0.

In this case our hypotheses can be stated rigorously as:

$$H_0 : r > n,$$
$$H_1 : r \leqslant n, \tag{3.21}$$

where r is the change point we are looking to locate and n is the length of our signal (or a sliding window on it).

Given the assumptions of this problem, we can define the likelihood ratio between the two hypotheses by:

$$\prod_{k=r}^{n} \frac{p_1\left(y_k\right)}{p_0\left(y_k\right)}, \tag{3.22}$$

where $p_j = \mathcal{N}\left(\mu_j, \sigma\right)$.

Now the log of this likelihood can be written as:

$$\mathcal{L}_n\left(r\right) = \frac{\mu_1 - \mu_0}{\sigma^2} \sum_{k=r}^{n} \left(y_k - \frac{\mu_1 + \mu_0}{2}\right) = \frac{1}{\sigma^2} S_r^n\left(\mu_0\right), \tag{3.23}$$

where

$$S_r^n\left(\mu\right) = \left(\mu_1 - \mu_0\right) \sum_{k=r}^{n} \left(y_k - \frac{\mu_1 + \mu_0}{2}\right). \tag{3.24}$$

This assumes that we actually know r but in reality that is what we are after (the change point). For this reason, we can replace r by its maximum likelihood estimate $\hat{r}$ which can be calculated as the value that maximizes S_r^n (notice that σ is independent of this value). This gives us the following estimate for $\hat{r}$:

$$\hat{r} = \underset{1 \leqslant r \leqslant n}{argmax} \left(\prod_{i=0}^{r-1} p_0\left(y_i\right) \prod_{j=r}^{n} p_1\left(y_j\right)\right) = \underset{1 \leqslant r \leqslant n}{argmax} \left(S_r^n\left(\mu_0\right)\right). \tag{3.25}$$

After calculating $\hat{r}$, we can then decide that a change happens within the n-point signal (or window) if $\underset{1 \leqslant r \leqslant n}{max} \left(S_r^n\left(\mu_0\right)\right)$ was higher than some predefined threshold τ.

This leads to the CUMSUM statistic which can be computed recursively using:

$$g_n = \left(g_{n-1} + y_n - \mu_0 - 0.5\left(\mu_1 + \mu_0\right)\right). \tag{3.26}$$

Whenever g_n exceeds the predefined threshold τ we announce a change at the point $\hat{r}$ estimated as described above.

Of course in real life we do not know μ_0 and μ_1 but we can use the same strategy used for r and use their maximum likelihood estimates from the data.

3.5   Singular Spectrum Analysis Based Methods

As the discussion of CUMSUM showed, most CPD algorithms require the specification of some threshold (e.g. τ in CUMSUM) that is used for localization. They also assume some predefined model of the signal (e.g. a constant value corrupted by a Gaussian noise in CUMSUM). In our application for motor-babbling (See Sect. 11.1), for example, the signals we deal with are very different ranging from speech to motor commands. It is difficult to have a model for each kind of these signals and it is also very difficult to decide apriori appropriate threshold values.

A promising approach that can overcome all of these problems is based on Singular Spectrum Analysis (Sect. 2.3.5). There are several SSA based CPD algorithms in literature (for a survey see, Mohammad and Nishida 2011). Here we discuss one of the simplest versions that was used in early implementations of our social robotics work called Robust Singular Spectrum Transform (RSST).

The essence of the $RSST$ is to find for every point $x(i)$ the difference between a representation of the dynamics of the few points before it (i.e. $x(i - p) : x(i)$ or M_p in Fig. 3.1) and the few points after it (i.e. $x(i + g) : x(i + f)$ or M_f in Fig. 3.1). This difference is normalized to have a value between zero and one and named $x_s(i)$.

Rather than relying on a predefined model (e.g. AR, ARMA, HMM, etc.), the system tries to directly model the *shape* of the time-series using Singular Spectrum Analysis (See Sect. 2.3.5).

The subsequences before and after the current point are represented using the Hankel matrix which is calculated in two steps:

1. A set of subsequences $seq(t)$ are calculated as:

$$seq(t - 1) = \{x(t - w), \ldots, x(t - 1)\}^T. \qquad (3.27)$$

2. The Hankel matrix is calculated as the concatenation of n overlapping subsequences:

$$H(t) = [seq(t - n), \ldots, seq(t - 1)]. \qquad (3.28)$$

Singular Value Decomposition (SVD) is then used to find the singular values and vectors of the Hankel Matrix by solving:

$$H(t) = U(t) S(t) V(t)^T, \qquad (3.29)$$

where $S(i - 1, i - 1) \leq S(i, i) \leq (i + 1, i + 1)$.

Only the first $l(t)$ left singular vectors $(U_l(t))$ are kept to represent the past change pattern as the hyperplane defined by them. Ide and Inoue (2005) showed that this hyperplane encodes the major directions of change in the signal. In RSST the value of $l(t)$ is allowed to change from point to point in the time series depending on the complexity of the signal before it. To calculate a sensible value for $l(t)$ we first sort the singular values of $H(t)$ and find the corner of the accumulated sum of them ($l_{inf}(t)$) (the point at which the tangent to the curve has an angle of $\pi/4$).

To find a first guess of the change score around every point, RSST tries to utilize as much information as possible from the future Hankel Matrix ($G(t)$) by using the $l_f(t)$ Eigen vectors of $G(t) G(t)^T$ with highest corresponding Eigen values ($\lambda_{1:l_f}$). The value of $l_f(t)$ is selected using the same algorithm for selecting $l(t)$.

$$G(t) G(t)^T u^g = \mu u^g, \tag{3.30}$$

$$\beta_i(t) = u_i^g, \; i \le l_f \; and \; \lambda_{j-1} \le \lambda_j \le \lambda_{j+1} \; for \; 1 \le j \le w. \tag{3.31}$$

Each one of these l_f directions are then projected onto the hyperplane defined by $U_l(t)$.

The projection of $\beta_i(t)$s and the hyperplane defined by $U_l(t)$ is then found using:

$$\alpha_i(t) = \frac{U_l^T \beta_i(t)}{\left\| U_l^T \beta_i(t) \right\|}, i \le l_f. \tag{3.32}$$

The change scores defined by $\beta_i(t)$s and $alpha_i(t)$s are then calculated as:

$$cs_i(t) = 1 - \alpha_i(t)^T \beta_i(t). \tag{3.33}$$

The first guess of the change score at the point t is then calculated as the weighted sum of these change point scores where the Eigen values of the matrix $G(t)$ are used as weights.

$$\hat{x}(t) = \frac{\sum\limits_{i=1}^{l_f} \lambda_i \times cs_i}{\sum\limits_{i=1}^{l_f} \lambda_i}. \tag{3.34}$$

After applying the aforementioned steps we get a first estimate $\hat{x}(t)$ of the change score at every point t of the time series. RSST then applies a filtering step to attenuate the effect of noise on the final scores. The filter first calculates the average and variance of the signal before and after each point using a subwindow of size w.

$$\mu_b(t) = \frac{\sum\limits_{i=0}^{w-1} \hat{x}(t-i)}{w}, \tag{3.35}$$

$$\sigma_b(t) = \frac{\sum\limits_{i=1}^{w} \left(\hat{x}(t-i) - \mu_b(t) \right)^2}{w-1}, \tag{3.36}$$

$$\mu_a(t) = \frac{\sum\limits_{i=1}^{w} \hat{x}(t+i)}{w}, \tag{3.37}$$

$$\sigma_a(t) = \frac{\sum\limits_{i=1}^{w} \left(\hat{x}(t+i) - \mu_a(t) \right)^2}{w-1}. \tag{3.38}$$

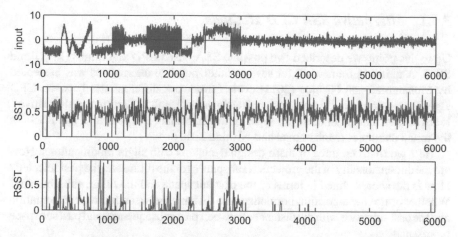

Fig. 3.3 Application of RSST and SST. The second half of the input consisted of pure Gaussian noise, yet SST is still generating many false positives while SST could generate a much more specific score

The guess of the change score at every point is then updated by:

$$\tilde{x}(t) = \hat{x}(t) \times |\mu_a(t) - \mu_b(t)| \times \left| \sqrt{\sigma_a(t)} - \sqrt{\sigma_b(t)} \right|, \qquad (3.39)$$

where $\tilde{x}(t)$ is then normalized to get $x(t)$ which represents the final change score of RSST.

This approach is implemented in the function $rsst()$ in the MC^2 toolbox. The original approach of (Ide and Inoue 2005) called SST is implemented in the function $sst()$ which lacks the final filtering step for removal of noisy segments and uses always a fixed number of Eigen vectors for representing the past Hankel matrix for score calculation instead of automatically calculating it as in RSST. Figure 3.3 shows an example of applying both algorithms to a time-series generated by concatenating different simple shapes. Notice that the second half of the input time-series (last 3000 points) was pure Gaussian noise. Nevertheless, SST was confused to generate many false positives. The reason for this behavior is that the Hankel matrices generated by this random noise contain random data. When converted into hyperplanes they will be oriented more or less randomly and the same is true for the Eigen vector representing the future. The comparison between these two randomly oriented constructs is expected to lead a random values which is why SST has so many false positives.

This problem is sidestepped in RSST through the use of the filtering step which simply detects these regions in the score signal and removes them.

3.5.1 Alternative SSA CPD Methods

Up to this point, we described two possible SSA based CPD algorithms (RSST and SST). A unifying framework for several other possible alternatives was presented by Mohammad and Nishida (2011). This framework differentiates between CPD algorithms using four different design decisions that result in 32 different alternatives. It also shows that six of these are superior in performance to both SST and RSST in the tested examples (Mohammad and Nishida 2011).

Here we focus on three of these design decisions with slight modifications: How are the dimensionality of the projection subspaces for the future and the past selected? How is the score defined in terms of the past subspace and the future subsequence? Whether or not the algorithm normalizes the distance between future representation and the past subspace with the distances between past subsequences and past subspace representation.

The toolbox contains the function $cpd()$ which allows the user to change all of these variables generating more than 64 different alternatives for SSA based change point discovery. The reader is encouraged to experiment with these variations for her dataset before settling on the variation to use. Guidelines can be found in our earlier work (see Mohammad and Nishida 2011).

3.6 Change Localization

The algorithms described in the previous sections use two stages for CPD. A score $\{\tilde{x}_t\}$ is first calculated then a localization step has to be conducted. In some cases—like the Brandt's GLR test presented in Sect. 3.3—the scoring algorithm suggests a localization algorithm. In other cases (e.g. SSA CPD described in Sect. 3.5), no such bias exists and the practitioner is free to choose among many possible localization systems.

One of the simplest possible localization approaches is thinning (Sect. 2.5.2) in which we just extract the local maxima of the score. This approach is bound to generate too many change points due to noise in score calculation.

Figure 3.4 shows two simple alternatives for thinning for change localization. In both cases a threshold is used and the score signal is scanned from beginning to

Fig. 3.4 Localization methods used in this book

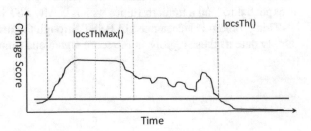

end. Once the threshold is exceeded a segment is started that continues until the score becomes less than the threshold once more. This segment is then converted to a single change point location. Several alternatives can be used to achieve that. The simplest is to use the point with the maximum score value within the segment or the ridge center if the maximum was a ridge. Another approach is to us the point in the middle of the segment. Both approaches are implemented in the toolbox with functions $locThMax()$ and $locTh()$ respectively.

Even though more accurate localization schemes can be devised (e.g. using a form of GLR for localization), these simple approaches proved accurate enough for our applications.

3.7 Comparing CPD Algorithms

Comparing change point discovery algorithms is not an easy task. Consider the case given in Fig. 3.5. A time-series is shown with the ground truth change points in blue. The outputs of three fictitious CPD algorithms are also shown. In one case (green), the algorithm finds two change points perfectly but gave smaller scores for them and failed to find the other two. In another case (red), it could find all of the four points but reported the change a little bit earlier than where it should have been reported. The third algorithm (dashed) also found all of the four change points but reported them later than when they happened. This variety of ways in which a CPD algorithm may fail (e.g. failure to find a change, false positives, delay in change reporting, too early announcement of change, inaccurate scoring of change, etc.) makes it difficult to quantify—using a single number—the relative quality of CPD algorithms.

Comparison between change detection algorithms is usually done using one of three approaches. The first approach is using the traditional confusion matrix based statistics including the F measure, MCC, Precision, Recall, and ROC curves (we call these confusion matrix measures). The second approach is using information theoretic measures like the Kullback–Leibler divergence or Jennsen–Shannon Divergence between the true change point locations and estimated locations (we call these

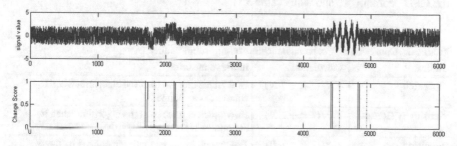

Fig. 3.5 (*Top*) A time-series. (*Bottom*) Ground truth change points (*blue*) and the output of three CPD algorithms

divergence methods). The third approach is using the delay between the change and
its discovery (we call these delay measures). In earlier work, we argued that all of
these approaches cannot effectively differentiate the performance of different CPD
algorithms and proposed a new approach that can solve this problem (Mohammad
and Nishida 2011).

3.7.1 Confusion Matrix Measures

All confusion matrix measures are based on evaluating four values: true positives
(TP), false positives (FP), true negatives (TN), and false negatives (FN). The total
number of positives (P) equals the sum of true positives and false negatives and the
total number of negatives (N) equals the sum of true negatives and false positives.
False positives are sometimes called Type I Errors or false alarms and false negatives
are sometimes called Type II errors or misses.

Consider true–positives. These are usually defined as the number of positives
(announced changes in CPD) that correspond to real positives (real change points
in CPD). This implies some form of localization of the changes. A general problem
of confusion matrix measures then is that they can only compare CPD algorithms
after localization and cannot directly be used to compare the scoring step which is
sometimes all what is needed (e.g. in constrained motif discovery as will be discussed
in the Chap. 4).

Even with a localization algorithm, the possible delay in CP detection may com-
plicate the matters. Consider an algorithm that announces a change at point 503
while a true change happens at point 500. Should we consider this a false positive
(because the point is different) or a true-positive because it is near enough (only 3
points difference). The simplest way to solve this problem is to use a predefined
delay threshold (which can even be different for early and late announcements of
changes). After a localization algorithm is specified and this threshold is fixed, it is
possible to calculate the four quantities needed for the confusion matrix.

Given the four quantities of the confusion matrix, it is easy to calculate several
indicators to compare CPD algorithms. Some of the most important indicators for
the CPD problem are shown in Table 3.1.

Table 3.1 Confusion matrix based indicators of algorithm quality

Indicator	Calculation	Intuitive meaning
Specificity	$TN/(FP + TN)$	How much can we trust the algorithm when it announces a change
Sensitivity (Recall)	$TP/(TP + FN)$	How much can we trust the algorithm when it announces no changes
Precision	$TP/(TP + FP)$	How much of the changes announced by the algorithm are correct

Notice that some of the most widely used confusion matrix based indicators are of little value for comparing CPD algorithms due to the large imbalance between P and N (positives are usually much rarer than negatives in the ground truth). For example accuracy (defined as $(TP + TN)/(P + N)$) is not a good indicator of algorithm quality for CPD despite its widespread use in comparing classification algorithms because $P \ll N$ and assuming for example that an algorithm will just output $\tilde{x}_t = 0$ for $1 \le t \le T$, it will has $TP = 0$ and $TN = N$ which leads to an accuracy of $N/(P + N) \approx 1.0$.

Usually we need two of the indicators given in Table 3.1 to compare CPD algorithms (e.g. precision vs. recall or specificity vs. sensitivity). Some other indicators try to use a single number to signal algorithm quality by combining these indicators. For example F_α combines precision and recall using:

$$F_\alpha = \frac{\left(1 + \alpha^2\right) Precision \times Recall}{\left(\alpha^2 \times Precision\right) + Recall} = \frac{\left(1 + \alpha^2\right) TP}{\left(a + \alpha^2\right) TP + \alpha^2 FN + FP}. \tag{3.40}$$

F_1 is the harmonic mean of precision and recall and puts the same emphasis on both of them. F_2 puts more emphasis on recall while $F_{0.5}$ puts more emphasis on precision. Depending on the application one of the other of these indicators may be more appropriate.

A similar approach is taken by Matthews correlation coefficient (MCC) which is defined as:

$$MCC = \frac{TP \times TN - FP \times FN}{\sqrt{P \times N \times (TP + FP) \times (TN + FN)}}. \tag{3.41}$$

MCC is specially useful for CPD problems because of the imbalance between the number of positives and negatives in the time-series. It represents a correlation coefficient ranging from -1 to 1 between the prediction and correct value of change point locations (after taking the acceptable delay into account).

These confusion matrix metrics can be computed in the MC^2 toolbox using the function $cpquality()$ which accepts the localized change points (or scores and localization function) and an acceptable delay threshold and calculates all the confusion matrix based indicators discussed in this section.

3.7.2 Divergence Measures

Given the problems of confusion matrix measures of CPD algorithm, another approach (that is more appropriate for comparing scores instead of localization results) is treating the output of the algorithm as a probability distribution and comparing the two probability distributions using one of many possible probability distribution distance functions. We call these methods divergence methods as they try to measure the divergence of the estimated change scores and ground truth change point locations.

Assuming that we have M changes at locations γ_m where $1 \le m \le M$ in a time-series of length T. Ground-truth distribution $p^{gt}()$ will then be defined as:

$$p^{gt}(t) = \begin{cases} 1/M & t \in \{\gamma_m\}, \\ 0 & otherwise. \end{cases} \tag{3.42}$$

Another possible approach is to locate a small Gaussian at every one of the change points with small variance σ leading to the following definition:

$$p^{gt}(t) = \sum_{m=1}^{M} \mathcal{N}\left(t; \gamma_m, \sigma^2\right). \tag{3.43}$$

The advantage of the ground truth distribution described by Eq. 3.43 over that described by Eq. 3.42 is that it does not penalize algorithms that fail to detect some change point the same way as algorithms that merely detect it slightly after (or even before) its occurrence.

Other possibilities can be considered. For example, we can us the idea of acceptable delay τ from the previous section and define p^{gt} as:

$$p^{gt}(t) = \begin{cases} 1/(\tau M) & \exists 0 \leqslant \delta \leqslant \tau \text{ where } t + \delta \in \{\gamma_m\} \vee t - \delta \in \{\gamma_m\} \\ 0 & otherwise \end{cases}. \tag{3.44}$$

There are several other possibilities for defining $p^{gt}(t)$ for all $1 \le t \le T$ but these three methods represented by Eqs. 3.42, and 3.43 can cover our needs for this chapter.

Now, given some output from a CPD algorithm $\{\tilde{x}_t\}$, we can define a probability distribution by just normalizing the output using:

$$p(t) = \frac{\tilde{x}_t}{\sum \tilde{x}_t}. \tag{3.45}$$

If we have instead the output of a localization algorithm in the form $\{\gamma_m\}$ for $1 \le m \le M$ and $\gamma \in [1, T]$, we can use any one of the three approaches used to generate the ground truth distribution (p^{gt}) to generate the distribution to be tested.

Now that we have both $\left\{p^{gt}(t)\right\}$ and $\{p(t)\}$, we can use any distance measure defined for probability distributions to compare them.

One of the most widely used such measure is the Kullback–Leibler divergence of two discrete probability distributions which is implemented in the function $KLDiv()$ in the toolbox. Other possibilities include Jennsen–Shannon Divergence implemented in $jsdiv()$.

A more appropriate distance measure for CPD comparison with ground truth is the Earth Mover's Distance (EMD) which is defined intuitively as follows: Assume that $p^{gt}(t)$ represents some dirt distributed over T bins corresponding to the T possible values of t. Moreover, assume that $p(t)$ represents the same number of holes but with depths corresponding to the values of $p(t)$. Assume moreover that moving dirt from bin i to bin j costs some cost c_{ij} which is usually taken to be $|i - j|^p$ where

$p \in \{1, 2\}$. Now the EMD is defined as the total cost needed to completely fill all the holes using all the dirt by transporting dirt into holes. Because we know that the volume of dirt is the same as the total volume of the holes (both are 1.0), we know that a solution exists. In general EMD is calculated by solving a linear programming problem.

For probability distributions though, it was shown by Levina and Bickel (2001) that EMD is equivalent to Mallow's distance which is defined as:

$$M_p (p, q) = \min_F E_F (\|X - Y\|^p)^{1/p},$$ (3.46)

where p and q are probability distributions of the random variables X and Y (respectively) and F represents a joint probability distribution where $(X, Y) \sim F$ subject to the two constraints:

$$X \sim p, Y \sim q.$$

Mallows distance is the expected difference between the two random variables X and Y taken over all joint probability distributions such that the marginal distribution of X and Y equal the compared p and q distributions.

It can be shown that $M_p (p^{gt}, p)$ (and so $EMD (p^{gt}, p)$) can be found as:

$$EMD (p^{gt}, p) = M_p (p^{gt}, p) = \left(\frac{1}{T} \sum_{t=0}^{T-1} |p^{gt} (t) - p (t)|^p \right)^{1/p}.$$ (3.47)

In our context, Mallows distance represents the expected difference between the ground-truth location of a change point and a sample from the distribution representing the CPD result. An immediate problem with this measure is that it does not take into account the direct correspondence between ground truth change points and estimated change points. A more subtle problem has to do with the fact that the scores generated from CPD algorithms are rarely uniform yet we do not care much about their value but their location. To make this point clear, consider a time series of length 100 with two change points at points 10 and 80. Now consider two change point algorithms: one generating an only two nonzero equal values at points 80 and 10 (call this Algorithm 1), and the other generates two nonzero but unequal values at the same points (call this Algorithm 2). Both algorithms detected the two changes perfectly and for most purposes they should be considered equal as long as both nonzero values are beyond the decision threshold, yet Algorithm 1 will have an EMD distance of zero while Algorithm 2 will have nonzero distance signaling inferior performance.

What we really care about in most of our applications is how many ground truth change points are discovered by the CPD algorithm. A more appropriate measure for this value will be presented in Sect. 3.7.3.

3.7.3 Equal Sampling Rate

The discussion of CPD quality measures so far leads to a set of desirable features in any such measure:

1. It should take into account acceptable delays and allows for penalizing delayed discoveries of changes while allowing also for rewarding early discovery.
2. It should be able to compare both raw change scores and localized change point estimates.
3. It should be bounded.
4. It should give maximum quality score for estimates that match the ground truth exactly and give minimum scores for random estimates.
5. It should correspond to our intuitive notion of change discovery accuracy. This means that algorithms declaring changes at every point or no changes at any points without considering the data should get minimum possible scores (similar to algorithms that just assign random estimates).

This section describes a form of a divergence measure that was developed explicitly for measuring the quality of CPD algorithms. It also treats the estimated and ground truth locations or scores as a probability distribution as we did in the previous section. The main idea of this measure (called Equal Sampling Rate) is to find the probability that a random sample from the distribution representing the estimate will be within some predefined *acceptable* distance from a real change point while discounting delayed discovery and possibly encouraging early (before change) prediction.

Consider again the example given by the end of the last section. We had a time series of length 100 with two change points at points 10 and 80. Now consider two change point algorithms: one generating an only two nonzero equal values at points 80 and 10 (call this Algorithm 1), and the other generates two nonzero but unequal values at the same points (call this Algorithm 2). According to ESR, both algorithms will have exactly the same score (which is 1.0 the maximum possible score) because no matter from where do we sample in the estimated change locations in both algorithms we get exactly one of the ground truth change points. This is in contrast to the other divergence methods discussed in the previous section that always will give different scores to these two algorithms.

Given two discrete probability distributions $p(t)$ and $q(t)$ defined on the range $1 \leq t \leq T$, a positive integer n and a scaling function $s(i)$ defined for $-n \leq i \leq n$, we define the equality of sampling from p to q with allowance n and distance scaling s (ES in short) as:

$$ES(p, q, n, s) = \sum_{t=1}^{T} q(t) \sum_{j=\max(1, t-n)}^{j=\min(T, t+n)} p(j) s(j-t). \tag{3.48}$$

If we treat p as the *true* distribution and q as the *measured* distribution, then $ES(p, q, n, 1)$ can be informally considered as a measure of how likely that a sample selected according to the measured distribution (q) will be within n points of a point with high probability in p. Notice that n is allowed to be infinite. The function s (i) representing another way to integrate delays within the evaluation by assigning a weight based on the distance between a sample from q and the real nonzero values in p. Usually one uses a specific value of n or a decreasing function s (i) but they can be combined. The function s (i) allows us in principle to treat change predictions before a real change differently than predictions on the same distance after that change.

Relating $ES\left(p^{gt}, p, n, s\right)$ to confusion matrix based measures, it can be seen as a generalization of the notion of *recall* while $ES\left(p, p^{gt}, n, s\right)$ can be related to a generalized form of *precision*. This suggests that a combination of them (inspired by F_α) can be defined as:

$$ESR\left(p^{pg}, p, n, s\right) = \alpha \times ESR\left(p, p^{pg}, n, s\right) + (1 - \alpha) \times ESR\left(p^{pg}, p, n, s\right).$$
(3.49)

In this book we always use $\alpha = 0.5$ and most of the results reported in this chapter and future applications of CPD will not depend much on the selection of this parameter. Mohammad and Nishida (2011) introduced the ESR measure of CPD quality and used it to compare different SAA based CPD algorithms (See Sect. 3.5).

3.8 CPD for Measuring Naturalness in HRI

This section presents one possible application of CPD in HRI. Future chapters will report many more applications of this core technology in achieving autonomous social behavior envisioned in this book.

How do we define naturalness in HRI? It is difficult to find an overarching definition of naturalness that is applicable to all situations. Nevertheless, one common thread of natural robot behavior would be a subjective evaluation of the behavior by human interlocutors of the robot to be *as expected* or to cause no anxiety in them.

This means that we can use human's subjective response to the behavior of the robot as some indicator of the naturalness of this behavior in a given social context. But, how can we measure this response to the robot's behavior? The most direct method is subjective evaluation and measuring it using questionnaires. The limitations of questionnaires are well known: they are cognitively mediated, they are prone to mistakes caused by priming effects resulting from the wording of the questionnaires and people in general are not very accurate in reporting their subjective responses after the fact. For all of these reasons, an objective measure of the human response to robot's behavior is needed.

As an operational definition, we define natural behavior as the behavior that reduces stress, cognitive load and other negative subjective experiences in the robot's interlocutors while maximizing positive responses like engagement and enjoyment. This definition is based on our earlier work (Mohammad and Nishida 2010). Several physiological signals have been shown to correlate with each of these subjective experiences and can be used as a basis for finding an objective measure of naturalness in HRI (Mohammad and Nishida 2010). For a detailed discussion of several of these signals see our earlier work (Nishida et al. 2014).

One problem of using these signals in general is that the correlations with sub-jective state are only clear for extreme situations and it is difficult to elicit these differences in natural interaction situations. This makes it difficult to compare dif-ferent robot behaviors based on them and here we can find a practical application of CPD (Mohammad and Nishida 2009).

Even though it may be difficult to find differences in these physiological signals between natural and unnatural interaction situations, we may be able to find clearer differences in the change scores resulting from applying CPD to these signals. The rationale for this hypothesis is that change point scores represent probable variations in the generating dynamics of the signal and even though that the unnatural day-to-day interaction situations of interest may not have enough effect on the measured physiological signals that can be measured by direct comparison of the signals them-selves, the instability of human perception of the interactions situation is expected to cause such changes.

This instability in situation perception will lead to changes in how the human responds to the robot leading to some changes in the generating dynamics. Even though these changes are not enough for detection using standard signal processing techniques, CPD focuses on the probability that such changes occur (without empha-sizing the degree of change in the signals themselves). In principle, this may lead to more accurate estimation of the subjective state. Mohammad and Nishida (2010) experimentally tested this hypothesis. We used the explanation scenario described in Sect. 1.4.

The participants (44 subjects) where randomly assigned either the listener or the instructor role. The instructor interacts in three sessions with two different human listeners and one humanoid robot (Robovie II). The instructor always explains the same assembly/disassembly task after being familiarized with it before the sessions. To reduce the effect of novelty, the instructor sees the robot and is familiarized with it before the interaction (Kidd and Breazeal 2005).

The listener listened to two different instructors explaining two different assem-bly/disassembly tasks. In one of these two sessions (s)he plays the *Good* listener role in which (s)he tries to listen as carefully as possible and use normal nonverbal interaction behavior. In the other session (s)he plays the *Bad* listener role in which (s)he tries to appear distracted, and use abnormal nonverbal interaction protocol. The listener is free to decide what is *normal* and what is *abnormal*.

We used the following set of features that were calculated from the smoothed signals: Heart Rate (HR), Heart Rate Variability (HRV) as well as raw pulse data (P) were calculated from BVP data. Skin Conductance Level (SCL) and Galvanic Skin Response (GSR) were calculated from skin-conductance sensor data. Respiration Rate (RR), Respiration Rate Variability (RRV), and raw respiration pattern (R) were calculated from the respiration sensor.

We then extracted the statistics usually used in estimating the psychological state of people from their physiological signals (see Shi et al. 2007; Lang 1995; Liao et al.

2005): mean (MEA), median (MED), standard deviation (STD), minimum (MIN) and maximum (MAX) leading to a 160 features for every session.

We employed RSST (See Sect. 3.5) to measure changes in the physiological signal and derived two features from the output of this algorithm: the number of local maxima per second (RSST LMD) and maximum number of local maxima per minute (RSST LMR) were calculated.

Analysis of the data revealed that these two RSST based features were more accurate in distinguishing between natural and unnatural interaction sessions. Moreover, the robot which just moved its head randomly was ranked as unnatural according to these features which accords with our expectations (Mohammad and Nishida 2010).

3.9 Summary

This chapter introduced the first of our three data mining pillars over which the second part of the book will be erected. Change point discovery is the problem of estimating the points at which generating dynamics of a given time-series are changing without access to these dynamics. The chapter introduced several methods applicable to stochastic time-series and model based CPD algorithms as well as shape based CPD algorithms like SSA CPD. The chapter also introduced several methods for comparing CPD algorithms in practical situations focusing on the Equal Sampling Rate (ESR) measure of discovery quality which is the main measure we utilize in our work. Finally, the chapter introduced briefly one practical application of change point discovery in assessing the naturalness of robot behavior based on human response to it using objective physiological signal analysis.

References

Andre-Obrecht R (1988) A new statistical approach for the automatic segmentation of continuous speech signals. IEEE Trans Acoust Speech Signal Process 36(1):29–40

Basseville M, Kikiforov I (1993) Detection of abrupt changes. Printice Hall, Englewood Cliffs, New Jersy

Fearnhead P, Liu Z (2007) On-line inference for multiple changepoint problems. J R Stat Soc: Ser B (Statistical Methodology) 69(4):589–605

Ide T, Inoue K (2005) Knowledge discovery from heterogeneous dynamic systems using change-point correlations. In: SDM'05: SIAM international conference on data mining, pp 571–575

Kidd CD, Breazeal C (2005) human–robot interaction experiments: lessons learned. In: Robot compaions: hard problems and open challenges, pp 141–142

Lang PJ (1995) The emotion probe: studies of motivation and attention. Am Psychologiest 50(5):285–372

Levina E, Bickel P (2001) The earth mover's distance is the mallows distance: some insights from statistics. In: ICCV'01: 8th IEEE international conference on computer vision, IEEE, vol 2, pp 251–256

Liao W, Zhang W, Zhu Z, Ji O (2005) A decision theoretic model for stress recognition and user assisstance. In: AAAI'05: the 20th national conference on artificial intelligence, pp 529–534

Mohammad Y, Nishida T (2009) Measuring naturalness during close encounters using physiological signal processing. In: International conference on industrial, engineering, and other applications of applied intelligence IEA/AIE, pp 281–290

Mohammad Y, Nishida T (2010) Using physiological signals to detect natural interactive behavior. Appl Intell 33:79–92. doi:10.1007/s10489-010-0241-4

Mohammad Y, Nishida T (2011) On comparing SSA-based change point discovery algorithms. In: SII'11: IEEE/SICE international symposium on system integration, pp 938–945

Nishida T, Nakazawa A, Ohmoto Y, Mohammad Y (2014) Measurement, analysis, and modeling, Springer, pp. 171–197 (Chap. 8)

Page E (1954) Continuous inspection schemes. Biometrika 41:100–115

Shi Y, Choi EHC, Ruiz N, Chen F, Taib R (2007) Galvanic skin respons (GSR) as an index of cognitive load. In: CHI'07: ACM conference on human factors in computing systems, ACM, pp 2651–2656

Willsky AS (1976) A survey of design methods for failure detection in dynamic systems. Automatica 12(6):601–611

Chapter 4
Motif Discovery

A recurring problem in the second part of this book is the problem of discovering recurrent patterns in long multidimensional time-series. This chapter introduces some of the algorithms that can be employed in solving this kind of problems for both discrete and continuous time-series. As usual the treatment does not target exhaustiveness but focuses on the algorithms most important for later developments in this book. We try to systematize the varied literature focusing on motif discovery into four categories: algorithms for discrete time-series that can be employed for continuous time-series after discretization, algorithms that find exact motifs with some predefined notion of similarity, algorithms that find approximate motifs using some form of random sampling and algorithms that utilize constraints on motif locations to speedup the processing.

4.1 Motif Discovery Problem(s)

The history of motif discovery is rich and dates back to the 1980s. The problem started in the bioinformatics research community focusing on discovering recurrent patterns in RNA, DNA and protein sequences leading to several algorithms that we will discuss later in this chapter. Since the 1990s, data mining researchers started to shift their attention to motif discovery in real-valued time-series and several formulations of the problems and solutions to these formulations were and are still being proposed. The motif discovery problem in real-valued time-series is harder to pose compared to the discrete case due to the complications arising from the fact that noise and inaccurate productions never allow the pattern to be repeated *exactly* and this forces the practitioner to decide upon allowed distortions to consider two time-series as realizations of the same pattern. Even in the discrete case, noise may cause repetition, deletion or letter-change of a part of the motif occurrences rendering the problem again more difficult and more interesting.

© Springer International Publishing Switzerland 2015
Y. Mohammad and T. Nishida, *Data Mining for Social Robotics*,
Advanced Information and Knowledge Processing,
DOI 10.1007/978-3-319-25232-2_4

Due to these problems, there is no single motif discovery problem in the literature but a set of related problems that all focus on discovering—probably approximately—repeated patterns in the time-series. This makes it much harder to compare different approaches to motif discovery and Sect. 4.7 will be dedicated to our approach for comparing the outputs of these algorithms.

4.2 Motif Discovery in Discrete Sequences

Motif discovery appeared first as the problem of discovering recurrent patterns in discrete sequences. It was inspired by research in bioinformatics which focused on analyzing long sequences of RNA, DNA and proteins (see for example Staden 1989). For our purposes, these sequences are just long strings of an alphabet of predefined letters. This set of letters (alphabet) is $\{A, T, C, G\}$ for DNA, $\{U, T, C, G\}$ for RNA and $\{A, C, D, E, F, G, H, I, K, L, M, N, P, Q, R, S, T, V, W, Y\}$ for proteins. An early use of the term *motif discovery* is attributed to Waterman et al. (1984). In that paper, the recurrent pattern was called a *consensus motif*.

The following definitions will be used throughout this section: An alphabet $\mathcal{A}$ is a finite set of items of length N_A. Each member of an alphabet is called a character c_k^A where the subscript identifies the character number and the superscript identifies the alphabet. We may skip indicating the alphabet when it is clear from the context. A string is a sequence of characters that follows our normal time-series terminology but is usually indexed by letters i, j, k instead of t. For example $S = \{s_i\}$ is string of characters. We will use T_s to indicate the length of string S but may skip the subscript if the string is known from the context. When multiple strings are being considered we will use superscripts to distinguish them (e.g. S^1, S^i, etc.). Substrings are subsequences using our standard time-series terminology and use the same naming conversions where, again, superscripts identify the string if multiple strings are implied. For example: $s_{i,l}^j$ is the subsequence starting at index i of length l of the string number j (called S^j). We usually reserve N_x to indicate the number of strings if a set of strings $\{X^1, X^2, \ldots, X^N\}$ is used. Notice that we will use the terms string, sequence, and time-series interchangeably in this section and as usual any subsequence is a sequence.

A distance function $D_l(p, q)$ is a mapping $|\mathcal{A}|^l \times |\mathcal{A}|^l \rightarrow \mathbb{R}$ that measures how different are the sequences p and q. Examples are the *Levinstein* distance which measures the number of edit operations (mutation, insertion and deletion) that are required to match the two sequences and the *Hamming* distance which measures the number of places in which the two sequences differ. Again when l is known from the context we will drop it and use $D(., .)$.

Given a distance function of specific length, it is possible to define a generalized distance function between strings of different lengths as the minimum distance between the shortest string and all subsequences of its length belonging to the longest string. This generalization can be represented mathematically as:

$$D_{\min(T_p,T_q)}(p,q) = D_{\min(T_p,T_q)}(q,p) = \begin{cases} \min_i D_{T_p}\left(p,q_{i,T_p}\right) & T_p \leqslant T_q, \\ \min_i D_{T_q}\left(q,p_{i,T_q}\right) & otherwise. \end{cases} \quad (4.1)$$

A sequence p is said to occur (or occur exactly) in another sequence q (where $T_p \leq T_q$) iff $D_{T_p}(p,q) = 0$ and is said to occur approximately within a range $r \in \mathbb{R}^+$ iff $D_{T_p}(p,q) \leq r$ and in both cases the shorter sequence p will be said to occur (exactly or approximately) at point (or index or location) τ ($0 \leq \tau \leq T_q - 1$) of q iff $D_{T_p}\left(p,q_\tau,T_p\right) = \min_i D_{T_p}\left(p,q_{i,T_p}\right)$. Notice that this definition means that the locations at which a short subsequence appears in a long subsequence may be the empty set, a singleton set of one item (single occurrence) or a multiple items set (multiple occurrences). In some cases, the locations at which a string p occurs in another may be separated by less than T_p points. In another word, the occurrences of p in q can overlap.

There are two general approaches to motif discovery in discrete time-series that do not only constitute different algorithms but different views of the problem. This means that some MD problems are best understood using the first and others using the later. We call these approaches: Pattern-First and Sequence-First approaches. In Pattern-First problems and solutions, the repeated pattern is not even required to occur in the time-series; only approximations of it may appear. In Sequence-First problems and solutions, the sought motifs MUST occur exactly at least once in the time-series. This means that the space of possible motifs considered for Pattern-first problems is much larger than Sequence-first problems. Consider an alphabet of n_a characters. The total number of possible sequences of length l that can occur is $(n_a)^l$ while considering a time-series of length T, the total number of sequences of the same length that actually occur in it is upper-bound by $T - l$ (this is usually an exaggerated count assuming that all subsequences are distinct). This shows that for realistic values of l and T, pattern-first approaches always consider more motifs than sequence-first approaches.

One of the earliest pattern-first MD problems to be formulated was the *common motif discovery problem* (Sagot 1998), which can be stated as follows:

Definition 4.1 Common Motif Discovery Problem *Given a set of N strings ($\{S^n\}$), a motif length l, a distance function D (of the same length) and a threshold value τ, find all sequences that occur approximately with range τ in at least M out of the N sequences.*

This problem was introduced in slightly different form by Pavesi et al. (2001) but in this case, the length l is considered as an upper limit and the problem is solved for all lengths up to it. Research in solving this problem was partially motivated by the need to discover binding sites in unaligned DNA sequences (e.g. ribosome binding site problem Tompa 1999).

The simplest pattern-driven method is to generate a list called P of all patterns of length l which will have a length of $|\mathcal{A}|^l = N_A^l$. For each member of this table p_i, we find the list of exact occurrence locations of p_i in each string S^n according to our earlier definition. Let's call this set $O_i = \{o_i\}$. We keep the length of this list n_i. These counts ($\{n_i\}$) are stored in a list called N with members n_i storing the count for sequence p_i. For each pattern $p_i \in P$, we then calculate a new count (c_i) which sums the counts of all other member of P that occur approximately within range τ from p_i (i.e. $c_i = \sum_{j|D_l(p_i,p_j) \leqslant \tau} n_j$).

Finally from each subset of P that is within range τ from each other, we keep only the member with maximum c_i value. This approach was used by Staden (1989).

It is evident that this simple algorithm is not scalable as it scales exponentially with motif length. It scales only linearly with the time-series length for the strings involved. This means that the approach is only beneficial for discovering short motifs (small l).

A more subtle problem with this algorithm (and the problem formulation of the common motif discovery problem in general) is that it depends solely on the number of repetitions of a subsequence. This may not be a good indicator of the importance of the subsequence specially for short lengths because short subsequences may occur due to random variations in the data. For most realistic noise models leading to these random variations, some sequences will be more likely to appear randomly due only to being near a larger number of sequences that are usually generated. This will give these sequences an undue advantage and bias the algorithm toward discovering them. For this reason another significance measure is usually used to keep only the patterns that are significant (i.e. having small probability of being generated randomly in the sequences). The most used such measure is the relative entropy between character frequencies in the pattern p_i and character frequencies in the whole sequence set s^k (Pavesi et al. 2001).

To illustrate the sequence-first approach, consider the following problem due to Sagot (1998):

Definition 4.2 Repeated Motif Discovery *Given a string S, a motif length l, a real number τ, and a positive integer $r > 1$, and a distance function D_l, find all sequences of length l that occur in S approximately within τ at r or more different locations.*

This problem can be solved using a pattern-first algorithm by simply finding all possible sequences of length l and counting the number of occurrences within τ distance units using D_l. We can then filter the ones that occur r or more times. This approach though seems extremely wasteful because we will have to consider too many sequences that will have zero occurrences.

A sequence-first solution can be devised based on the idea presented in Fig. 4.1 which represents a metric space showing the ball (circle for 2D) containing all possible sequences that are τ distance units or less from M (the red circle in the center). The maximum possible distance between any of the points within this ball is $2 \times \tau$. This is true under the assumption that the triangular inequality is true for the distance function D_l. Given this assumption a simple sequence-first algorithm for solving the repeated motif discovery problem defined above.

Fig. 4.1 Example of a
metric space showing the ball
(*circle* for 2D) containing all
possible sequences that are τ
distance units or less from M
(the *red circle* in the center)

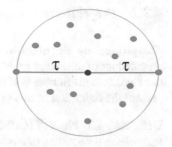

The algorithm firstly scans the input string and builds a list L of all subsequences
of length l occurring exactly in the string. The algorithm then creates a graph $\mathcal{G}$ with
all members of Γ as vertices. An edge is connected between any two vertices γ_1 and
γ_2 iff $D_l (\gamma_1, \gamma_2) \leq 2 \times \tau$. Now the problem is reduced to the standard clique finding
problem. Each fully connected clique in this graph will represent a set of occurrences
that are all within the desired $2 \times \tau$ distance limit from each other. The cliques with r
or more members can then be returned to represent the occurrences of the motif and
the mean of these members can then be used as the motif M. This is a modified version
of the WINNOWER algorithm introduced by Pevzner et al. (2000) for solving the
slightly simpler planted motif problem which fixes the distance function to counting
point replacements and finds motifs that occur exactly τ distance units apart.

An important point about this sequence-first algorithm is that it is quadratic in the
length of the string. This is a common feature of exact MD algorithms for both discrete
and continuous domain time-series and is the main reason why MD is a challenging
problem because quadratic time is not good enough for processing long time-series.
Most of the efforts that will be described later in this chapter to come up with more
advanced algorithms for motif discovery (specially in the continuous domain) will
be directed toward finding subquadratic solutions to the problem. Usually quadratic
solutions are as easy to find as the algorithms we just described.

The aforementioned discussion of the pattern-first and sequence-first approaches
highlights the fuzzy line between them. Even though the *repeated motif discovery*
problem tries to find a motif that may never occur exactly in the time-series, it is still
a sequence-first problem which can be seen by the extreme difference between the
exponential time needed to solve it using a pattern-first algorithm and the quadratic
time needed to solve it by the sequence-first simple alternative. In fact we can say
that this problem is not strictly a sequence-first problem if we limit the definition of
sequence first problems to finding motifs that occur exactly within the time-series at
least once. We will see later that restricted sense of sequence-first problems leads to
the approach commonly known as *exact motif discovery* for real-valued time-series.

We will now turn to one of the most important discrete time-series motif discovery
algorithms which will be the basis of a whole set of algorithms for discovering motifs
in real-valued time-series later.

4.2.1 Projections Algorithm

The projections algorithm tries to solve the planted(n, l) problem related to the repeated motif discovery problem discussed earlier in this section. The exposition of the algorithm given here is based on our earlier book (Nishida et al. 2014). We will present the definition of this problem here again for concreteness:

Definition 4.3 Planted (l, d) **Motif Discovery Problem (PMD)**: *Let* M *be a fixed but unknown sequence (the motif) of length l. Suppose that* M *occurs once in each of N strings* S^n *of common length T, but that each occurrence of* M *is corrupted by exactly d point substitutions in positions chosen independently at random. Given the N sequences, recover the motif occurrences and the motif* M.

The WINNOWER algorithm described earlier can be used to solve this problem by simply using all subsequences from all strings as edges to the graph and only considering subsequences from different strings when generating the edges. This algorithm is clearly quadratic in T which makes it applicable only to short strings.

A more scalable algorithm was proposed by Buhler and Tompa (2002) for solving the special case of Planted(l, 2) problem.

Projections is a seeding step that aims at discovering probable candidates for the hidden motif M. The main idea of the algorithm is to select several random hash functions $f_j \left(s_i^t \right)$ and use them to hash the input sequences. Occurrences of the hidden motif are expected to hash frequently to the same value (called *bucket*) with a small proportion of background noise. Noise sequences on the other hand are not expected to hash to similar values. If the hash functions are different enough (and complex enough), we can use the buckets with largest hits as representing occurrences of the motif. This is an initialization step that can then be refined using the EM algorithm in order to recover the full motif.

There are several internal parameters to Projections that need to be set. Firstly, the hashing functions f_j must be determined and their number J must be decided. Secondly, it is not expected that all occurrences of the same motif will hash to the same bucket for every function so we need an estimate of the number of hits to count as significant for each bucket (h).

Projections use hashing functions that balance ease of computation and versatility. Each function is a mapping from sequences of n characters (the input size) to sequences of K characters where $K < n$. The hashing function $f_j(s; \langle l_1, l_2, \ldots, l_K \rangle)$ simply selects the characters of s at the K positions l_k and concatenates them to create a sequence. This leads to $|\Sigma|^K$ buckets. Notice that we need not store all of these buckets as most of them will be empty which makes the bucket list a sparse matrix for large enough values of K.

Now that we have decided the form of the hashing functions, we need to decide how to select them from all possible hashing functions of that form (there are $K!$ possible such functions). We simply select J random functions of this function set.

Two questions now remain concerning the hashing functions: How to select K and J. A simple way to select K is to select a value that minimizes the probability that two random sequences of length n will hash to the same bucket. This can be satisfied by selecting $K \geqslant \log_{|\Sigma|} \left((T(n-l+1)) / E \right)$ where E is the allowed expected number of random sequences that hash into some bucket (usually selected as some number less than 1). The guiding principle in selecting J is that we want to continue hashing until the number of occurrences of the hidden motif that hash to the same bucket are expected to be greater than some cutoff value s that can then be used to decide which buckets contain possible occurrences of the motif we are after. Buhler and Tompa (2002) have shown that we can select J such that:

$$J = \left\lceil \frac{\log(1-q)}{\log \left(B_{l,p}(s) \right)} \right\rceil, \tag{4.2}$$

where $B_{a,p}(c)$ is the probability that there are fewer than c success at a independent Bernoulli trials each having probability p of success, and p is the probability that an occurrence will hash to a specific bucket which can be calculated as:

$$p = \binom{l-d}{T} \Big/ \binom{l}{T}. \tag{4.3}$$

The final piece of the puzzle is how to select the cutoff number of hits to a bucket to be considered as a candidate occurrence set of a motif (s) which played an important rule in selecting the value of J. Unfortunately, there is no agreed upon method to select s but practical tests have shown that as long as it is near the value of d usually good results can be achieved (e.g. 3–4 for the case of the (20,4)-motif case Buhler and Tompa 2002). The projections algorithm is implemented in the $projections()$ function in MC^2.

Even though, all of the formulations and algorithms discussed so far focus on discrete sequences, it can be directly extended to real world time series by first discretizing the time series then just using one of these algorithms.

4.2.2 GEMODA Algorithm

Gemoda is another MD algorithm that stemmed from bioinformatics research (Jensen et al. 2006). This algorithm though works perfectly with real-valued time-series (even multidimensional time-series) given an appropriately defined distance function. Even though the algorithm is clearly quadratic in the length of the time-series, it exemplifies several basic ideas that will appear again and again in this chapter.

The problem for which Gemoda is an exact solution is another variant of the motif discovery problem that requires slightly different definitions.

Definition 4.4 Gemoda Motif Given a time-series $X = \{x_t\}$, a distance function D, a minimum length L, and a threshold real value τ; a Gemoda motif $\mathcal{GM}$ is a set of subsequences $\{O^i\}$ of the same length W that satisfy the following properties:

- O^i occurs exactly in X.
- $D_L\left(O^i_{k,L}, O^j_{l,L}\right) < \tau$ for all $1 \leq i \leq |\mathcal{GM}|, 1 \leq j \leq |\mathcal{GM}|, 1 \leq k \leq W - L + 1$, and $1 \leq l \leq W - L + 1$.
- There exists no W point subsequences in X ($x_{i,w}$ for $1 \leq i \leq T_x - W + 1$) that can be added to $\mathcal{GM}$ without conflicting with one the previous property. This is called *maximal occurrence coverage*.
- There can be no integers $\hat{W} > W$ such that the second property is still satisfied if each member of $\mathcal{GM}$ is extended from right or left by adding the corresponding values from X by $\hat{W} - W$ points. This property is called *maximal extensibility*.

Notice that Gemoda motifs are maximal in the sense that they satisfy both maximal occurrence coverage and maximal extensibility.

Discovering Gemoda motifs amounts to finding all possible such motifs in an input set of time-series. In the rest of this exposition we will assume that the input is a single time-series, yet extension to multiple input time-series is trivial.

The problem definition of Gemoda motif discovery is the first in this chapter that does not assume apriori a length for motifs but only a minimal motif length L. These types of motifs will be called variable length motifs and their discovery will be of special interest to us because the length of recurring patterns in social robotics datasets is usually impossible to guess apriori. Moreover, the same kind of activity (e.g. gestures) may contain different recurrent patterns of different lengths. A variable length MD algorithm can discover them all in one call.

Gemoda runs in three stages (comparison, clustering, and convolution). We will explain each of them in turn.

During the comparison stage, a similarity matrix $\mathbb{M}_{T-L+1 \times T-L+1}$ is built by scanning the input time-series and calculating all pairwise distances between all subsequences of length L. All entries of $\mathbb{M}$ that are greater than the threshold τ are set to zero and the rest of the elements are set to 1. Depending on the clustering algorithm, the diagonal may be set to 1 or to zero. This results in a new binary matrix $\mathbb{B}$.

Clustering and convolution stages only reference thresholded binary matrix $\mathbb{B}$ and never reference the input time-series X again. This means that once this similarity matrix is created, the algorithm is agnostic about the datatype of the input time-series. For this reason, Gemoda is applicable without any modifications (at least in principle) to real-valued time-series as well as discrete strings.

The clustering stage, clusters the subsequences in the time-series using any clustering algorithm like K-means. Jensen et al. (2006) proposed two simple clustering algorithms based on graph operations. Considering $\mathbb{B}$ as a connectivity matrix representing a graph with $T - L + 1$ vertices and $\mathbb{B}_{ij} = 1$ represents a directed edge from node i to node j, we can either find cliques or connected components in this graph and use them as our clusters. The main advantage of this graph based approach is that it requires no specification of the number of clusters. Moreover, connected components can be found efficiently (in linear time). Clique finding is in general a *NP-complete* problem but for small time-series it can be carried out sufficiently efficiently because connectivity is usually low in $\mathbb{B}$. This step results in a set of clusters $\{c_i^L\}$ where $c_i^L = \{c_{i1}^L, c_{i2}^L, \ldots, c_{in_i^L}^L\}$ and $1 \leq c_{ij}^Z \leq T - L + 1$ represents the starting position of a subsequence of length Z belonging to cluster c_i^Z and occupying index j in this cluster. n_i^L represents the number of subsequences in cluster c_i^L. For convenience, we define a function

$$S^Z (i, j) = x_{c_{ij}^Z, Z}.$$

The function S^Z maps from cluster and index values to the corresponding subsequence. We also define its inverse that reverses that mapping S^{-1} for any value of Z as:

$$S^{-1}\left(x_{c_{ij}^L, L}\right) = c_{ij}^L.$$

After Jensen et al. (2006), we will define the following shortcut notation: $x_{i,l} + 1 \equiv x_{i+1,l+1}$. This means that $S^Z (i, j) + 1 \equiv x_{c_{ij}^Z + 1, Z + 1}$.

If clique finding is used for clustering, the set of resulting clusters all satisfy the definition of Gemoda motifs of length L. These short motifs are called motif *stems*. Given these stems, the convolution stage tries to extend them as far as possible while preserving the definition of Gemoda motifs and in the same time removing any redundant motifs when longer motifs are discovered. Many algorithms for real-valued time-series that will be considered later in this chapter are based on this same idea.

The final convolution stage is the heart of the Gemoda algorithm and it tries to extend the motif stems found in the clustering stage. The algorithm uses only the cluster set $\{c_i^L\}$ and does not access $\mathbb{B}$, $\mathbb{M}$, or X.

Given the set $\{c_i^Z\}$ for some value of L, the algorithm generates a new set $\{c_i^{Z+1}\}$ of one point longer motifs and also discovers all maximal motifs of length Z from the set $\{c_i^Z\}$ that we call $\{m_i^Z\}$ hereafter. The final maximal motif set is then found as the union of all of these maximal motif sets:

$$\{\mathcal{GM}\} = \cup_{z=L}^{\infty} \{m_i^Z\}.$$

To understand the convolution step, we need to define two more operations: directed intersection ($\cap$), and motif inclusion ($\ddot{\subset}$). Directed intersection finds the set of stems in its right hand that can *extend* the any stems on its left hand side and generates the extended stems. Formally:

$$c_i^Z \cap c_j^Z \equiv c_k^{Z+1},$$

where c_k^{Z+1} is the set of all indices q such that

$$\mathcal{S}^{-1}\left(\mathcal{S}^Z\left(c_{iq}^Z\right)+1\right) \in c_j^Z.$$

Directed inclusion allows us to find the stems that can be extended one point (even partially by extending a subset of its occurrences) without violating Gemoda motif definition. If no such extension is possible (signaled by all directed inclusions of c_i^Z with all other members of $\left\{c_j^Z\right\}$ resulting in the empty set), we can simply add c_i^Z to $\left\{m_i^Z\right\}$ because it is now *proved* to be a maximal motif.

If $c^Z \cap c_j^Z = c_{new} \neq \phi$, then c_{new} represents a proposed motif of length $Z + 1$. Before adding this motif to the list $\left\{c_k^{Z+1}\right\}$, we first have to make sure that no previously added motif to this set includes this proposed new motif. This is where we need the motif inclusion operator ($\ddot{\subset}$). The motif inclusion operator returns true only if its left operand is either a superset of a subset of the motif represented by its right operand.

We only add c_{new} to $\left\{c_k^{Z+1}\right\}$ if $c_{new} \ddot{\subset} c_k^{Z+1}$ returns false for all k. If $c_{new} \ddot{\subset} c_k^{Z+1}$ was true for some value k, we combine c_{new} and c_k^{Z+1} using a standard union operation.

The convolution step repeats the aforementioned steps from $Z = L$ until for some value L_{max}, the set $\left\{c^{L_{max}}\right\}$ is the empty set which can be shown to always occur for some value of $L_{max} \leq T$. Actually L_{max} will be much less than T.

It can be shown that this convolution operation when seeded with Gemoda motifs will only return Gemoda motifs at all lengths. It can also be shown that it is an exhaustive operation which means that if the Gemoda motifs of the first length passed to it were representing all possible Gemoda motifs at that length, then by the end of the convolution operation, all Gemoda motifs at all lengths will be discovered.

Gemoda is implemented in the MC^2 toolbox by the function *gemoda*(). In our implementation, it is possible to restrict the allowable overlap between different motifs and different motif occurrences. It is also possible to build the similarity matrix from a predefined set of locations in the time-series rather than using exhaustive search. These options render Gemoda an approximate algorithm but they can significantly increase its speed. This is the simplest way to utilize knowledge about probable motif locations in Gemoda (See Sect. 4.6).

4.3 Discretization Algorithms

The first algorithms for discovering real-valued motifs in real-valued time-series to be considered in this chapters are discretization algorithms. These algorithms were among the first to be introduced and tried to take advantage of the wealth of research in motif discovery for discrete sequences in the bioinformatics literature. Many of these early algorithms depended on the projections algorithm described in the previous section.

For real valued time-series, distances can rarely (if ever) be exactly zero. This requires us to use a softer definition of a match between two sequences in this case. We say that two time-series X and Y match up to $\tau \in \mathbb{R}$ given a distance function $D(.,.)$ iff $D(X, Y) \leq \tau$.

Another problem that we face usually in real-valued time-series analysis is the problem of trivial matches. A trivial match is a match that happens between two subsequences of a time-series $x_{i,l}$ and $x_{j,l}$ just because i and j are very near making the two subsequences overlapping. For a smooth time-series X, it is expected that x_i and $x_{i+\delta}$ will be small for small values of δ. This is the underlying cause of trivial matches. For all algorithms considered in this section and the following ones, we will always assume that trivial matches are ignored. The simplest method to ignore trivial matches is to set $D\left(x_{i,l}, x_{j,l}\right) = \infty$ for all $j \leq i + l$. This forces all occurrences of all motifs discovered to have no overlaps.

One of the earliest discussions of motif discovery in real-valued time-series and the paper that first used the term explicitly was by Lin et al. (2002). In that paper motifs were defined as follows:

Definition 4.5 *K-Motifs*: (Lin et al. 2002) Given a time series x, a subsequence length l and a range R, the most significant motif in x (called thereafter 1st-*Motif*) is the subsequence C_1 that has the highest count of non-trivial matches (ties are broken by choosing the motif whose matches have the lower variance). The Kth most significant motif in x (called thereafter K-*Motif*) is the subsequence C_K that has the highest count of non-trivial matches, and satisfies $D(C_K, C_i) > 2R$, for all $1 \leq i < K$.

Note that a K-*Motif* according to this definition is a subsequence that exactly occurs within the time-series. This is clearly a sequence-first problem and, as we will see, the sequence-first approach dominates real-valued time-series motif discovery in most of its forms with the exception of the newly suggested optimal motif discovery problem which forms a return to the earlier pattern-first approach to motif discovery.

One of the simplest discretization based algorithms that were suggested in literature for discovering the *1-Motif* problem is to discretized the input time-series using SAX (See Sect. 2.3.2) then just apply the projections algorithm (See Sect. 4.2.1). This was the approach taken in Chiu et al. (2003). Following this milestone paper, several other researchers used variations of this approach for motif discovery (see of example Minnen et al. 2007; Tang and Liao 2008). This approach can be implemented in the MC^2 toolbox in two lines by calling $sax()$ followed by $projections()$.

Because projections is only a stemming operation, it only finds pairs of probable motif candidates. After recovering these pairs from the projections matrix, we need to scan the original real-valued time-series and find all subsequences that are no more than R distance units far from either of the two subsequences corresponding to the two members of the stem. This is a linear time operation.

Several extensions have been suggested over the years to this general framework (Minnen et al. 2007; Tang and Liao 2008; Tanaka et al. 2005). We here describe one such extension that represents this body of work and present a complete related approach in Sect. 4.3.1.

One limitation of the method proposed by Chiu et al. (2003) and discussed earlier is the need to specify R (the range). Minnen et al. (2007) proposed a method for automatically selecting this range dynamically for each proposed motif stem found through projections. This approach bears some similarity to the approach discussed in Sect. 3.5 while introducing RSST for the selection of the appropriate dimensionality of the past Hankel matrix.

Given a motif stem from the projection step consisting of two subsequences of length L ($x_{i,L}^1$ and $x_{j,L}^2$), we start by scanning X to find the distance from each subsequence of length L in it and this stem which is defined as:

$$D_k = \min\left(D_L\left(x_{k,L}, x_{i,L}\right), D_L\left(x_{k,L}, x_{j,L}\right)\right).   \tag{4.4}$$

The vector $\{D_k\}$ will have length $T - L + 1$. We then calculate the α's order statistic where usually α is taken as $T/10$ and keep only the α smallest distances in D_k leading to the α long vector $\hat{D}_k$. We now sort $\hat{D}_k$ in ascending order generating $\tilde{D}_k$ and find the knee in the curve by finding the first difference $\delta D_k = \tilde{D}_{k+1} - \tilde{D}_k$ and find the estimated range as the expected value of this derivative:

$$\hat{R} = \mathbb{E}\left(\delta D\right) = \frac{\sum\limits_{k}^{\alpha-1} \tilde{D}_k \times \left(k + \frac{1}{2}\right)}{\sum\limits_{k}^{\alpha-1} \tilde{D}_k}.   \tag{4.5}$$

This value of the range can then be used as usual to find other occurrences of the motif represented by the stem.

Li and Lin (2010) proposed using the Sequitur (Nevill-Manning and Witten 1997) algorithm for discovering motifs of variable length using grammar inference after discretization using SAX (Lin et al. 2007). This technique can discover variable length motifs but a small burst of outliers in a single motif occurrence will result in dividing this motif into two disjoint motifs.

4.3.1   MDL Extended Motif Discovery

This algorithm was proposed by Tanaka et al. (2005) to discover motifs of any length starting from some minimal length L_{min}. The main new features of this algorithm compared with the projections based approach of Chiu et al. (2003) are the following:

- The input is assumed to be multidimensional time-series instead of a single-dimensional time-series.
- The user needs to provide a minimal motif length but the algorithm can still find motifs of longer lengths.
- The algorithm uses a principled MDL based approach for selecting the importance of different motifs.

The algorithm receives a multidimensional input time-series $X^{md} = \{x_t^{md}\}$ of length T. The first step is to convert it to a single-dimensional time-series. PCA can be used to achieve this as previously discussed in Sect. 2.5.5. This leads to a single-dimensional time-series X that will be used for the rest of the algorithm.

The second step of the algorithm is discretization using SAX as described in details in Sect. 2.3.2 resulting in a $T - w + 1 \times w$ matrix of symbols called S where w is the length of the window used for SAX. Notice that in this case, we convert ALL subsequences of X to symbols rather than just converting X (which would have resulted in a $T/w \times w$ symbols matrix instead). Each row of this matrix is called a word hereafter and represent the SAX transformation of a subsequence of length w starting at the same index as the word's row index. Notice that application of SAX requires other than the window size w, the specification of alphabet size and number of time-series points per symbol.

The third step is to build a dictionary of *behavior symbols* that represents the shape of each one of these words. This has two advantages: firstly, we reduce the storage space required by storing a single symbol for each word and secondly, we detect words of similar behavior by giving them the same behavior symbol. Each word is then replaced by a unique symbol from the set $\mathcal{B}$ that represent all possible behavior symbols found in the time-series. This results in a discrete time-series of behavior symbols BS of length $T - w + 1$.

The string of behavior symbols BS is then scanned to generate all substrings (subsequences) of length $L_{min} - w + 1$ which corresponds to L_{min} points of the original time-series (X or X^{md}). This results in another matrix of length $(T - L_{min} + 1) \times L_{min} - w + 1$ words representing the behavior of subsequences of length L_{min} in the original time-series. This matrix is called M hereafter.

Now we start digging for motifs in M by finding repeated words (rows). The rows are sorted from the one with maximum repeat count to the one with minimum repeat count in a sorted matrix SB.

For each element of SB (in order), we find all of its appearances in M and consider them a motif candidate. Let the number of these occurrences be K. Even though these K words are the same after discretization, the time-series subsequences on X^{md} corresponding to them may not satisfy the required maximum distance threshold (τ). Let the function $\mathcal{S}$ map between the index of the behavior symbol word in M and the corresponding time-series sequence in X^{md} (or X). Call the set of indices of the behavior symbol words in the current motif candidate $\{L_i^{sb}\}$. We find the corresponding subsequences in X^{md} as $L_i^{ts} = \mathcal{S}\left(L_i^{sb}\right)$. A similarity matrix $\mathcal{M}$ is then created where:

$$\mathcal{M}_{ij} = D_{L_{min}}\left(x_{L_i^{ts}, L_{min}}, x_{L_j^{ts}, L_{min}}\right). \tag{4.6}$$

A thresholded version of $\mathcal{M}$ called $\mathcal{TM}$ is then created by setting the diagonal using:

$$\mathcal{TM}_{ij} = \begin{cases} 0 & i = j \vee \mathcal{M}_{ij} > \tau \\ 1 & otherwise \end{cases}. \tag{4.7}$$

Now counting the number of nonzero elements in any row (or column for symmetric distance functions) of $\mathcal{T M}$ gives the number of time-series subsequences that are within the acceptable range defined by τ of every candidate occurrence. This is called *connectedness*. This indicates how *central* this occurrence is given this motif. For this reason the subsequence with maximum connectedness is called the *center* of the motif. Usually, multiple occurrences will share the maximum value of connectedness and in this case we select the *center* as the subsequence of them with the minimum total distance to all other occurrences in $\mathcal{M}$. Once the center is selected, all subsequences connected to it according to $\mathcal{T M}$ are combined with it to form the final motif candidate MC.

This process can be repeated for all words in SB and then for all lengths from L_{min} to $T/2$. This results in a large set of motif candidates. We need a way to select the most significant motifs from this large list. Firstly, we can remove candidates that are completely covered by longer candidates. Still this gives us an indication of motif lengths but not their significance. To calculate motif significance, we use an MDL approach proposed first by Tanaka et al. (2005) which is explained for the rest of this section.

Assume that we have the behavior symbols representing a motif candidate represented by $\mathcal{MC}$ with n_p behavior symbols in each. Assume that we have n_s different symbols in this $\mathcal{BS}$. We can then calculate the description length (e.g. the number of bits required to describe this motif candidate) as:

$$DL\left(\mathcal{MC}\right) = \log_2 n_p + n_p \log_2 n_s. \tag{4.8}$$

We would like also to know the description length of the time-series given that we are going to replace all of the occurrences of this motif with a symbol. This gives us $DL\left(BS|\mathcal{MC}\right)$. Assume that the total length of X^{ml} was T. We know that $|BS| = T - L_{min} + 1$. This means that the length of the time-series after replacing all occurrences of $\mathcal{MC}$ with a single symbol will be: $n_T = T - L_{min} + 1 - (n_o) \times (L_{MC} - 1)$ where L_{MC} is the length of the subsequences in X^{ml} corresponding to any occurrence of $\mathcal{MC}$ and n_o is the number of occurrences of $\mathcal{MC}$ in BS. Now assume that there are m_s unique symbols in BS. Given these definitions, we can find the description length of BS given $\mathcal{MC}$ as:

$$DL\left(BS|\mathcal{MC}\right) = \log_2 n_T + n_T \log_2 \left(n_o + m_s\right). \tag{4.9}$$

Given Eqs. 4.8 and 4.9, we can find the minimum description length corresponding to accepting the motif $\mathcal{MC}$ as:

$$MDL\left(BS, \mathcal{MC}\right) = DL\left(BS|\mathcal{MC}\right) + MDL\left(\mathcal{MC}\right), \tag{4.10}$$

leading to the final estimation Equation:

$$MDL\left(BS, \mathcal{MC}\right) = \log_2 n_T + n_T \log_2 \left(n_o + m_s\right) + \log_2 n_p + n_p \log_2 n_s. \tag{4.11}$$

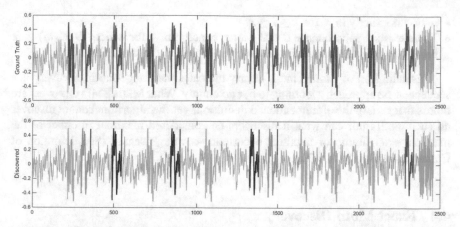

Fig. 4.2 An example time-series with implanted motifs (*top*) and the results of discovery using MDL extended motif discovery algorithm

This MDL value is calculated for all candidate motifs discovered and used to sort them ascendingly. A cut-off value can then be used to select the motifs to be returned to the user. The toolbox implements this algorithm by the function $tanaka()$. Figure 4.2 shows an example application of it to a time-series with implanted motifs.

This algorithm (MDL EMD) can discover motifs of a range of lengths (in that it is similar to GEMODA and different from Projections) but it is an approximate algorithm in the sense that it has no guarantees of discovering maximal motifs or even to discover any motifs. That is in direct contrast with GEMODA which has good theoretical guarantees (at least for discrete sequences). One main cause of this problem is the reliance on the discretization step which introduces noise in distance calculations. The algorithm introduced in the following sections work directly with the real-valued time-series to avoid this problem.

An important idea that was introduced by MDL EMD is the importance of relying on a significance measure other than the length of the motif. By basing the decision of motif importance on MDL, this algorithm is biased toward longer more complex patterns that are usually of interest for the practitioner.

The algorithm also exemplifies a general method that we will use again and again to avoid the generation of the full similarity matrix used in GEMODA like algorithms. It tries first to generate candidate motifs using a heuristic (the discretized behavior symbols in this case) and builds a similarity matrix only of these candidate motifs. It is important to make a clearer definition of what is discovered by this algorithm.

Definition 4.6 Range Motif Given a time-series X (possibly multidimensional), a symmetric distance function $D(.,.)$ and a real-valued threshold τ called the range, A range motif $\mathcal{RM}$ is a set of subsequences $\{o_i\}$ of cardinality N that satisfy the following properties:

- o_i occurs exactly in X for $1 \leq i \leq N$.
- There exists a subsequence C that occurs exactly in X for which: $D(C, o_i) \leq \tau$ for $1 \leq i \leq N$. C is called the *center* or *exemplar* of the motif $\mathcal{RM}$.

This definition of a range motif clearly indicates that range-motif discovery is a sequence-first approach according to our taxonomy. What MDL EMD discovers are motifs that satisfy the aforementioned definition, yet the algorithm cannot discover all such motifs nor can what it discovers be considered maximal in either of the maximal extensibility or maximal occurrence coverage senses (See definition of Gemoda motifs in Sect. 4.2.2).

4.4 Exact Motif Discovery

The methods discussed in Sect. 4.3 discover motif stems in the discretized version of the time-series then solves a motif detection rather than a motif discovery in the original time-series which can be solved efficiently.

Another possible approach is to generate the stems directly from the time-series. Any exact solution to this problem will be quadratic in the worst case but some tricks can be used to speed up this operation significantly achieving amortized linear speeds as will be discussed in this section. This is currently one of the leading approaches for solving motif discovery problems and we will present a series of algorithms all based on the groundbreaking work of Mueen et al. (2009b).

Definition 4.7 *Exact Motif*: (Mueen et al. 2009a) An Exact Motif is a pair of L-point subsequences $x_{i,L}$, $x_{j,L}$ of a time series x that are most similar. More formally, $\forall_{a,b,i,j}$ the pair $\{x_{i,L}, x_{j,L}\}$ is the exact motif iff $D(x_{i,L}, x_{j,L}) \leq D(x_{a,L}, x_{b,L})$, $|i - j| \geq w$ and $|a - b| \geq w$ for $w > 0$.

This definition defines exactly a single *exact motif* with exactly two occurrences which is simply the pair most similar in the time-series. What we care about though are recurrent motifs and we care about finding more than a single motif. For this purpose we make the following extended definition:

Definition 4.8 *Top K Pair-Motifs*: The top K Pair Motifs is an ordered list of K pairs of L-point subsequences $x_{i,L}^k$, $x_{j,L}^k$ of a time series x that are most similar. More formally, $\forall_{a,b,i,j}$ the pair $\{x_{i,L}^1, x_{j,L}^1\}$ is the first pair-motif iff $D\left(x_{i,L}^1, x_{j,L}^1\right) \leq D(x_{a,L}, x_{b,L})$, $|i - j| \geq w$ and $|a - b| \geq w$ for $w > 0$. The kth pair-motif is the pair that satisfy the aforementioned definition after ignoring the first to the $(k - 1)$th pair motifs.

The distance function of choice for pair-motif discovery is the Euclidean distance between z-score normalized time-series defined as:

$$D_{z-score}(x, y) = \sum_{k=0}^{L} \left(\frac{x(k) - \mu_x}{\sigma_x} - \frac{y(k) - \mu_y}{\sigma_y} \right)^2, \qquad (4.12)$$

where μ_i and σ_i are the mean and standard deviation of time-series i.

4.4.1 MK Algorithm

The most basic algorithm discussed in this section is the MK algorithm used to solve the exact motif discovery problem (See Definition 4.7) (Mueen et al. 2009b).

The main idea of the MK algorithm is to use two features of the Euclidean distance to speedup calculation:

- Early abandoning of distance calculation. The summation in Eq. 4.12 can be abandoned earlier if we know that it already exceeded the current minimum subsequence distance because summed elements are always positive.
- The triangular inequality can be used to ignore certain distance calculations completely because it gives a lower bound on them that is higher than the current minimum subsequence distance.

To understand the role of triangular inequity, consider the following situation where a, b, and c are all time-series and D is the Euclidean distance: $D(a, b) = 50$, and $D(a, c) = 200$. The triangular inequality in this case can be stated as:

$$D(b, c) \geqslant D(b, c) - D(a, c), \qquad (4.13)$$

$$\therefore D(b, c) \geqslant 150.$$

Now, we can immediately conclude that $D(b, c) \geq D(a, b)$ without bothering to calculate it. The MK algorithm is a clever way to exploit this basic idea combined with early abandoning of distance calculations that leads practically to linear performance for the discovery of exact motifs.

The MK algorithm works on two stages. The first stage (called the referencing stage hereafter), calculates the z-score normalized distances between all subsequences of length L and a set of R randomly chosen subsequences of the same length (called $\{ref_r\}$). This parameter (R) is usually selected something around 8 based on the recommendation of Mueen et al. (2009b). This results in R vectors of length $T - L + 1$ each (call these D^r where $1 \leq r \leq R$). We calculate the variance of these vectors and find the one with maximum variance (let the corresponding best random subsequence $x_{b,L}$) and order it ascendingly (call this list $\mathcal{D} \equiv \{D_i^b\}$ for $1 \leq i \leq T - L + 1$). We assume that the function $\mathcal{F}(i)$ gives the subsequence in x corresponding to item t in $\mathcal{D}$ and the function $\mathcal{G}_r(i)$ that returns the distance $D_{z-score}(ref_r, x_{i,L})$ where ref_r is the reference subsequence at index r. These functions can easily be implemented as constant-time lookup operations by trivial book-keeping during the referencing stage.

The justification of this decision is that the distances to the corresponding random subsequence ref_r will have maximum variability which allows the triangular inequality criterion discussed before to remove more pairs from consideration. This stage is clearly linear in the length of the time-series. Notice that during this stage and for the rest of the algorithm we always keep the pair with minimum distance $\{x_{m_1,L}, x_{m_2,L}\}$ and the value of this minimum distance d_{min}. During the second stage of the algorithm, any distance calculation that exceeds the current value of d_{min} will be abandoned.

The referencing stage gives us an ordered list of distances between one selected random subsequence of length L (ref_r) and all other subsequences in the time-series $(\mathcal{D})$. The second stage (called the scanning stage) utilizes this list to remove as many pairs as possible from consideration without distance calculation. When distance calculation is deemed necessary, we use early abandoning of the summation operation as explained earlier to further speedup the operation.

The scanning stage uses $\mathcal{D}$ to order the pairs of subsequences considered for distance calculation. To understand this stage of the algorithm consider the first three elements of $\mathcal{D}$.

$$D_{z-score}(r, \mathcal{F}(1)) - D_{z-score}(r, \mathcal{F}(2)) \geqslant D_{z-score}(ref_b, \mathcal{F}(1)) - D_{z-score}(ref_b, \mathcal{F}(3)).$$
(4.14)

From the triangular inequality, we know that:

$$D_{z-score}(\mathcal{F}(2), \mathcal{F}(1)) \geqslant D_{z-score}(ref_b, \mathcal{F}(2)) - D_{z-score}(ref_b, \mathcal{F}(1)),$$
$$D_{z-score}(\mathcal{F}(3), \mathcal{F}(1)) \geqslant D_{z-score}(ref_b, \mathcal{F}(3)) - D_{z-score}(ref_b, \mathcal{F}(1)).$$
(4.15)

Considering Eqs. 4.14 and 4.15, it is trivial to show that there exists two non-negative values γ and δ such that:

$$\therefore D_{z-score}(\mathcal{F}(1), \mathcal{F}(2)) \geqslant \gamma,$$
$$D_{z-score}(\mathcal{F}(1), \mathcal{F}(3)) \geqslant \gamma + \delta.$$
(4.16)

Equation 4.16 suggest that the lower bound on the distance between the two subsequences corresponding to the first two elements of $\mathcal{D}$ is lower than that for the first and third subsequences which leads the following heuristic. Consider the pair $\{\mathcal{F}(1), \mathcal{F}(2)\}$ before the pair $\{\mathcal{F}(1), \mathcal{F}(3)\}$. The same reasoning can be extended to lead to the following general heuristic:

Given three integers $1 \leq i \leq j \leq k \leq T - L + 1$, consider pair $\{\mathcal{F}(i), \mathcal{F}(j)\}$ before considering pair $\{\mathcal{F}(i), \mathcal{F}(k)\}$ for being the exact motif.

This suggests that we scan $\mathcal{D}$ taking first consecutive indices, then ones with one index in the middle, and one with two indices in the middle, etc. This can be done in $T - L$ loops each with $T - L + 1$ distance calculations in the worst case. It seems that this will be quadratic. In practice though, only few of these runs will be enough and the rest of the calculations will not be done by virtue of another application of the triangular inequality as will be explained later. We call the offset between considered indices α and it takes the values from 1 to $T - L$.

Because this is just a heuristic, in some cases the pair considered later will have less distance that the pair considered earlier. This means that we still need to calculate these distances.

The triangular inequality can be utilized once more to make this operation as efficient as possible. We can do that by first calculating R lower-bounds for the distance between the candidate subsequence starting locations and only calculate the Euclidean distance when any of these lower-bounds is lower than the current minimum distance d_{min}. This can be put formally as:

$$\forall 1 \leqslant r \leqslant R \wedge 1 \leqslant i \leqslant T - L + 1 \wedge 1 \leqslant j \leqslant T - L + 1,$$
$$D_{z-score}\left(\mathcal{F}(i), \mathcal{F}(j)\right) \geqslant \Delta D_{ij}^{r} \equiv D^{r}\left(\mathcal{F}(i)\right) - D^{r}\left(\mathcal{F}(j)\right). \qquad (4.17)$$

We simply calculate $\Delta D_{i,j}^{r}$ for all r and only calculate $D_{z-score}\left(\mathcal{F}(i), \mathcal{F}(i)\right)$ when at least one of these values is less than d_{min}.

When a full scan with some value of α results in no change in the best pairs considered, then it can be shown that all other distance calculations need not be considered as the triangular inequality (once more) will rule them out (Mueen et al. 2009b).

The MK algorithm is implemented in the function $mk()$ in the MC^2 toolbox which calls a slightly modified version of the MK utility written in C and provided by Mueen et al. (2009b).

4.4.2 MK+ Algorithm

A naive way to extend the MK algorithm to discover top K pair-motifs rather than only the first is to apply the algorithm K times making sure in each step (i) to ignore the $(i - 1)$ pair-motifs discovered in the previous calls. To ignore trivial matches, a maximum allowable overlap between any two motifs can be provided and all candidates overlapping with an already discovered motif with more than this allowed maximum are ignored. A similar idea was used in (Mueen et al. 2009b) to discover an appropriate ordering in anytime nearest neighbor applications. This is called *naive MK* hereafter.

MK+ was proposed to overcome the slow performance of naive MK for discovering top K pair motifs (Mohammad and Nishida 2012b). The main idea of MK+ is to keep a sorted list of pairs and corresponding distances instead of d_{min} used in MK. By keeping this list updated appropriately, the algorithm can discover the top K pair-motifs with a single referencing and scanning stages. Mohammad and Nishida (2012b) showed empirically that this approach can achieve an order of magnitude speedup compared with naive MK even for medium sized time-series.

The algorithm keeps an ordered list $D^{best} \equiv \{d_k^{best}\}$ of K elements at most containing the lowest distances obtained at any point from all subsequence distance calculations so far. We assume that the pair of functions $\mathcal{I}_1(k)$ and $\mathcal{I}_2(k)$ can return the starting positions of the two subsequences of length L corresponding to the distance recorded in d_k^{best}. These two functions can be implemented as constant time operations by appropriate book-keeping. To simplify the exposition of the algorithm we will only keep track of updates to D^{best} and will ignore all of this book-keeping. The source code of the implementation provided in the MC^2 toolbox ($MK+.cpp$) provides detailed explanation of this book-keeping. A MATLAB binding is also available in $mkp()$.

In contrast to MK, MK+ needs a way to avoid trivial matches because it generates multiple pair-motifs as its output. This is achieved by considering the distance between any two subsequences overlapping at more than $w_{external} \times L$ points as infinite. Hereafter, we call two subsequences overlapping only if this condition is met.

During referencing stage and when a new distance d_{ij} between a pair $\langle i, j \rangle$ is larger than $d_{K_-}^{best}$ where K_- is the number of elements currently in D^{best} and $K_- < K$, we just insert d_{ij} at location $K_- + 1$. When d_{ij} is less than $d_{K_-}^{best}$ where $K_- \leq K$, the list is updated by looping over D^{best} elements executing the following rules: If the new pair is overlapping the corresponding $\langle \mathcal{I}_1(k), \mathcal{I}_s(k) \rangle$ pair, then $d_k^{best} \leftarrow min\left(d_{ij}, d_k^{best}\right)$. If they are not overlapping then, again, $d_k^{best} \leftarrow min\left(d_{ij}, d_k^{best}\right)$ but in this case, we will need to remove any elements of D^{best} that are overlapping the pair $\langle i, j \rangle$ to avoid trivial matches. A more detailed discussion of this process and its justification is provided by Mohammad and Nishida (2012b). This same procedure will be used during the scanning stage whenever the best distance so far needs to be checked and updated.

One complicating factor of the simple story we told so far is that it is possible that a pair of subsequences gets removed because of an overlap with another pair with smaller distance which in turn gets removed because of another overlap. To make sure of the algorithm exactness, we keep a linked list attached to each position of the D^{best} list and use it to save such removed pairs on the first overlap event and allow them to be recovered when the second overlap event happens.

It can be shown easily that with appropriate book-keeping, MK+ provides an exact solution to the top K pair motifs discovery problem. Nevertheless, that is not enough for most applications to find recurring patterns.

A simple idea to use the top K pair motifs for general motif discovery is to find them for a large value of K. A similarity matrix can then be used to compare them and GEMODA like convolution process can then be used to recover approximate Gemoda motifs at variable lengths (See Sect. 4.2.2). To recover range motifs from these pairs, we can keep increasing K until the maximum distance in D^{best} is larger than the user-selected threshold (τ). At this point, we know that these K motifs contain all subsequences that are no more than τ distance unites apart. A similarity matrix can then be built and clusters in this matrix can be used to recover range motifs.

Another approach is to combine pair-motifs into clusters using connected components directly from D^{best} and then use an extension mechanism like GSteXB that will be discussed in Sect. 4.5 to extend discovered motifs to higher lengths (Mohammad and Nishida 2012b).

A third approach that will be pursuit in the following sections for extending MK+ to higher lengths is to find top K pair motifs at multiple lengths at the same time, then just use clustering to recover approximate range motifs without the need for motif extension.

4.4.3 MK++ Algorithm

If the distance function we use is the standard Euclidean distance, or if it is even mean normalized Euclidean distance (defined by applying the Euclidean distance after subtracting the means), then we can use the distances at one length to lower-bound distances at higher lengths and this can provide another heuristic to be combined with the triangular inequality to speedup calculations at higher lengths. This is the main idea behind the MK++ algorithm that finds top K pair motifs at all lengths starting from some minimum length L.

To use lower length distances to bound higher length distances, the distance function must satisfy the following condition:

$$D_l(i, j) \leq D_{l+1}(i, j) \ \forall l > 0 \land 1 \leq i; j \leq T - l + 1. \tag{4.18}$$

If subsequences are not normalized and the distance measure used is the Euclidean is used for $D(., .)$, it is trivial to show that Eq. 4.18 holds. Subtracting the mean of the whole time-series does not change this result. It is not immediately obvious that normalizing each subsequences by subtracting its own mean will result in a distance function that respects Eq. 4.18. Nevertheless, we can prove the following theorem:

Theorem 4.1 *Given that the distance function $D(., .)$ is defined for subsequences of length l as:*

$$D_l(i, j) = \sum_{k=0}^{l-1} \left[(x_{i+k} - \mu_i^l) - (x_{j+k} - \mu_j^l) \right]^2, \tag{4.19}$$

and $\mu_i^n = \frac{1}{n} \sum_{k=0}^{n-1} x_{i+k}$; then

$$D_l(i, j) < D_{l+1}(i, j) \ \forall l > 0 \land 1 \leq i; j \leq T - l + 1.$$

A sketch of the proof was given by Mohammad and Nishida (2015a) and is given below for concreteness: Let $\Delta_k \equiv \Delta_{j+l}^{i+k} \equiv x_{i+k} - x_{j+k}$ and $\mu_{ij}^l \equiv \mu_j^l - \mu_j^l$, then the definition of $D_l(i, j)$ can be written as:

$$D_l(i, j) = \sum_{k=0}^{l-1} \left(\Delta_k - \mu_{ij}^l\right)^2 = \sum_{k=0}^{l-1} (\Delta_k)^2 - 2 \sum_{k=0}^{l-1} \left(\Delta_k \mu_{ij}^l\right) + \sum_{k=0}^{l-1} \left(\mu_{ij}^l\right)^2,$$

$$D_l(i, j) = -l\left(\mu_{ij}^l\right)^2 + \sum_{k=0}^{l-1} (\Delta_k)^2.$$

We used the definition of μ_{ij}^l in arriving at the last result. Using the same steps with $D_{l+1}(i, j)$, subtracting and with few manipulations we arrive at:

$$D_{l+1}(i, j) - D_l(i, j) = \Delta_l^2 + l\left(\mu_{ij}^l\right)^2 - (l + 1)\left(\mu_{ij}^{l+1}\right)^2. \qquad (4.20)$$

Using the definition of μ_{ij}^l, it is straight forward to show that:

$$(l + 1)\,\mu_{ij}^{l+1} = l\mu_{ij}^l + \Delta_l. \qquad (4.21)$$

Substituting in Eq. 4.20 and rearranging terms we get:

$$D_{l+1}(i, j) - D_l(i, j) = \Delta_l^2 + \left(\mu_{ij}^l\right)^2 - 2\mu_{ij}^l \Delta_l = \left(\Delta_l - \mu_{ij}^l\right)^2 \geqslant 0.$$

This proves Theorem 4.1.

The MK++ algorithm starts by applying MK+ as explained in Sect. 4.4.2 to find top K pair motifs at length L. The algorithm then progressively uses Theorem 1 to lower-bound the distances at higher lengths during the referencing phase. The same D^{best} list from every length can then be used to seed the search at higher lengths. Updating the distances in D^{best} to the higher length may take them out of order. We then have to re-sort D^{best} in ascending order and update the *best-so-far* accordingly. This gives us a—usually—tight upper bound on the possible distances for the top K pair-motifs at the higher length. This is specially true if the increment in motif length is small and the time-series is smooth. This is the source of the speedup achieved by MK++ over repeated application of MK+ which has to find these lists from scratch at every length. Moreover, the referencing step will not need to be completely recalculated at every length because of the lower-bound property.

The exactness of MK++ depends on the applicability of the condition in Eq. 4.18 to the distance function. z-score Normalized Euclidean distance (the distance function of choice in may applications because of its relative scale independence), does not satisfy this condition. The following section provides a solution to this problem and another algorithm that can discover top K motifs using scale normalized distance functions. MK++ is implemented in C++ with a Matlab interface provided by the function $mkpp()$ in the MC^2 toolbox. Figure 4.3 shows the results of applying this algorithm to the same time-series used to test MDL EMD (Sect. 4.3.1). Comparing the results with the results reported in Fig. 4.2, it is clear that the *exact* solution discovers only a single motif in the data but completely, while MDL EMD discovered partially both embedded motifs but in smaller segments.

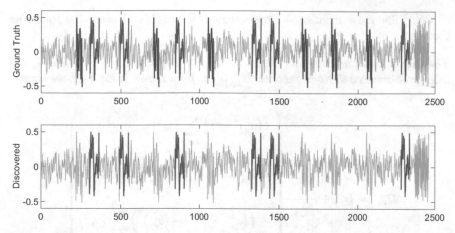

Fig. 4.3 Results of applying MK++ on the same time-series shown in Fig. 4.2

4.4.4 Motif Discovery Using Scale Normalized Distance Function (MN)

We utilize the following notation in this section: The symbols μ_x^l, σ_x^l, mx_x^l, mn_x^l stand for the mean, standard deviation, maximum and minimum of x^l where we use superscript to indicate the lengths of time-series. The normalization constant r_x^l is assumed to be a real number calculated from x^l and is used to achieve scale-invariance by either letting $r_x^l = \sigma_x^l$ (z-score normalization), or $r_x^l = mx_x^l - mn_x^l$ (range normalization).

The distance function (between any two time-series x and y) used in this section has the general form:

$$D_{xy}^l = \sum_{k=0}^{l-1} \left(\frac{x_k - \mu_x^l}{r_x^l} - \frac{y_k - \mu_y^l}{r_y^l} \right)^2. \qquad (4.22)$$

This is an Euclidean distance between two subsequences $\bar{x}$ and $\bar{y}$, where

$$\bar{z}_k = \left(z_k^l - \mu_z^l \right) / r_z^l.$$

This means that it satisfies the triangular inequality which allows us to use the speedup strategy described in Sect. 4.4.1. Nevertheless, because of the dependence of r_x^l and r_y^l on data and length, it is no longer true that $D_{xy}^{l+1} \geq D_{xy}^l$. Moreover, once any of these two values change, we can no longer use any catched values of $\bar{x}$ and $\bar{y}$.

We need few more definitions: $\alpha_x^l \equiv r_x^{l-1}/r_x^l$, $\theta_{xy}^l \equiv r_x^l/r_y^l$, $\Delta_k^l \equiv x_k - \theta_{xy}^l y_k$, $^n\mu_{xy}^m \equiv \mu_x^m - \theta_{xy}^n \mu_y^m$ and $\mu_{xy}^l \equiv {}^l\mu_{xy}^l$. Notice that it is trivial to prove that the mean of the sequence $\langle \Delta_{xy}{}^l \rangle$ is equal to μ_{xy}^l.

Using the above mentioned terminology, it is possible to prove the following theorem:

Theorem 4.2 *For any two time-series x and y of lengths $L_x > l$ and $L_y > l$, and using a normalized distance function D_{xy}^l of the form shown in Eq. 4.22, we have:*

$$D_{xy}^{l+1} = D_{xy}^l + \frac{1}{(r_x^l)^2} \left(\begin{array}{l} \left((\alpha_x^l)^2 - 1\right) \sum_{k=0}^{l-1} (x_k^2) \\[2mm] + 2\left(\theta_{xy}^l - (\alpha_x^l)^2 \theta_{xy}^{l+1}\right) \sum_{k=0}^{l-1} x_k y_k \\[2mm] + \left((\alpha_x^{l+1}\theta_{xy}^{l+1})^2 - (\theta_{xy}^l)^2\right) \sum_{k=0}^{l-1} (y_k^2) \\[2mm] + l(\mu_{xy}^l)^2 + (\alpha_x^{l+1}\Delta_l^{l+1})^2 \\[2mm] - (l+1)(\alpha_x^{l+1}\mu_{xy}^{l+1})^2 \end{array} \right).$$

A sketch of the proof for this theorem is given by Mohammad and Nishida (2015a). The important point about this theorem, is that it shows that by having a running sum of x_k, y_k, $(x_k)^2$, $(y_k)^2$, and $x_k y_k$, we can incrementally calculate the scale invariant distance function for any length l given its value for the previous length $l-1$. This allows us to extend the MK+ algorithm directly to handle *all* motif lengths required in parallel rather than solving the problem for each length serially as was done in MK++.

The form of D_{xy}^{l+1} as a function of D_{xy}^l is quite complicated but it can be simplified tremendously if we have another assumption:

Lemma 4.1 *For any two time-series x and y of lengths $L_x > l$ and $L_y > l$, and using a normalized distance function D_{xy}^l of the form shown in Eq. 4.22, and assuming that $r_x^{l+1} = r_x^l$ and $r_y^{l+1} = r_y^l$, we have:*

$$D_{xy}^{l+1} = D_{xy}^l + \frac{1}{(r_x^l)^2}\left(\frac{l}{l+1}\right)(\mu_{xy}^l - \Delta_l^l)^2.$$

Lemma 4.1 can be proved by substituting in Theorem 4.2 noticing that given the assumptions about r_x^l and r_y^l, we have $\Delta_k^{l+1} = \Delta_k^l$ and $^{l+1}\mu_{xy}^{l+1} = {}^l\mu_{xy}^{l+1}$.

What Lemma 4.1 shows is that if the normalization constant did not change with increased length, we need only to use the running sum of x_k and y_k for calculating the distance function incrementally and using a much simpler formula. This suggests that the normalization constant should be selected to change as infrequently as possible while keeping the scale invariance nature of the distance function. The most used normalization method to achieve scale invariance is z-score normalization in which $r_x^l = \sigma_x^l$. Mohammad and Nishida (2015a) proposed using the—less frequently used—range normalization ($r_x^l = mx_x^l - mn_x^l$) because the normalization constants change much less frequently. The formulas for incremental evaluation of the distance function given in this section assume that the change in length is a single point. Both formulas can be extended to the case of any difference in the length but proofs are much more involved.

The main idea of the Motif discovery using Normalized distance metrics (MN) algorithm is to utilize Theorem 4.2 and run the two phases of the MK algorithm in parallel for all lengths. The algorithm starts similarly to MK+ by calculating the distance between all subsequences of the minimum length and a randomly selected set of reference points. These distances will be used later to find lower bounds during the scanning phase. Based on the variance of the distances associated with reference points, these points are ordered. The subsequences of the time-series are then ordered according to their distances to the reference point with maximum variance. These steps can be achieved in $O(nlogn)$ operations. The distance function used in these steps (D_{full}) uses Eq. 4.22 for distance calculation but in the same time keeps the five running summation (x_k, y_k, $(x_k)^2$, $(y_k)^2$, and $x_k y_k$) needed for future incremental distance calculations as well as the maximum and minimum of each subsequence. After each distance calculation the structure S_{bests}^l is updated to keep the top K motifs at this length with associated running summations using the same rules of MK++.

The next step is to calculate the S_{bests}^l list storing distances and running summations for all lengths above the minimum length using the function D_{inc} which utilized Theorem 4.1 to find the distances at longer lengths. The list is then sorted at every length. Both D_{inc} and D_{full} update the bsf variable which contains the best-so-far distance at all lengths and is used if the run is approximate to further prune out distance calculations during the scanning phase.

The scanning phase is then started in which the subsequences as ordered in the previous phase are taken in order and compared with increasing offset between them. If a complete run at a specific length did not pass the lower-bound test, we can safely ignore all future distance calculations at that length because by the triangular inequality we know that these distances can never be lower than the ones we have in S_{bests}^l. Scanning stops when all lengths are fully scanned.

During scanning we make use of Theorem 4.1 once more by using an incremental distance calculation to find the distances to reference points and between currently tested subsequences. If we accept approximate results based on Lemma 4.1, we can speed things up even more by not calculating the distances to reference points during the evaluation of the lower bound and by avoiding this step all-together if the distance at lower length was more than the current maximum distance in S_{bests}^l (Mohammad and Nishida 2015a).

This algorithm is implemented in the function $mn()$ in the MC^2 toolbox. Figure 4.4 shows an example application of this algorithm.

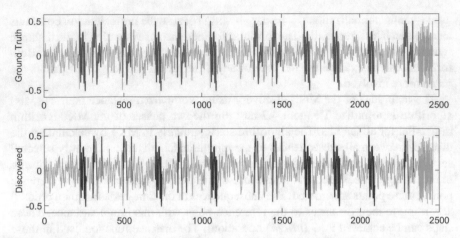

Fig. 4.4 Application of the MN algorithm to the same time-series used in Fig. 4.2

4.5   Stochastic Motif Discovery

The algorithms discussed so far try to be as exhaustive as possible in their search for motifs in the time-series. This reduces their scalability to very long time-series (in the range of millions or billions of points). When these extremely long time-series are considered, even linear time performance may prove too slow for practical application. In such case we can resort to stochastic discovery algorithm that sample the time-series more or less randomly instead of scanning it during the search for motifs.

4.5.1   Catalano's Algorithm

Catalano et al. (2006) provided one of the earliest stochastic algorithms for motif discovery. The algorithm is designed to find motifs with frequent repetitions and with known maximum length.

The main idea behind the algorithm is to randomly sample a user specified number of large candidate windows of data, sub-window and store them, then these sub-windows are compared to the sub-windows of comparison windows sampled from the time series.

If the candidate window contains a recurring pattern (a motif occurrence) then the sampling process will yield comparison sub-windows with similar properties. To distinguish patterns from noise, the mean similarity of the best matches for each candidate sub-window is compared to a distribution of mean similarities constructed so as to enforce the null hypothesis that the matching sub-windows do not contain an instance of the same pattern. When the null hypothesis is rejected the candidate sub-window probably contains a part of a pattern instance and adjacent sub-windows with this property are concatenated together to obtain the entire pattern in a process similar to GEMODA's convolution.

The steps of the algorithm can be summarized as:

1. Select a subsequence s^w of length $w \geq l_{max}$.
2. Select w values randomly from X and concatenate them to form a noise sequence n^w.
3. Select a set of n_c comparison subsequences of length w from X ($\{c^w_i\}$).
4. Find the set $S^{\hat{w}}$ of subsequences of X of length $\hat{w}$ where $\hat{w} \leq l_{min}$ for s^w (call this $\{s^{\hat{w}}_k\}$), and n^w (call this $\{n^{\hat{w}}_k\}$) where $1 \leq k \leq w - \hat{w} - 1$. Then normalize all of the resulting subsequences to have unit mean square.
5. For the candidate subsequence of s^w ($s^{\hat{w}}_k$) do the following:

   a. Randomly select $w - \hat{w} - 1$ subsequences from the set of all subsequences of the comparison windows (c^w_i). Call this set the comparison set $\hat{c}^{\hat{w}}_j$.
   b. Find the distances $d_{kj} = d\left(s^{\hat{w}}_k, \hat{c}^{\hat{w}}_j\right)$.
   c. Group the set $\hat{c}^{\hat{w}}_j$ with their parent subsequence c^w_i and for every group select $\hat{c}^{\hat{w}}_j$ that has the minimum distance d_{kj}. This leads to a set of R subsequences $\tilde{c}^{\hat{w}}_r$ where $R \leq n_c$.
   d. Keep only the $\hat{R}$ subsequences of $\tilde{c}^{\hat{w}}_r$ with least d_{kr}.
   e. Repeat the previous three steps for subsequences of the noise subsequence n^w. This leads to another set of $\hat{R}$ subsequences called $\tilde{n}^{\hat{w}}_r$.
6. Remove all candidate subsequences $s^{\hat{w}}_k$ that have average d_{kr} less than a threshold γ. To calculate γ, sort $\tilde{n}^{\hat{w}}_r$ by their distances and call these distances $dn(u)$. γ is selected as $dn(\lfloor(\alpha n)\rfloor)$ where α is a parameter of the algorithm. Repeat the steps above using this reduced set. α is a parameter of the algorithm. Repeat this reduction for n_r times.
7. If the final set $s^{\hat{w}}_k$ is not empty output each of them as a motif seed M_s after concatenating any continuous subset of them.

There are six parameters of this algorithm: w, $\hat{w}$, $\hat{R}$, n_r, n_c, α. Catalano et al. (2006) suggested that w should be selected larger than l_{max} (we select it as $l_{max} + 1$). They have also shown that a choice of α between 0.05 and 0.2 gives good results. n_r is decided based on the time available to run (the more the better). n_c is decided based on the space available. We select $\hat{w}$ to be less than $l_{min}/2$ in order to be sure that the smallest pattern has more than one subwindow. $\hat{R}$ is difficult to select but we can use $\hat{R} = n_c/2$ which gave good results in all our evaluations (Mohammad and Nishida 2009). In case motif length limits l_{min} and l_{max} are not available, it is necessary to estimate them by visualizing parts of the data or trying different values for them.

In this form, Catalano's algorithm can discover motifs only if they are frequent enough for two of its occurrences to fit perfectly within the candidate window and at least one comparison window. Several approaches have been proposed to increase the probability of this happening leading to the stream of constrained motif discovery algorithms to be discussed in Sect. 4.6.

As stated, the algorithm does not discover range motifs but something akin to GEMODA motifs because the criteria for selecting similarity is having a distance less than a threshold *for every short subsequence of length* $\hat{w}$ instead of satisfying a range constraint. It is easy to convert the algorithm to a range motif discovery system by filtering discovered motifs and running motif detection over the complete time-series for every discovered motif.

4.6   Constrained Motif Discovery

Constrained motif discovery algorithms receive not only the time-series X to be searched for motifs but another time-series P of the same length representing some estimate of the odds that a motif occurrence exists near every point in X. In a sense, this is a simpler problem than standard unconstrained motif discovery and the main challenge is to utilize this information efficiently for motif discovery.

Where can we get this constraint P? It can contain some domain knowledge but more interestingly for us, it can be found efficiently in linear time by a single scan of the time-series using a change point discovery algorithm (See Chap. 3). The basic idea is that motif occurrences are assumed to be disjoint and do not repeat continuously. In such cases, the beginning and ending of motif occurrences are likely to appear as change points to the CPD algorithm. Here CPD is used to focus the search around points of high interest at which we expect motifs to start and end.

One simple approach to solve this kind of problem is to build a similarity matrix for short subsequences around these change points and use it for motif discovery. Another simple approach that will be explored here is to use this estimate to bias the sampling process of a stochastic motif discovery algorithm like Catalano's algorithm.

4.6.1   MCFull and MCInc

If calculating the constraint P is not computationally expensive or if it cannot be calculated for any required single point of the series X using only local information, we can speed up Catalano's algorithm by modifying the first three steps as follows (Mohammad and Nishida 2009):

1. Apply a Gaussian smoothing filter ($N(0, \sigma^2)$) to the original P constraint which results on the smoothed constraint $\tilde{P}$, then normalize $\tilde{P}$ so that $\sum_{t=1}^{n} \hat{p}(t) = 1$ and $0 \le \hat{p}(t) \le 1$.
2. Randomly select a subsequence s^w of length $w \ge l_{max}$ using $\hat{P}$ as the probability distribution.

3. Randomly select a set of w values from X by concatenating short subsequences of length $\hat{w}$ using $1 - \hat{P}^m$ where $m \geq 1$. This is the noise sequence n^w.
4. Randomly select a set of n_c comparison subsequences of X ($\{c^w{}_i\}$) each of length w using $\hat{P}$ as the probability distribution.

The rest of the algorithm goes exactly as the original. The smoothing step is required to account for any inaccuracy of the constraint. This way we understand the constraint as specifying the probability that a motif occurrence ends *near* and not necessary *at* the corresponding time step.

This algorithm is called MCFull (standing for Modified Catalano Full) because the P constraint must be calculated before the algorithm is run for the complete input time-series.

The only parameter added to the set of Catalano's parameters is the smoothing variance σ. If an estimation of the locality variance of the constraint is available (how accurate is it in finding the exact starting/ending point of motifs) then it can be used.

If calculating the constraint P is computationally expensive and if $p(t)$ can be calculated using only local information around t, then an incremental version of MCFull can be constructed (named MCInc hereafter) by modifying the MCFull steps as follows:

1. Randomly select a time step $1 \leq \tau \leq n$ using a uniform distribution.
2. Calculate $p(t)$ for $\tau - \delta \leq t \leq \tau + \delta$ and calculate $\hat{p}(\tau) = \frac{1}{2 \times \delta + 1} \sum_{t=\tau-\delta}^{\tau+\delta} p(t)$.
3. Repeat steps 1 and 2 as long as $\hat{p}(\tau) \leq th$.
4. Select the subsequence s^w of length $w \geq l_{max}$ that ends at τ.
5. Randomly select a set of $\hat{w}$ values from X and concatenate them to form a noise sequence n^w.
6. Repeat steps 1, 2, 3 for n_c times to select the comparison subsequences of X ($\{c^w{}_i\}$) each of length w.

The only added parameters by this algorithm are δ and τ. δ is selected based on the locality variance of the constraint (similar to σ for MCFull). τ should be decided experimentally by finding a tradeoff between increasing the number of iterations (low τ) and increasing the number of random samplings (high τ) (Mohammad and Nishida 2009).

The MC^2 toolbox accompanying this book provides implementations of both MCFull and MCInc algorithms.

Mohammad and Nishida (2009) analyzed the theoretical benefits of using MCFull instead of relying on the basic Catalano's algorithm and proved the following relationship:

$$\frac{\overset{mc}{\Pr}}{\overset{c}{\Pr}} = \left(\frac{T}{m \log_2 m} \times \frac{D\left(C_r^{\hat{P}} || C_r^{P_g}\right) + H\left(C_r^{\hat{P}}\right)}{w - \mathbb{E}(L) + 1} \right)^2, \tag{4.23}$$

where $C_r^P = \sum_{j=1}^{r(i)} P\,(j + (i - 1)\,r\,(i))$, $\sum r\,(i) = T$, m is the actual number of occurrences of the motif, $\mathcal{E}\,(L)$ is the expected length of these occurrences, $D\,(x\|y)$ is the relative entropy of x given y, and $H\,(x)$ is the entropy of x. P_g is the ground truth probability distribution of motif occurrences (having the value $1/m$ at the endings of motif occurrences and zero everywhere else), and P is the constraint after being normalized to have a sum of 1. $\frac{\overset{mc}{\text{Pr}}}{\underset{\text{Pr}}{c}}$ is the ratio between the probability of finding a motif using MCFull to the probability of finding it using the basic Catalano's algorithm.

Several points of Eq. 4.23 need to be highlighted. Firstly, the odds of discovering a motif using random sampling improves with the square of the time-series length not linearly. This suggests that for longer time-series using constraints to guide the search for motifs is necessary. Moreover, the ratio depends also on $m^2 log^2 m$ which shows that the smaller the number of occurrences, the more focusing the search will be beneficial.

4.6.2   Real-Valued GEMODA

Having found a constraint P (usually from a change point discovery algorithm), we can assume that motifs will be found around points with relatively high values for p_t. This suggests a simple approach for motif discovery. Firstly, we sample random points from P (after proper normalization), and then we can collect subsequences at these random locations and build a complete similarity matrix for them.

The similarity matrix can then be fed to the GEMODA algorithm (Sect. 4.2.2) to generate GEMODA motifs. This option is implemented in the function $gemoda()$ by passing candidate locations to use when building the similarity matrix. The results of using this approach is shown in Fig. 4.5. It is clear that the approach is approximate in the sense that it is no longer having the maximal discovery guarantees of GEMODA when using the full similarity matrix. This is the price we pay for having a scalable algorithm.

4.6.3   Greedy Motif Extension

Mohammad et al. (2012) proposed a different approach for extending the motif stems discovered through comparison of distances at candidate locations sampled randomly from the constraint called greedy motif extension (GSteX) which can either work sequentially (GSteXS) or using bisection (GSteXB).

Given a list $candLoc$ of candidate motif locations found for example by an application of CPD, a set of stem locations is generated from $candLoc$ by collecting all subsequences of length l_{min} around each member of $candLoc$. This set is called $sequences$. Notice that l_{min} is a small integer that is used for initialization and

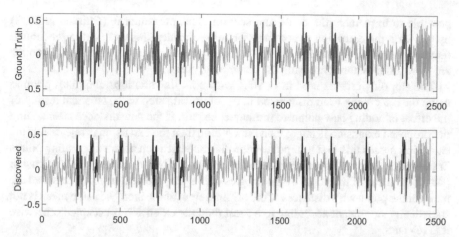

Fig. 4.5 Application of the Gemoda algorithm to the same time-series used in Fig. 4.2 using RSST for change point discovery to focus the search for motifs. All occurrences of the same motif are shown with the same color and occurrences of different motifs are shown in different colors

the stems will then be allowed to grow. In our experiments, we selected l_{min} to be $\max\left(10, 10^{-5}T\right)$, where T is the length of the input time series. The results presented were not dependant on the choice of this parameter as long as it was less than half the discovered motif lengths.

The distances between all members of *sequences* is then calculated and clustered into four clusters using K-means. The largest distance of the cluster containing shortest distances is used as an upper limit of distances between *similar* subsequences (*maxNear*) and is used to prune future distance calculations.

The sequences generating the first distance clusters (*nearSequences*) are then used for finding motif stems. Each one of these distances corresponds to a pair of sequences in the *sequences* set. This is the core step in G-SteX. For each pair of sequences in the *nearSequences* set, the first of them is slid until minimum distance is reached between the pair and then a stem is generated from the pair using one of the two following two techniques.

The first extension technique is called G-SteXB (for binary/bisection) and it starts by trying to extend the motif to the nearest end of the time series and if the stopping criteria is not met, this is accepted as the motif stem. If the stopping criteria is met, the extension is tried with half of this distance (hence the name binary/bisection) until the stopping criteria is not met. At this point, the extension continues by sequentially adding half the last extension length until the stopping criteria is met again. This is done in both directions of the original sequences. By the end of G-SteXB, we have a pair of motif occurrences that cannot be extended from any direction without meeting the stopping criteria.

The second extension technique is called G-SteXS (for sequential) and it extends the sequences from both sizes by incrementally adding l_{min} points to the current pair from one direction then the other until the stopping criteria is met. At this step it is possible to implement the don't care section ideas presented in (Chiu et al. 2003) by allowing the extension to continue for subsequences shorter than the predefined maximum outlier region length (don't care length) as long as at least one l_{min} points are then added without breaking the stopping criteria.

The stopping criteria used in G-SteX with both its variations combines pruning using the $maxNear$ limit discovered in the clustering step with statistical testing of the effect of adding new points to the sequence pair. If the new distance after adding the proposed extension is larger than $maxNear$, then the extension is rejected. If the extension passes this first test, point-wise distance between all corresponding points in the motif stem and between all corresponding points in the proposed extension part. If the mean of the point-wise distances of the proposed extension is less than the mean of the point-wise distances of the original stem or the increase in the mean is not statistically significant according to a t-test then the extension is accepted otherwise it is rejected.

This approach is implemented by the function $dgr()$ in the MC^2 toolbox with the actual extension logic in the function $extendStem()$. Figure 4.6 shows the results of applying this approach to the same running example of this chapter. We can see that the algorithm has relatively high false negatives rate due to the stringent criteria it uses for combining the stems and extending them.

4.6.4 Shift-Density Constrained Motif Discovery

The last stochastic motif discovery algorithm we will consider is called shift density constrained motif discovery (sdCMD) (Mohammad and Nishida 2015b). The

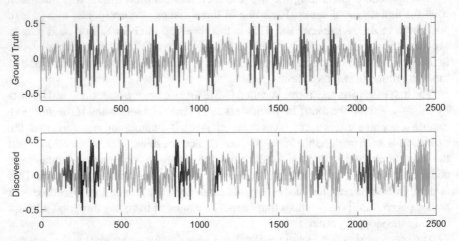

Fig. 4.6 Application of the GSteXB algorithm to the same time-series used in Fig. 4.2 using RSST for change point discovery to focus the search for motifs. All occurrences of the same motif are shown with the same color and occurrences of different motifs are shown in different colors

algorithm starts by finding candidate locations for motifs using the same change point discovery technique and random sampling used by all other constrained motif discovery algorithms discussed so far.

The goal of sdCMD is to discover GEMODA-like motifs but without requiring the complete occurrences to fit within the candidate and comparison windows (in contrast to Catalno's and related algorithms like MCFull). This is achieved by converting the problem of motif discovery in multidimensional data to the problem of shift-clustering.

To understand the idea of this algorithm, consider Fig. 4.7. The subsequences of a time-series with two motif occurrences. The top subsequence contains a partial occurrence, the middle subsequence contains the complete occurrence, while the bottom subsequence contains no occurrences. We now consider 10-points short windows of the middle subsequence at indices 31, 46, and 73. The first two of these windows are within the partial motif occurrence while the third is outside it. Now we find the subsequence that has the minimum distance to each of these three windows in the remaining two subsequences. The top subsequence with the full motif occurrence gives 1, 16 and 80 in order. Notice that the two windows that correspond to parts of the motif gave the same *shift* between the top and middle subsequences of -30 while the window outside the motif occurrence gave a shift of 7. This result is expected because windows inside the middle motif occurrence is expected to show maximum similarity (minimum distance) with the corresponding windows of the top subsequence which will be shifted by exactly the same amount (-30 points in this case). The parts outside the motif occurrences in the middle and top subsequences will show no such pattern.

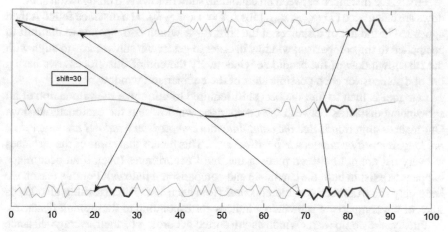

Fig. 4.7 Three subsequences of a time-series with two motif occurrences. The *top* subsequence contains a partial occurrence, the *middle* subsequence contains the complete occurrence, while the *bottom* subsequence contains no occurrences

Now consider the bottom subsequence and repeat the same procedure. Now the windows corresponding to windows 31, 46, and 73 of the middle subsequence will be 20, 69, 20 with shifts of -11, 23, and -53. There is no pattern to these shifts as expected.

This simple example suggests that the shifts between the windows with minimum distance to parts of a subsequence contain information about not only whether two motif occurrences happen to be inside the two compared subsequences but furthermore about the needed shifting of one of them to align these occurrences if they exist and the limits of these occurrences. If we have a streak of shift values of the same (or similar) values we can conclude that the two subsequences have at least partial occurrences of a motif and furthermore, we know how to align them to get the full motif. Moreover, we know that the parts that show no such behavior in the shifts are outside the motif occurrence.

The sdCMD algorithm capitalizes on this feature of shifts between most similar short subwindows of candidate and comparison windows to discover motif occurrences even if they only appear partially within the windows and complete them (an advantage over Catalano's and similar algorithms).

Assuming that the sampling based on the constraint generated a set C of windows. They are considered a candidate window in turn and compared with the rest of the windows in C.

The current candidate window (c) is divided into $w - \bar{w}$ ordered overlapping subwindows ($x_{c(i)}$ where w is the length of the window, $\bar{w} = \eta l_{min}$, $1 \leq i \leq w - \bar{w}$ and $0 < \eta < 1$). The same process is applied to every window in the current comparison set. The following steps are then applied for each comparison window (j) for the same candidate window (c).

Firstly, the distances between all candidate subwindows and comparison subwindows are calculated ($D\left(x_{c(i)}, x_{j(k)}\right)$ for $i, k = 1 : w - \bar{w}$). The distance found is then appended to the list of distances at the shift $i - k$ which corresponds to the shift to be applied to the comparison window in order to get its kth subwinodw to align with the ith subwindow of the candidate window. By the end of this process, we have a list of distances for each possible shift of the comparison window.

Our goal is then to find the best shift required to minimize the summation of all subwindow distances between the comparison window and the candidate window. Our main assumption is that *the candidate and comparison windows are* larger *than the longest motif occurrence to be discovered*. This means that some of the distances in every list are not between parts of the motif occurrences (even if an occurrence happens to exist in both the candidate and comparison windows). For this reason we keep only the distances considered *small* from each list. The algorithm also keeps track of the comparison subwindow indices corresponding to these *small* distances.

Finally, the comparison windows are sorted according to their average distance to the candidate window, with the best shift of each of them recorded.

At the end of this process and after applying it to all candidate windows, we have for each member of C a set of best matching members with the appropriate shifts required to align the motif occurrences in them (if any).

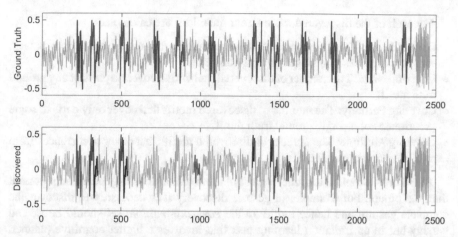

Fig. 4.8 Application of the Shift Density Constrained Motif Discovery algorithm to the same time-series used in Fig. 4.2 using RSST for change point discovery to focus the search for motifs. All occurrences of the same motif are shown with the same color and occurrences of different motifs are shown in different colors

Given this list, we can create a *matching matrix* showing for each subwindow of the candidate window the shift required to match it with the comparison window. We search this matrix for streaks of the same (or similar) values at every row and these streaks define motif stems containing one occurrence in the candidate window and another in a comparison window. These stems can then be extended by GEMODA like convolution process (Mohammad and Nishida 2015b) leading to complete motifs.

This approach is implemented in the function $sdCMD()$ in the MC^2 toolbox and its results are exemplified in Fig. 4.8.

4.7 Comparing Motif Discovery Algorithms

Comparing motif discovery algorithms is not a trivial task. For example, a MD algorithm that finds a hundred occurrences of a 5-points motif with a single point shift will have 200 errors if compared directly with the ground truth. Another algorithm that finds exactly only 61 occurrences and fails to find the rest may be considered better if only false positives and negatives are counted. This may or may not be appropriate depending on the application.

Mohammad et al. (2012) proposed to use a multi-dimensional criterion for evaluating MD algorithms assuming that ground-truth information about the motifs and their occurrences is available.

For each of the discovered motifs, four quantities are calculated:

- Correct Motifs: The number of discovered motifs that completely cover at least some occurrences of a single ground truth motif.
- Covering None: The number of discovered motifs that cover no parts of any ground truth motif.
- Covering Partially: The number of discovered motifs that cover only parts of some occurrences of a single ground truth motif.
- Covering Multiple: The number of discovered motifs that cover occurrences from multiple ground truth motifs.

Based on the application, one or more of these dimensions may be more important than the others. For example, in gesture discovery it is necessary to discover the complete gesture and hence motifs in the *covering-multiple* set should be treated harshly but in an imitation learning task that involves a higher-cognitive planner, these motifs may be counted as correct motifs because the boundaries between acts are not important in this task.

These four criteria are calculated from the view-point of the discovered motifs. We also calculate two criteria from the ground-truth motifs view point: *Fraction Covered* is the fraction of occurrences of each one of the ground-truth motifs that is covered by discovered motifs. *Extra Fraction* is the fraction of the discovered motif occurrences used in calculating the fraction covered that are not covering a part of the ground truth occurrence.

This approach to comparison of motifs is implemented in the function $mdquality()$ in the MC^2 toolbox.

Another simpler approach to motif discovery algorithm comparison is to collect the output motif locations in a single list and compare it to ground truth using the same techniques suggested for comparing change point discovery algorithms. This is done in the MC^2 toolbox using the function $mdq2cpq()$.

4.8   Real World Applications

This section presents three applications of motif discovery to real-world datasets related to robotics and social robotics. The second part of the book will contain other cases.

4.8.1   *Gesture Discovery from Accelerometer Data*

The first application of motif discovery to real world data that we will consider is in the area of gesture discovery (Mohammad and Nishida 2015b). Our task is to build a robot that can be operated with free hand-gestures without any predefined protocol. The way to achieve that is to have the robot *watch* as a human subject is guiding another robot/human using hand gestures. The learner then discovers the gestures related to the task by running a motif discovery algorithm on the data collected from an accelerometer attached to the tip of the middle finger of the operator's dominant hand. We collected only 13 min of data during which seven gestures were used. The data was sampled 100 times/s leading to a 78000 points 3D time-series. The time-series was converted into a single space time series using PCA as proposed in Mohammad and Nishida (2011). sdCMD as well as GSteXS were applied to this projected time-series. sdCMD discovered 9 gestures, the top seven of them corresponded to the true gestures (with a discovery rate of 100 %) while GSteXS discovered 16 gestures and the longest six of them corresponded to six of the seven gestures embedded in the data (with a discovery rate of 85.7 %) and five of them corresponded to partial and multiple coverings of these gestures.

4.8.2 Differential Drive Motion Pattern Discovery

The second application we will consider is discovery of motion patterns of a differential drive robot (Mohammad and Nishida 2012a). In this experiment, we employed a simulated differential drive robot moving in an empty arena of area $4m^2$. The robot had the same dimensions as an e-puck robot (Fig. 1.4d) and executed one of three different motions at random times (a circle, a triangle and a square). At every step, the robot selected either one of these patterns or a random point in the arena and moved toward it. The robot had a reactive process to avoided the boundaries of the arena.

Ten sessions with four occurrences of each pattern within each session were collected and the proposed algorithm was applied to each session after projecting the 2D time-series representing motor commands sent to the robot's two motors into a 1-D time-series as in previous applications.

MK++ discovered 3 motifs corresponding to the three motion patterns. In this case there were no partial motifs or false positives and discovery rate was 100 %. These results were obtained in a simulated environment and are not expected to be the same in a real world situation but they support the claim that motif discovery can in principle be applied to motion pattern discovery.

4.8.3 Basic Motions Discovery from Skeletal Tracking Data

The final real world application to be considered here is basic motion discovery from skeletal tracking data (Mohammad and Nishida 2015b). Discovering basic motions (sometimes called motion primitives or motor primitives) from observation of human behavior is a heavily researched problem in robotics (Kulic and Nakamura 2008; Pantic et al. 2007; Minnen et al. 2007; Vahdatpour et al. 2009). A recurrent motion primitive is a special case of a motif but in a multi-dimensional time-series.

This work uses a recording of the joint angles of a human skeletal's two arms $(\Theta_{left}(t), \Theta_{right}(t))$ and the goal is to discover motion primitives (e.g. recurring patterns of motion) using a MD algorithm. As discussed in Sect. 4.3.1, it is possible to apply a dimensionality reduction technique to the data before applying the proposed method. The problem of this approach in our case is that sometimes the motion primitive does not span all of the dimensions but involves a subset of the joints measured. For this reason, we applied MD (using sdCMD) to each dimension of the input and found recurrent patterns in a single dimension.

The experiment was designed to make it easy to find a quantitative measure of performance by having accurate ground truth data. The robot used in this experiment is a NAO robot (See Fig. 1.4c). The head was fixed and the legs DoFs were fixed to a supporting pose. The experimental setup was designed as a game called *follow-the-leader* where NAO was the leader in the beginning. The subject faced the NAO and a Kinect sensor. The NAO repeatedly selected one of 4 different predefined motions for each arm and executed it while the subject watched the motion. The subject then imitated the motion of NAO. Each motion was conducted 5 times (in random order). This gives a total of 20 motions. After each five of these motions, the leader was switched to the subject who was allowed to make 5 motions that were imitated by the NAO. The subject was instructed to move freely/randomly and try not to repeat his motion. This gave a total of 40 motions from both partners in the game for each session (Mohammad and Nishida 2015b). sdCMD could discover 75 % of the motifs in both streams and with a correct discovery rate of 71 % (human data) and 83 % (robot data).

4.9   Summary

Discovering recurrent patterns in long multidimensional time-series is a core problem for our approach to autonomous learning of social behavior. This chapter introduced several definitions of the patterns we are after (e.g. GEMODA motifs, K-Motifs, Pair-Motifs, and Range Motifs) and reported several algorithms that are adequate for discovering each of them. The algorithms covered in this chapter range from algorithms appropriate mostly for discrete time-series like projections and GEMODA to algorithms designed for multidimensional real-valued time series like MDL-Extended Motif Discovery algorithm. We reported also discretization algorithms for solving real-valued motif discovery problems as well as other approaches that do not require any discretization. Three applications of motif discovery to HRI related situations were also presented: gesture discovery, differential drive motion pattern discovery, and motion discovery from skeletal data. In the second part of this book, motif discovery will be used extensively for several applications and will prove to be a useful tool for social robotics.

References

Buhler J, Tompa M (2002) Finding motifs using random projections. J Comput Biol 9(2):225–242

Catalano J, Armstrong T, Oates T (2006) Discovering patterns in real-valued time series. In: PKDD'06: knowledge discovery in databases, pp 462–469

Chiu B, Keogh E, Lonardi S (2003) Probabilistic discovery of time series motifs. In: KDD'03: the 9th ACM SIGKDD international conference on knowledge discovery and data mining, ACM, New York, USA, pp 493–498. http://doi.acm.org/10.1145/956750.956808

Jensen KL, Styczynxki MP, Rigoutsos I, Stephanopoulos GN (2006) A generic motif discovery algorithm for sequenctial data. BioInformatics 22(1):21–28

Kulic D, Nakamura Y (2008) Incremental learning and memory consolidation of whole body motion patterns. In: International conference on epigenetic robotics, pp 61–68

Li Y, Lin J (2010) Approximate variable-length time series motif discovery using grammar inference. In: The 10th international workshop on multimedia data mining, ACM, pp 101–109

Lin J, Keogh E, Lonardi S, Patel P (2002) Finding motifs in time series. In: The 2nd workshop on temporal data mining, at the 8th ACM SIGKDD, pp 53–68

Lin J, Keogh E, Wei L, Lonardi S (2007) Experiencing SAX: a novel symbolic representation of time series. In: KDD'07: international conference on data mining and knowledge discovery, vol 15(2), pp 107–144. doi:10.1007/s10618-007-0064-z

Minnen D, Starner T, Essa IA, Isbell Jr CL (2007) Improving activity discovery with automatic neighborhood estimation. In: IJCAI'07: 16th international joint conference on artificial intelligence, vol 7, pp 2814–2819

Mohammad Y, Nishida T (2011) On comparing SSA-based change point discovery algorithms. In: SII'11: IEEE/SICE international symposium on system integration, pp 938–945

Mohammad Y, Nishida T (2012a) Fluid imitation: discovering what to imitate. Int J Social Robot 4(4):369–382

Mohammad Y, Nishida T (2012b) Unsupervised discovery of basic human actions from activity recording datasets. In: SII'12: IEEE/SICE international symposium on system integration, IEEE, pp 402–409

Mohammad Y, Nishida T (2009) Constrained motif discovery in time series. New Gener Comput 27(4):319–346

Mohammad Y, Ohmoto Y, Nishida T (2012) GSteX: greedy stem extension for free-length constrained motif discovery. In: IEA/AIE'12: the international conference on industrial, engineering, and other applications of applied intelligence, pp 417–426

Mohammad Y, Nishida T (2015a) Exact multi-length scale and mean invariant motif discovery. Appl Intell 1–18. doi:10.1007/s10489-015-0684-8

Mohammad Y, Nishida T (2015b) Shift density estimation based approximately recurring motif discovery. Appl Intell 42(1):112–134

Mueen A, Keogh E, Bigdely-Shamlo N (2009a) Finding time series motifs in disk-resident data. In: ICDM'09: IEEE international conference on data mining, IEEE, pp 367–376

Mueen A, Keogh E, Zhu Q, Cash S, Westover B (2009b) Exact discovery of time series motifs. In: SDM'09: SIAM international conference on data mining, pp 473–484

Nevill-Manning CG, Witten IH (1997) Identifying hierarchical strcture in sequences: a linear-time algorithm. J Artif Intell Res 7:67–82

Nishida T, Nakazawa A, Ohmoto Y, Mohammad Y (2014) Conversation quantization. Springer, chap 4, pp 103–130

Pantic M, Pentland A, Nijholt A, Huang T (2007) Machine understanding of human behavior. In: AI4HC'07: IJCAI 2007 workshop on artificail intelligence for human computing. University of Twente, Centre for Telematics and Information Technology (CTIT), pp 13–24

Pavesi G, Mauri G, Pesole G (2001) Methods for pattern discovery in unaligned biological sequences. Brief Bioinform 2(4):417–431

Pevzner PA, Sze SH et al (2000) Combinatorial approaches to finding subtle signals in dna sequences. In: ISMB'00: the 8th international conference on intelligent systems for molecular biology, vol 8, pp 269–278

Sagot MF (1998) Spelling approximate repeated or common motifs using a suffix tree. In: Theoretical informatics. Springer, pp 374–390

Staden R (1989) Methods for discovering novel motifs in nucleic acid sequences. Comput Appl Biosci 5(4):293–298

Tanaka Y, Iwamoto K, Uehara K (2005) Discovery of time-series motif from multi-dimensional data based on mdl principle. Mach Learn 58(2/3):269–300

Tang H, Liao SS (2008) Discovering original motifs with different lengths from time series. Knowl-Based Syst 21(7):666–671. http://dx.doi.org/10.1016/j.knosys.2008.03.022

Tompa M (1999) An exact method for finding short motifs in sequences, with application to the ribosome binding site problem. In: ISMB'99: the 7th international conference on intelligent systems for molecular biology, vol 99, pp 262–271

Vahdatpour A, Amini N, Sarrafzadeh M (2009) Toward unsupervised activity discovery using multi-dimensional motif detection in time series. In: IJCAI'09: the international joint conference on artificial intelligence, pp 1261–1266

Waterman M, Arratia R, Galas D (1984) Pattern recognition in several sequences: consensus and alignment. Bull Math Biol 46(4):515–527

Chapter 5
Causality Analysis

The study of causality can be traced back to Aristotle who defined four types of causal relations (material, formal, efficient and final causes). From these four types, only efficient causality is still considered a form of *causality* today. In his *Treatise of Human Nature* (1739–1740), David Hume described causality in terms of regular succession. For Hume, causality is a regular succession of event-types: one thing invariably following another. In his words: *We may define a CAUSE to be 'an object precedent and contiguous to another, and where all the objects resembling the former are placed in like relations of precedence and contiguity to those objects, that resemble the latter'.*

Since Hume, there were many definitions of causation (Glymour 2003). There are two ingredients of the concept that are widely accepted by *most* definitions: causation is related to *predictability* of effects from knowing causes and causation involves time asymmetry between causes and effects. Simply put, causes *precede* effects or at least cannot *follow* them in time. In fact this is the basis of the definition of Causal Linear Systems as used in electrical engineering literature.

There are three main definitions of causation that can be used to devise computational models. Some theorists equated causality with increased predictability leading to Granger-causality tests (Granger 1969; Gelper and Croux 2007). Some theorists have equated causality with manipulability (Menzies and Price 1993). Under these theories, x causes y if and only if one change y by changing x. This is the root of causation from perturbation models (Hoover 1990). Finally, yet other theorists defined causality as a counterfactual which means that the statement "x causes y" is equivalent to "y would not have happened if x did not". This kind of definition of causality is the basis of Pearl's formalization of the Structure Equation Model (SEM) (Pearl 2000). All of the three definition agree on the time asymmetric nature of causality.

There are two problems that most of these systems (taken alone) run into: causality cycles and common causes. Causality cycles (sometimes called feedbacks) happen

© Springer International Publishing Switzerland 2015
Y. Mohammad and T. Nishida, *Data Mining for Social Robotics*,
Advanced Information and Knowledge Processing,
DOI 10.1007/978-3-319-25232-2_5

when an event directly or indirectly causes itself to happen again in the future. Linear Granger causality can detect these cycles only if two independent null-hypothesis were rejected (Bai et al. 2010). Common causes happen when event A causes both B and C but with different time delays. most techniques analyzing the behavior of B and C alone would result on the false conclusion that one of them is causing the other. When representing causality using graphs, causality cycles correspond to cycles in the graphs and common causes appear as bifurcations in these graphs (Mohammad and Nishida 2010). This chapter will introduce three algorithms for discovering causal relations between time-series that will be of use for both interaction protocol learning (Chap. 11) and fluid imitation (Chap. 12). This chapter presents three algorithms for causality discovery.

5.1 Causality Discovery

The central problem in causality discovery can be stated as follows: Given a set of n_t time-series of length T characterizing the performance of n processes (where $n_t \geq n$), build a n-nodes directed graph representing the causal relations between these processes. Each edge in the DAG should be labeled by the expected delay between events in the source processes and their causal reflection in the destination node.

We assume that every process P_i generates a time series of dimension $n_i \geq 1$. This means that $n_t = \sum n_i$. Our goal is to build a graph that represents each process (not each time series) by a node and connects processes with causal relations using directed edge (cause as a source and effect as a destination). Each edge should be labeled by the expected delay time ($E(\tau)$) between a change in the cause processes and the resulting change in the effect process.

5.2 Correlation and Causation

It is common to hear that *correlation does not imply causation* which is true in most cases but may hide the more subtle fact that causation may not imply correlation. This means that neither of these two concepts can in general be used to infer the other. Examples of correlations that do not imply causation can be found everywhere. To show the other side of this coin consider the following system:

$$x(t+1) = x(t)(3.8 - 3.8x(t) - 0.02y(t)),$$
$$y(t+1) = \alpha y(t)(3.5 - 3.5y(t) - 0.1x(t)). \tag{5.1}$$

It is clear that the two variables X and Y are causally connected because we need each of them to determine the other. Figure 5.1 shows the first 100 samples of

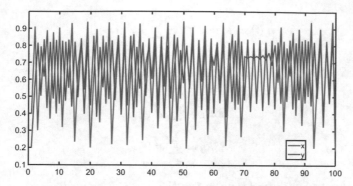

Fig. 5.1 Example of two causally connected time-series

these two time-series with initial conditions $(x(0) = 0.4, y(0) = 0.2)$. Calculating the correlation between the two time-series between samples 50 and 100 gives a correlation coefficient of 0.0537 with a p-value of 0.7084. Calculating the correlation for 3000 samples gives a correlation coefficient of only 0.0481 now with a p-value of 0.0084. The correlation coefficient of the samples between 200 and 300 is -0.45 with a p-value of less than 0.0001. These conflicting results ranging from no correlation to strong negative correlation signify the difficulty in relying on correlations for inferring causation.

5.3 Granger-Causality and Its Extensions

The main idea of Granger causality is to base the decision on causality on predictability instead of correlation. This may sidestep the problem highlighted in the previous section for systems basing causality on correlation.

Granger causality is defined between two stationary stochastic processes X and Y within a universe U as follows: X Granger-causes Y with respect to U (or X g-causes Y with respect to U) iff $\sigma^2(Y_t | U_{-\infty:t-1}) < \sigma^2(Y_t | U_{-\infty:t-1} - X_{-\infty:t-1})$ where $\sigma^2(A|B)$ is the variance of the residuals for predicting A from B.

There are many ways in which to implement a test of Granger causality. One particularly simple approach uses the autoregressive specification of a bivariate vector autoregression. First we assume a particular autoregressive lag length for every time series (ρ_a, ρ_b) and estimate the following unrestricted equation by ordinary least squares (OLS):

$$\hat{A}(t) = \varepsilon_1 + u(t) + \sum_{i=1}^{\rho_a} \alpha_i \hat{A}(t-i) + \sum_{i=1}^{\rho_b} \beta_i \hat{B}(t-i). \tag{5.2}$$

Second, we estimate the following restricted equation also by OLS:

$$\hat{A}(t) = \varepsilon_2 + e(t) + \sum_{i=1}^{\rho_a} \lambda_i \hat{A}(t-i). \tag{5.3}$$

We then calculate the sum of squared residuals (SSR) in both cases:

$$
\begin{aligned}
SSR_1 &= \sum_{i=1}^{T} u^2(t), \\
SSR_0 &= \sum_{i=1}^{T} e^2(t).
\end{aligned}
\tag{5.4}
$$

We then calculate the test statistic S_{ρ_b} as:

$$S_{\rho_b} = \frac{(SSR_0 - SSR_1)/\rho_b}{SSR_1/(T - 2\rho_b - 1)}. \tag{5.5}$$

If S_{ρ_b} is larger than the specified critical value then reject the null hypothesis that A does not g–cause B. The p-value in this case is $1 - F_{\rho_b, T-2\rho_b-1}\left(S_{\rho_b}\right)$. As presented here Ganger Causality can be used to deduce causal relations between two variables assuming a linear regression model. The test was extended to multiple variables and nonlinear relations (e.g. using radial basis functions) (Ding et al. 2006).

From this analysis of Granger causality, we can see that it is essential that the variable tested for being a cause must be separable from other possible causes. This means that it should be possible to remove the to-be-a-cause variable from the universe of all possible causation variables. In a linear system with a superposition guarantee, this can be assumed safely but in general it may not be possible to achieve this separability because other variables may carry information about the tested variable and this may affect the analysis outlined above. Separability is a technical term for decomposability which means that the system can be decomposed into pieces and understood by understanding each of these pieces and their relations. Even though linear systems and purely stochastic systems may be decomposable, this is usually lacking for deterministic dynamical systems with weak to moderate coupling (Granger 1969; Sugihara et al. 2012.)

To see this, consider the following example a modified version of the one given in Sugihara et al. (2012):

$$
\begin{aligned}
x(t+1) &= x(t)\left(r_x - r_x x(t) - \beta_{xy} y(t)\right), \\
y(t+1) &= \alpha y(t)\left(r_y - r_y y(t) - \beta_{yx} x(t)\right).
\end{aligned}
\tag{5.6}
$$

It is easy to solve the first equation for $x(t)$ which will be:

$$x(t) = \left(r_y - r_y y(t) - y(t+1)/\alpha y(t)\right)/\beta_{yx}. \tag{5.7}$$

From Eq. 5.7, we can see that $x(t)$ can be determined from $y(t)$ and $y(t+1)$ which means that y carries information about x. Moreover, increasing the value of

α reduces the importance of $y(t+1)$ in determining $x(t)$ and makes it dependent only on $y(t)$. If the system is noisy with large value of α, the effect of $y(t+1)$ on $x(t)$ will become smaller than the noise level rendering it impossible to discover the effect of $x(t)$ on $y(t+1)$.

In (Mohammad and Nishida 2010), we utilized an extended version of the Granger causality test combined with a constrained motif discovery (CMD) algorithm (Sect. 4.6) to discover causal relations. The main problem with this approach is that a large amount of data is required in order for the CMD algorithm to discover meaningful recurrent patterns. Also CMD requires the specification of an upper limit on the motif length which may be difficult to give in some contexts.

Because the optimal lag is not usually known in advance, we use BIC (Sect. 2.3) for selecting these lags in our evaluations. This approach is implemented in the function $detectGC()$ in MC^2.

5.4 Convergent Cross Mapping

The previous section discussed Granger causality which is applicable to stationary stochastic processes for which the condition of separability is satisfied. In many robotics applications, the time-series we are interested on are not stationary stochastic processes but dynamical systems with corrupting noise. This regime is different in-principle from the regime in which Granger causality excels and a new approach is needed to handle it.

A new development in this area was the introduction of convergent cross mapping (CCM) by Sugihara et al. (2012). The system is also based on predictability analysis rather than correlation but is both theoretically and practically different from Granger-causality.

For dynamical systems, we can say that two time-series have a causal link if they are both parts of the same dynamical system. Equation 5.6 gives one example of two causally linked time-series. Figure 5.2 shows the first 50 points of these two time-series. The primary structure in this figure is the manifold presented in the middle which depicts the two time-series together. The system moves over this manifold with time and the two time series (left and right) can be though of as representing projections of this main manifold onto the dimensions of the two time-series.

Figure 5.3 shows the first 100 points of these two time-series in the middle. Compared with the middle of Fig. 5.2, we can see that the lines traced by the time-series are nearer to each other. This is expected as longer time-series are not expected to trace the same path traced in the early samples leading to denser lines. Figure 5.3 shows also two other manifolds that are generated by drawing $x(t)$ against $x(t-1)$ (right) and $y(t)$ against $y(t-1)$. These two projection manifolds are called M_y (left) and M_x (right) while the common manifold is simply called M.

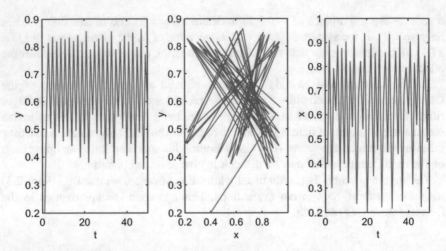

Fig. 5.2 Example of two causally connected time-series showing the common manifold (*middle*) and its projection on *Y* (*left*) and *X* (*right*)

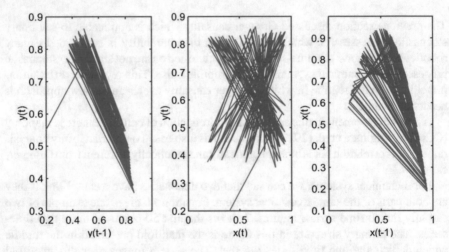

Fig. 5.3 Example of two causally connected time-series showing the common manifold (*middle*) and the manifold of lagged values of *Y* (*left*) and *X* (*right*)

The idea behind CCM is to study the prediction of each variable based on the manifold of the other variable (i.e. $\hat{Y}|M_x$ and $\hat{X}|M_y$). If the two time-series belong to the same dynamical system (i.e. *M* exists) then the longer the time-series used for prediction the more accurate will these prediction be compared with the original time-series. By studying the convergence of this cross mapping, we can quantify the causal relation between the two time-series and decide its direction.

A simple algorithm for achieving this analysis can be devised as follows: Firstly, we convert the two time-series into state-representations by concatenating lagged vectors (as we have done in Singular Spectrum Analysis in Sect. 2.3.5).

Given two time-series $X = \{x(t) \,|\, 0 \le t \le T - 1\}$ and $Y = \{y(t) \,|\, 0 \le t \le T - 1\}$, and a lag value N, we create the lagged state matrices $\tilde{X} = [X_N, X_2, ..., X_T]$ and $\tilde{Y} = [Y_N, Y_2, ..., Y_T]$ where $\tilde{X}_t = x(t - N + 1 : t)$ and $\tilde{Y}_t = y(t - N + 1 : t)$. The two matrices ($\tilde{X}$ and $\tilde{Y}$) are two Hankel matrices the same as the ones used for SSA (Sect. 2.3.5).

Now Each column of these matrices represent a point on the manifolds M_x and M_y. The next step is to calculate $\hat{Y}|M_x$. To achieve that, we calculate the nearby points to each vector of $\tilde{X}$ using K-nearest neighbors where $K = N + 1$. The indices of these neighbors will be called I for the rest of this section. We then use the points vectors of $\tilde{Y}$ that appear in the same indices (I) to predict the value of Y. This is done using simple weighted averaging:

$$\check{Y}|M_x = \sum_{k=1}^{N+1} w_k \tilde{Y}_{I(k)}, \tag{5.8}$$

where $\check{Y}|M_x$ is the prediction of $\tilde{Y}$ given the manifold M_x and the weights are calculated as:

$$w_i = \frac{e^{-d_i/d_1}}{\sum_{j=1}^{N+1} e^{-d_j/d_1}}. \tag{5.9}$$

Finally, we can find $\hat{Y}|M_x$ from $\check{Y}|M_x$ by the same procedure used to find the final time series in SSA from group approximations of the Hankel matrix (See Sect. 2.3.5) by simply averaging the values corresponding to each point of the time-series.

The same procedure can be used to find $\hat{X}|M_y$. If past values of Y affect X, we expect the projection $\hat{X}|M_y$ to predict X accurately. Moreover, comparing the middle panel of Figs. 5.2 and 5.3 we noticed that the larger the number of data points used (T) the more dense will the manifold lines be which will make the nearest neighbors better approximators of X. This means that if Y is a cause of X, we expect this approximation to converge to a high value. We can measure this approximation by the correlation coefficient (ρ) between $\hat{X}|M_y$ and X and the same for $\hat{Y}|M_x$ and Y.

Figure 5.4 shows the results of applying this algorithm to the system modeled by Eq. 5.6 with initial conditions 0.4, 0.2, $r_x = 3.8$, $r_y = 3.5$, $\beta_{xy} = 0.02$ and $\beta_{yx} = 0.1$. Figure 5.4a shows the correlation coefficients as a function of time-series length. It is clear that both coefficients converge but to different values (both near 1.0). Because $\beta_{yx} > \beta_{xy}$, the effect of X on Y is higher than the effect of Y on X. This is reflected in better approximation for $\hat{Y}|M_x$ compared with $\hat{X}|M_y$. This can be seen by comparing Fig. 5.4a, b which shows higher error in predicting Y. The correlation coefficient graph (Fig. 5.4a) shows this by faster convergence of $\rho_x = corr\left(X, \hat{X}|M_y\right)$ compared with $\rho_y = corr\left(Y, \hat{Y}|M_x\right)$ and having $\rho_x > \rho_y$ for all lengths.

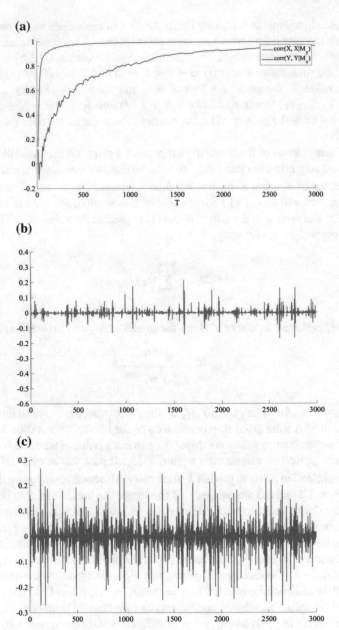

Fig. 5.4 Result of CCM analysis of the system of equation 5.6 with $\beta_{xy} = 0.02$ and $\beta_{yx} = 0.1$. **a** Correlation coefficient as a function of time-series length, **b** error in predicting X, **c** error in predicting Y

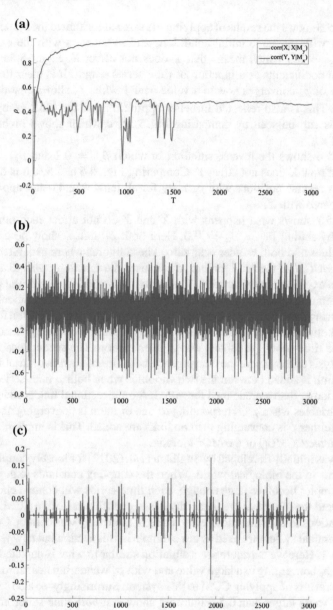

Fig. 5.5 Result of CCM analysis of the system of equation 5.6 with $\beta_{xy} = 0.0$ and $\beta_{yx} = 0.1$. **a** Correlation coefficient as a function of time-series length, **b** error in predicting X, **c** error in predicting Y

Figure 5.5 shows the results of applying the same algorithm to the system modeled by Eq. 5.6 with the same initial conditions and parameters with the exception of having $\beta_{xy} = 0.0$ which means that Y does not affect X. Figure 5.5a shows the correlation coefficients as a function of time-series length. It is clear that only the coefficient of ρ_x converges now to a value near 1 while ρ_y shows chaotic behavior near zero. This is also reflected in better approximation for $\hat{Y}|M_x$ compared with $\hat{X}|M_y$. This can be seen by comparing Fig. 5.5b, c which shows higher error in predicting Y.

Figure 5.6 shows the reverse situation in which $\beta_{xy} = 0.1$ and $\beta_{yx} = 0$ which means that now X does not affect Y. Comparing Figs. 5.5 and 5.6, it is clear that X and Y now reverse positions with $\rho_y > \rho_x$ for all lengths and better approximation of Y compared with X.

Figure 5.7 shows what happens with X and Y do not affect each other. This is achieved by setting $\beta_{xy} = \beta_{yx} = 0.0$. Here both ρ_x and ρ_y show no convergence to high values and both wander near zero. These figures where generated using the script *demoCCM* of the MC^2 toolbox which uses the function *ccm()* to apply CCM analysis as explained in this section. The reader is advised to modify the parameters of the generating dynamics and try different dynamical generation models.

Comparing these four cases together suggests a very simple algorithm for deciding causality between any N variables. We start by a graph containing N nodes representing the time-series (variables) under investigation. The second step is to apply CCM analysis and compare the convergence behavior of each two variables. A bidirectional link is added between the two variables when both ρ time-series converge to a value larger than a predefined threshold. A unidirectional link is added between the two variables when ρ corresponding to one of them is converging and the other is not. If neither ρ is converging then no links are added. This is implemented in the function *detectCCMC()* of the MC^2 toolbox.

CCM was initially developed by Sugihara et al. (2012) for loosely coupled dynamical systems in the biological world. When the coupling constants (e.g. β_{xy} and β_{yx} in our example) becomes large enough, each time-series will contain enough information about the other that predictability based causality tests like CCM will assign causal relations to both directions even if one of them only was correct. Consider the same dynamical system we used as our example in this section but now with $\beta_{xy} = 0$ and $\beta_{yx} = 1$. Here we expect to see a situation similar to what is depicted in Fig. 5.5 with only ρ_y converging to a large value and with ρ_x wandering near zero. Figure 5.8 shows the results of applying CCM to this system. Surprisingly, ρ_y also seem to converge to a value larger than 0.5. Figure 5.4 show that *both* time-series are perfectly predictable give the manifold of the other after a transient short period. This singular case, would be considered as a case of bidirectinoal causation even though $\beta_{xy} = 0$. To see how did this happen, consider the actual time-series generated as shown in Fig. 5.9. The large coupling constant caused each time-series to be predictable from the other which confused CCM.

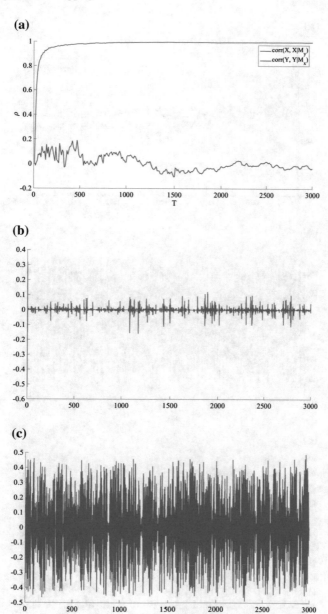

Fig. 5.6 Result of CCM analysis of the system of equation 5.6 with $\beta_{xy} = 0.1$ and $\beta_{yx} = 0.0$. **a** Correlation coefficient as a function of time-series length, **b** error in predicting X, **c** error in predicting Y

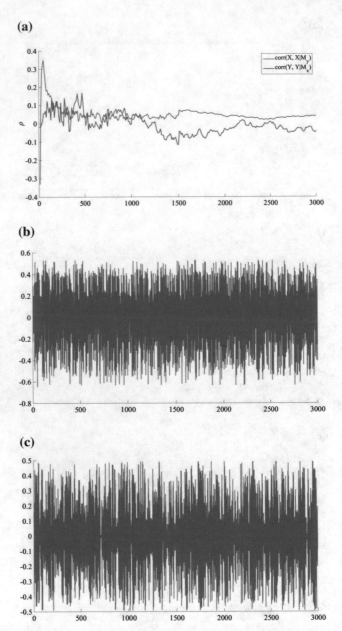

Fig. 5.7 Result of CCM analysis of the system of equation 5.6 with $\beta_{xy} = 0.0$ and $\beta_{yx} = 0.0$. **a** Correlation coefficient as a function of time-series length, **b** error in predicting X, **c** error in predicting Y

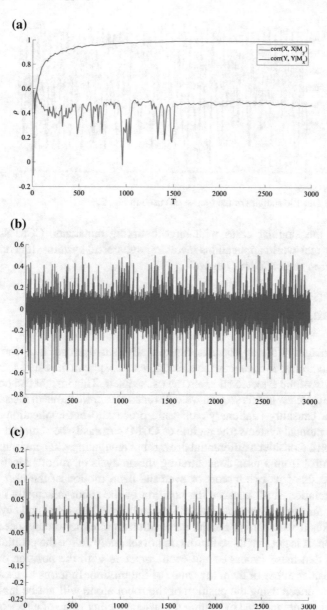

Fig. 5.8 Result of CCM analysis of the system of equation 5.6 with $\beta_{xy} = 0.0$ and $\beta_{yx} = 1$. **a** Correlation coefficient as a function of time-series length, **b** error in predicting X, **c** error in predicting Y

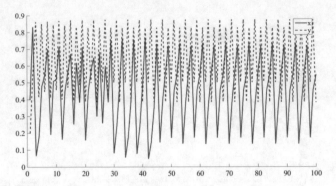

Fig. 5.9 The first 100 samples of the time-series used in Fig. 5.8

Despite the singular cases with large coupling constants, CCM shows great promise for real-world applications involving dynamical systems. It is implemented in the function *detectCCMC()* in MC^2.

5.5 Change Causality

The main insight of our approach to causal analysis is that causation can be defined as the predictability of change in the effect given a change in the cause rather than the predictability of the effect itself given the cause itself. This insight can be applied to most causality tests. In this section we will focus on applications in the same regime for Granger causality (stationary stochastic processes) but applications to weakly coupled dynamical systems (the regime of CCM) can easily be achieved with minor modifications. Consider a differential drive robot navigating a 2D environment under gesture control from a user. concentrating the analysis on robot's location and the rotation speeds of its two motors or even the hand motion of the user will reveal no causal relation at all because the position of the robot is actually not drivable from the commands given to it. The initial condition of the robot and environmental factors makes it very hard to predict its position from either motor speeds or user's hand motion. It is not that these factors do not *uniquely* define the position, the main problem is that these factors do not even *correlate* with the position even though we can consider either of them the cause of the motion. In terms of g-causality the AR model induced using the position of the robot alone will not be less predictive than the AR model induced by adding the instantaneous value of either of these two factors. The fact that both motor speed and hand motion are multidimensional further complicates the problem as either dimension alone gives no information whatsoever about the next location of the robot even if we know the current position (except in the very special cases of zero or maximum motor speed). Figure 5.10 shows the path of the robot, its horizontal (x), vertical (y), and orientation (θ) state as well as the commands given to the left and right motors in radians/second. From the figure, no clear causation can be inferred. Applying g-causality test as described in Sect. 5.3

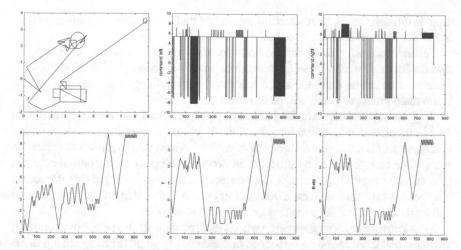

Fig. 5.10 Simulated differential drive Robot's path in a 2D environment

revealed no causality for AR models of orders 1–50 (ρ_b in Sect. 5.3 and corresponding to maximum delay before a change is propagated).

The main lesson of this example is that causation may not be readily inferred from predictability of the time series themselves. The concept of causation by perturbation comes here to rescue. If we can show that perturbing motor commands perturbs the behavior of the robot, then we may be able to infer the causality relationship. From this, we hypothesize that causation may be inferred not from the predictability of the time-series themselves as in the standard Granger causality test but from the predictability of the *change* in them.

For this approach to work, we need a general change detection algorithm that can provide us with the ground-truth with which we compare our predictions. We simply use RSST for change detection (Sect. 3.5). For multidimensional data we employ PCA as described in Sect. 2.5.

We apply multidimensional RSST (PCA+RSST) to get the set of $\tilde{P}_i$ processes representing the n processes in the system and use these signals as the inputs of our algorithm.

Our main assumption is that if Process i causes j then $\tilde{P}_j$ will *most of the time if not always* has major changes near τ_{ij} time-steps after $\tilde{P}_i$ where τ_{ij} is a constant representing the delay of the causation. Because in the real world, many factors will affect the actual delay (add to this inaccuracies in the change point detection algorithm), we expect that in reality the delays between these change points will be well approximated with a Gaussian distribution with a mean of $\hat{\tau}_{ij}$ where $\left| \hat{\tau}_{ij} - \tau_{ij} \right| < \varepsilon$ for some small value ε. The main idea of our algorithm is to check for the normality of the delays and to use the normality statistic as a measure of causality between the two processes.

We start with a causality graph with n nodes representing the processes and no edges. First, we scan all the $\tilde{P}_i$ time-series to find locations of change (L_i). This can be done in different ways. In this section we simply find the midpoints of subsequences in $\tilde{P}_i$ over some predefined threshold.

Second, for each pair of change point location vectors (L_i and L_j in order); we find the list of all delays between changes in processes i and j ($\{\tau_{ij}{}^k\}$). Notice that in general (if no causal loops exist) the sets $\{\tau_{ij}{}^k\}$ have nothing to do with $\{\tau_{ji}{}^k\}$. This is the reason that our graph will be directed.

To guard against inaccuracies in change point detection we remove from the set $\{\tau_{ij}{}^k\}$ all points that are more than 4 standard deviations from its median assuming that these points are outliers. Removal of outliers serves also to allow the system to work if the change caused by process i in process j happens with a probability less than one (but is still high enough to be detected). If it was expected that the causal changes are probabilistic then a different approach to this step would have been to cluster the delays and keep the cluster with largest number.

Prior knowledge about the system can be incorporated at this stage. For example if we know that there can be no self-loops in the causality graph (a change in a process does no directly cause another change later) we can restrict the pairs of L_i, L_j by having $i \neq j$. If we know that processes with higher index can never cause process with lower index, then we can restrict i to be less than or equal to j. Extension to more complex constraints is fairly straightforward.

We then calculate a causality score from set $\{\tau_{ij}{}^k\}$ using its mean (μ_{ij}) and standard deviation (σ_{ij}) (after removing the outliers as explained before) by:

$$score = 1 - \exp\left(-\sigma_{ij}/\mu_{ij}\right). \tag{5.10}$$

This score is always between zero and one. The larger this score is, the more probable is it that there is a causal relation between processes i and j. To construct the causality graph we accept the causal relation if this score was over some predefined threshold. In the causality graph we add an edge from i to j and associate with it the mean and variance of the delays (μ_{ij}, σ_{ij}) calculated from $\{\tau_{ij}{}^k\}$. It is also possible to add to this edge's label a *confidence* measure by dividing the number of times L_j has a change after $\mu_{ij} \pm \delta$ from a change in L_j to the number of changes of process i ($|L_i|$). This confidence measure characterizes the predicting power of this causal relationship. We write this information as:

$$i \xrightarrow{\mu_{ij},\sigma_{ij},c_{ij}} j.$$

After completing this operation for all L_i and L_j pairs (an n^2 order of operations), we have a directed graphs that represents the causal relationships between the series involved. The problem now is that this graph may have some redundancies. In this case, we resolve this ambiguity by simply removing the causal relation with the longest time delay. A better approach would be keep multiple possible graphs and then use a causal hypothesis testing system like Hoover's technique (Hoover 1990) for selecting one of them. We call this algorithm Delay Consistency-Discrete (DCD) and the MC^2 toolbox implements it in the function *detectCC()*.

5.6 Application to Guided Navigation

5.6.1 Robot Guided Navigation

This section presents a feasibility study to assess the applicability of causality discovery in learning the causal structure in the guided navigation task explained in Sect. 1.4. This task was selected because it has a known causal structure that we can compare quantitatively to the results of applying our proposed algorithm.

The evaluation experiment was designed as a Wizard of Ooz (WOZ) experiment in which an untrained novice human operator is asked to use hand gestures to guide the robot along the two paths in two consecutive sessions. The subject is told that the robot is autonomous and can understand any gesture (s)he will do. A hidden human operator was sitting behind a magic mirror and was translating the gestures of the operator into the basic primitive actions of the WOZ robot that were decided based on an earlier study of the gestures used during navigation guidance (Mohammad and Nishida 2008, 2009).

In this design the movement of the robot is known to be *caused* by the commands sent by the WOZ operator which in turn is partially *caused* by the gestures of the participant. This can be formally explained as: $G_i \xrightarrow{T_1,0,1} W_j$ and $W_j \xrightarrow{T_2,0,1} M_k$ where G_i represent some gesture done by the participant, W_j represent some action done by the WOZ operator (e.g. pressing a button on the GUI of the control software), and M_k represents some pattern in the movement of the robot (e.g. moving toward the participant, stopping, etc.).

The total number of sessions conducted was 16 sessions with durations ranging from 5:34 min to 16:53 min.

The motion of the subject's hands (G) was measured by six B-Pack (Ohmura et al. 2006) sensors attached to both hands generating 18 channels of data. The PhaseSpace motion capture system was also used to capture the location and direction of the robot using eight infrared markers. The location and direction of the subject was also captured by the motion capture system using six markers attached to the head of the subject (three on the forehead and three on the back). Eight more motion capture markers were attached to the thumb and index of the right hand of the operator.

The time of button presses in the GUI used by the WOZ operator was collected in synchrony with both participant gestures and robot actions. The interface had seven buttons (related to robot motion) each of which can be toggled on and off and a single button can be on at any point of time. The WOZ operator's actions were represented by a single input dimension giving the ID of the currently active button (from 1 to 7).

This leads to a total of 67 input dimensions. The generating causal model (ground truth) consists of seven gestures, seven button press configuration and corresponding seven robot actions (21 total patterns).

There are four processes in this system representing the user, the operator, the motors of the robot, and the configuration of the robot (location/orientation). The causal structure of this problem as set by the controlled experiment setup is very simple *user → operator → robotMotors → robotConfiguration*. The true value of the delays are not controlled in the experiment.

We applied the CDC algorithm first to the collection of all interactions and it was able to discover the exact causal structure of the problem with any threshold value larger than 0.1.

We also applied the algorithm to each interaction alone and calculated the number of times each causal relation was discovered correctly. The relation *user → operator* was found 83.4% of the time and the relation *operator → robot* was discovered 91.6% of the time. The system had two single false positives in two sessions and both where *robot → operator*. In fact these errors can be explained because the hidden operator in our experiment had some hard time dealing with rotation commands and repeatedly did the wrong rotation then he had to correct it which appeared as if these corrections are *caused* by robot's behavior. These results show that the proposed algorithm can successfully discover the causal structure of this real world experiment with known causal structure.

5.7 Summary

This chapter discussed the problem of causality discovery. Three approaches to discover causal structure from time-series were discussed in detail: Granger causality test, convergent cross mapping and change causality. Causality analysis is important for social robots because it allows them to discover the causal structure of human's behavior during interaction which helps them in deciding what to imitate (Chap. 12) and how to relate the behavior of different partners in interaction situations (Chap. 11). The chapter also reported a simple experiment to test change causality discovery in the context of the guided navigation scenario.

References

Bai Z, Wong WK, Zhang B (2010) Multivariate linear and nonlinear causality tests. Math Comput Simul 81(1):5–17
Ding M, Chen Y, Bressler S (2006) Granger causality: Basic theory and application to neuroscience, Wiley. http://arxiv.org/abs/q-bio/0608035
Gelper S, Croux C (2007) Multivariate out-of-sample tests for granger causality. Comput Stat Data Anal 51(7):3319–3329. doi:10.1016/j.csda.2006.09.021
Glymour C (2003) Learning, prediction and causal Bayes nets. Trends Cogn Sci 7(1):43–48
Granger CW (1969) Investigating causal relations by econometric models and cross-spectral methods. Econometrica: J Econometric Soc:424–438
Hoover K (1990) The logic of causal inference. Econ Philos 6:207–234
Menzies P, Price H (1993) Causation as a secondary quality. Br J Philos Sci 4:187–203

Mohammad Y, Nishida T (2008) Human adaptation to a miniature robot: precursors of mutual adaptation. In: RO-MAN'08: IEEE 17th international symposium on robot and human interactive communication, pp 124–129

Mohammad Y, Nishida T (2009) Toward combining autonomy and interactivity for social robots. AI & Society 24:35–49. doi:10.1007/s00146-009-0196-3

Mohammad Y, Nishida T (2010) Mining causal relationships in multidimensional time series. In: Szczerbicki E, Nguyen NT (eds) Smart information and knowledge management: advances, challenges, and critical issues. Studies in computational intelligence, Springer, pp 309–338

Ohmura R, Naya F, Noma H, Kogure K (2006) B-Pack: a bluetooth-based wearable sensing device for nursing activity recognition. The 1st international symposium on wireless pervasive computing, pp 6–11. doi:10.1109/ISWPC.2006.1613628

Pearl J (2000) Causality: models, reasoning, and inference. Cambridge University Press, Cambridge

Sugihara G, May R, Ye H, Hsieh Ch, Deyle E, Fogarty M, Munch S (2012) Detecting causality in complex ecosystems. Science 338(6106):496–500

Part II
Autonomously Social Robots

Chapter 6
Introduction to Social Robotics

Social robotics is an exciting field with too many research threads within which interesting new developments appear every year. It is very hard to summarize what a field as varied and interdisciplinary as social robotics is targeting but we can distinguish two main research directions within the field. The first direction (the engineering one) targets the design of robots that can interact with people in a social manner following human-like interaction patterns while the second direction (the scientific one) tries to evaluate the responses of people to robots and understand the expectations, and subjective responses to engineered social robots. This chapter tries to highlight some of the main findings in each of these two directions related to our quest for autonomous sociality. A third direction of research is concerned with using robots for understanding human's social behavior but we do not interest ourselves much with this research agenda in this book.

6.1 Engineering Social Robots

Social robotics is a multifaceted research field with many intertwined threads that makes it hard to summarize. This section provides an introduction to the field that does not claim exhaustiveness but only points to research ideas that are most related to our goal of achieving autonomous sociality.

Fascination with robots that can interact in a human-like manner dates back to the early days of biologically inspired robotics. One of the earliest examples are the robotic tortoises built by Walter in 1940s. The two robots involved (Elmar and Elsie) used lamps attached to their bodies and light sensitive sensors driving a phototaxis mechanism to produce what appeared as social behavior. This very early example has some lessons for us. Firstly, the phototaxis mechanism existed in the robots not primarily for social purposes but to allow them to find their charging station when they run low on battery in order to preserve their autonomy. This reuse of an autonomy enhancing mechanism to produce socially acceptable (or even socially describable)

© Springer International Publishing Switzerland 2015   171
Y. Mohammad and T. Nishida, *Data Mining for Social Robotics*,
Advanced Information and Knowledge Processing,
DOI 10.1007/978-3-319-25232-2_6

behavior is one of this book's main focusing points. Secondly, this kind of social-like behavior did not require complex computations but relied on a set of predefined reflexes that are easy to implement yet can produce complex behavior in line with ideas in behavioral robotics and reactive architectures that we discussed in Chap. 1. Thirdly, the robots did not actually produce social behavior in the sense that they did not really interact with people. Nevertheless, it is usually asserted that it is hard to see the videos of these robots without attributing some sense of agency to them (Fong et al. 2003). This mostly unconscious tendency to attribute agency to machines that produce certain kinds of behavior is the key human trait that social robotics relies upon and tries to exploit. Finally, Elmar and Elsie exemplify what we call later in the book the *engineering approach* to social robotics. The interaction between the agents was engineered carefully to give the illusion of sociability. The location of the light source and the photo-sensitive sensors as well as the internal wiring were all adjusted to produce the required behavior. 70 years later this approach is still the predominant approach in Human–Robot Interaction design either for social or industrial robots.

This book is in some sense a break from this tradition because it tries to rely less on careful engineering of the interaction and more on learning interaction protocols as discussed in full details in this part of the book.

Early research in robotics focused on the mechanical and control problems of industrial robots. After all, the first robot that left the researchers' imaginations and laboratories was unimate which started working on a General Motors assembly line at the Inland Fisher Guide Plant in Ewing Township, New Jersey in 1961.

The first attempts at *social* robotics focused on creating robots that interacted more or less like social insects. A common communication mechanism that was used in the early days was stigmergy which is a form of indirect communication through modifications of the environment. Stigmergy based ant-like robots started to appear in the 1990s followed by an avalanche of research in multi-robot systems that sometimes used modeling of social phenomena (e.g. interference, aggressive competition, and collaboration) to enhance the behavior of the robot group. This kind of research based on social animals is valuable in better understanding these animal societies as well as in solving genuine engineering problems including disaster response, collaborative map building etc. Nevertheless, this book focuses exclusively on robots that interact with people and other robots using individualized historical embodiment.

Dautenhahn and Billard (1999) defined social robots as:

> ... embodied agents that are part of a heterogeneous group: a society of robots or humans. They are able to recognize each other and engage in social interactions, they possess histories (perceive and interpret the world in terms of their own experience), and they explicitly communicate with and learn from each other.

This definition highlights the interplay between social robots and humans. It is not the case that the robot simply learns from the human *experts* but both kinds of agents are adapting to each other. This mutual adaptation will be one of the two bases of our proposed computational model in Chap. 8.

The interaction between the social robot and humans living in its social space shapes the development of its individuality. For example Kozima and Yano (2001) propose an epigenetic approach to social robot development based on three stages:

> (1) the acquisition of intentionality, which enables the robot to intentionally use certain methods for obtaining goals, (2) identification with others, which enables it to indirectly experience other people's behavior, and (3) social communication, in which the robot empathetically understands other people's behavior by ascribing to the intention that best explains the behavior.

It is instructive to consider in more details the design procedure for empathetic social robots proposed by Kozima and Yano (2001). The procedure starts by modeling how infant humans develop their ability to empathize. Based on research in developmental psychology, three stages are proposed: In the first 3 months of life, the infant shares gaze and exchanges voice and facial expressions with the caregiver. During this stage, the caregiver is responsible of structuring the interaction into a form of elementary turn-taking. The second stage starts at 3 months of age and lasts for around 6 months. During this stage, the caregiver starts to interpret the behavior of the infant in an intentional manner and later the infants starts to learn to interpret the behavior of the caregiver also in an intentional manner leading to a more symmetric interaction. The final stage the follow during which the infant learns to engage in joint attention (attention to the same object) with the caregiver. Two robotic platforms (infanoid and keepon) were then programmed to execute exactly these behaviors in that order (mutual gaze followed by mutual attention).

As another example, Kanda et al. (2007) developed a robot that *pretends* to listen to route guides given by a human subject using natural interaction protocol in such situations. The design procedure is similar to the one shown in the previous example. Human–human interaction in the same situation were analyzed and, based on them, different situated behaviors (See Sect. 6.3.2) were designed to capture each primitive action found in the human–human interaction trials and the relation between these situated behaviors were carefully engineered to produce the required *natural* listening behavior.

These two examples show success stories for the engineering approach (Fig. 6.1[upper]) which involves careful engineering of interaction rules learned from research in human–human interactions.

Given the large number of possible situations that the robot may find itself in and the complexity and context dependence of human behavior, this approach is not expected to build a complete system for a social robot even in a limited interaction context (e.g. within an office or a hospital room). Another approach to social robotics that we advocate in this book is shown in the lower panel of Fig. 6.1 which we call the autonomous approach. In this case, human–human interaction data is not modeled manually by a researcher but is used to automatically build a model of the interaction

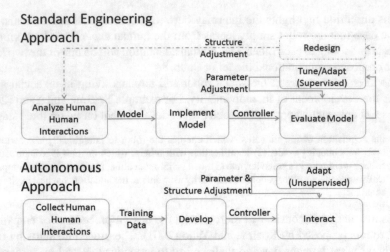

Fig. 6.1 The two main approaches for designing social robots

protocol involved using machine learning techniques. This model is then tested in actual HRI trials and the robot adapts the model based on the outcomes of these trials.

This machine learning approach is starting to gain ground in social robotics community only in the past decade and only slowly. The development approach we detail in Chap. 11 is based on this autonomous approach to sociality as well as the fluid imitation architecture described in Chap. 12. Other researchers are also working to advance this agenda. For example, Liu et al. (2014), proposed a system for teaching the robot how to interact in a shop keeper role in a simplified interaction with a customer. Interactions between people in the same situation were captured using networked sensors and both verbal and nonverbal behaviors were used to learn a set of semantically meaningful elementary *behavior elements* that are represented by joint behavior states highlighting the exchange of behaviors between two partners (customer and shop keeper). These basic elements are then clustered and the clusters are used in real-time to generate appropriate social behavior based on prediction of the current joint behavior state.

6.2   Human Social Response to Robots

How would humans respond to intelligent robots? This question lies at the heart of social robotics research, yet, it is not easy to answer or even to understand. Answering this question is the focus of the second *scientific* direction of research in social robotics.

Humans respond to robots differently based on the qualities of the robot, the context and the human. Most importantly, it depends on the expectations that the human brings to the situation. For example, Komatsu and Yamada (2007) compared the ten university students' ability to distinguish between the positive and negative attitudes of a PC, AIBO robot, and Mindstorms robot based on sound. They found that their subjects were best at classifying the attitudes of the PC. This result can be expected if we assume that the robot's embodiment affected the connection between sound and attitude attributed to it by people. The beeping sounds used in the experiments were *expected* from the PC but not the robots which made it more cognitively demanding to recognize the attitude of the robot from the sound.

A related hypothesis is the uncanny valley hypothesis suggested by Mori et al. (2012) which suggests that increased robotic anthropomorphism increases the positivity of humans' emotional response to them up to a point at which the similarity between the robot and a human becomes too close that further increases in human-likeness leads to decreased likability (to level even worse than clearly mechanical robotic forms) followed by another increase in likability when the robot becomes indistinguishable from humans and further on. If human-likeness is plotted against likability, according to this hypothesis, a valley appears which is what gives the hypothesis its name. The hypothesis further predicts that robotic motion exaggerates the whole curve leading to even faster increase and decrease of likability with human-likeness (See Fig. 6.2).

Experimental evaluations of the uncanny valley for robots can be conducted using pictures, videos or actual interactions with robots of variable *human-likeness*.

The uncanny valley hypothesis is widely cited in HRI and agents research. What causes it? A study by Bartneck et al. (2009) suggests one possible cause. In this study 16 subjects interacted with either a human (Prof. Ishiguro) or an android that was designed to resemble (as perfectly as technologically possible) that human called Geminoid HI-1. The android's eyes had either normal glasses or a visor (See Fig. 6.3). In each of these three cases the robot either moved normally or had limited movement. This lead to a 3 × 2 experimental design. Statistical analysis of the participants'

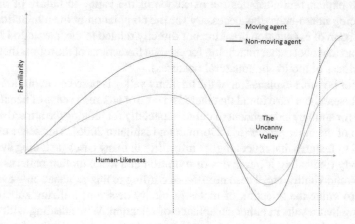

Fig. 6.2 The uncanny valley

Fig. 6.3 (*Left*) Prof. Ishiguro (*right*) Germinoid HI-1 during interaction. Reproduced with permission from Bartneck et al. (2009)

responses to questionnaires showed that they perceived the robot (both with the eye-glasses and the visor) as less human-like than the human which is expected. It also showed that this led to no effect on likability. Movement also had no effect on likability for the robot. These results seem to directly contradict the uncanny valley hypothesis but it can be explained in two different ways. It is possible to argue that the similarity of the human and the robot's appearance either did not approach the valley or passed it. It is difficult to maintain this position though as Fig. 6.3 shows that the similarity is high enough to confuse the two from a distance but is not perfect enough that participants who were seated in front of the robot could be deceived. Another, more plausible, explanation that was advocated by Bartneck et al. (2009) is that likability is not a single-dimensional factor. Humans do not just compare robots and other humans on the same scale but they use different scales to compare each of them. This again shows the importance of expectations in judgment of robot's appearance and behavior.

There can be other explanations for the uncanny valley that we will refer to here. Our discussion follows closely that provided by MacDorman and Ishiguro (2006). One such explanation attributes the existence of the valley to failure of the robot in producing micro-behaviors necessary for the completion of the interaction cycle. This is a form of expectation breaking but directly related to the interaction between the self and the robot rather than being focused on the actions of the robots themselves (e.g. jerkiness of motion or unnatural postures).

Another possible explanation of the uncanny valley is based on evolutionary aesthetics. Researchers have found that there is a core of common sense of beauty that is the same (or similar) independent of culture specially for judging the attractiveness of members of the other gender (MacDorman and Ishiguro 2006). This sense of beauty is related to features that reflect higher infertility in many cases including symmetry of the body (indicating lower levels of mutation and tear), motion patterns indicating youth and vitality, etc. These norms—according to this explanation—evolved in humans to guide the selection of mates (a binary decision) utilizing subconscious processing that results in either acceptance of rejection. When dealing with a robot, these same subconscious processes leads to a subconscious *rejection* for the robot based on this aesthetic criterion which leads in turn to it falling into the uncanny valley. The degree of uncanniness of a robot would then be proportional to its failure in achieving higher levels of aesthetic quality.

Ramey (2005) provides a different explanation which keeps that the similarity between the robot and the self forces the self to contemplate the existence of a form of an intermediate between its humanity and unhumanity of machines. This phenomenological confrontation leaves the sense of uncanniness. A related hypothesis is that the robot may elicit an implicit fear of death by showing a machine with human facade that forces us to contemplate that we are just soulless machines.

Even though all of these explanations may have some role in explaining the uncanny valley phenomenon, expectation violation continues to be the simplest explanation. A supportive study used fMRI to measure the difference in perception of a robot, an android and a human. Using neural adaptation analysis, it was found that watching a video of an android was associated with significant increase in activity of the anterior intraparietal cortex. This suggests that the uncanny valley has roots in perception conflicts within the brain's action perception system (Saygin et al. 2010).

This discussion of the uncanny valley highlights the importance of perception of the robot's behavior in acceptance by humans. This extends in our opinion to behavior not only appearance and this means that naturalness is not always a synonym of human-likeness when thinking about social behavior. A natural behavior of a robotic pet is not to be like humans but like pets.

6.3 Social Robot Architectures

Section 1.7 discussed briefly different architectures for robotics in general focusing on reactive and hybrid behavioral architectures. Here we try to give the reader a taste of HRI specific architectures that were designed with social interaction aspects in mind. We will introduce three of them each with its own strengths and weaknesses. Chapter 10 will provide detailed explanation of our proposed social architecture which relies completely on unsupervised learning techniques to approach our goal of autonomous sociality.

6.3.1 C4 Cognitive Architecture

The first architecture that we will consider here was not originally designed for social robots or for HRI but for synthetic creatures living in virtual worlds that can though interact with human subjects based on ideas from ethology and animal training (Blumberg and Galyean 1995). The later iterations of the architecture (C4 and C5m) were utilized is several research projects with the animal-like robot called Leonardo shown in Fig. 6.4 (Lockerd and Breazeal 2004; Breazeal et al. 2004, 2005; Breazeal and Aryananda 2002).

Fig. 6.4 Leonardo an animal—like robot controlled using the C4 architecture. Reproduced with permission from (Breazeal et al. 2006)

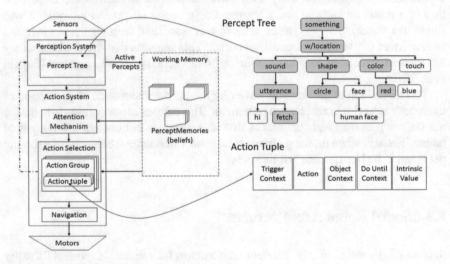

Fig. 6.5 Main components of the C4 architecture

Figure 6.5 shows the main component of this architecture (Isla et al. 2001). Perceptual processes in this architecture are organized in a tree structure with a single root (*something*). Each layer in this structure represent percepts of lower abstraction and more specificity. Figure 6.5 shows a part of a percept tree. The root node represents any thing and is always active as long as the robot is turned on. In this tree, all percepts are localized in a specific location in the perceptual field of the robot. This is represented by the single percept in the second level of the tree *w/location*. As long as a location can be assigned to any sensory input this percept becomes active. The third layer of the percept tree differentiates percepts based on their sensational modality and higher layers represent ever more specific percepts leading for example for a specific node for the word "fetch". The figure shows active nodes in shaded boxes and the whole tree represents that case when the robot perceives the word "fetch" with the scene of a red circle (ball). The architecture gives no specific constraints on the internal design of perception behaviors that generate the percept except the implicit assumption that it should work fast enough to give an honest picture of the world around the agent/robot in a timely fashion. The tree structure can be used to reduce the computational complexity of perception by activating the behaviors testing for every percept only if its parent percept is active. Each percept is also accompanied by a confidence level.

The basic percepts generated from the perception system are sent to the working memory. The working memory in C4 is not a passive container of information but an active entity that was designed to achieve some form of perception stability. The percepts active at any moment are combined together to form *percept memories*. A PerceptMemory is a bound representation of the perceptual state of the agent. When the working memory receives a PerceptMemory, it compares it with existing PerceptMemories and if it matches one of them, the confidences accompanying each percept are added to the existing PerceptMemory. This means that a PerceptMemory is not just a single snapshot but represents a history of bound perceptual states. The confidences in each PerceptMemory are automatically decreased at every timestep if not boosted by a matching percept.

When no match is found for percept bound inside a PerceptMemory at any timestep, the expected value of this percept is predicted in light of its previous values (e.g. through a regression mechanism) and this predicted value is added instead of the missing percept. This prediction when combined with natural decay of confidences, provide a simple form of perceptual stability for the agent.

Whenever a percept is available again that was predicted previously in Percept-Memory, the difference between its true value and its predicted value provide a measure of the *surprise* in that percept that can be used to drive the attentional system of the agent.

The PerceptMemory object provides a good supporting structure for grounded learning of perceptual concepts and objects. When localized perceptions of multiple modalities co-occur in percept memories enough times, a concept representing them can be abstracted which will be—by definition—grounded in the perceptual capacities of the robot.

The third major part of C4 (other than the perceptual system and working memory), is the action system. The most important component of the actions system is the action selection mechanism. Actions are represented in C4 by sets of action tuples. Each action tuple consists of five parts:

- Trigger context: this is the pre-conditions component that must be activated to execute the action. The trigger context relies on specific percepts (for example "red" and "circle") within the context represented by the current contents of percept memories.
- Action: that is the actual action code that generate commands to the robot motors/agent actuators.
- Object context: this part of the action tuple defines the objects that are to be used to execute the commands in the action part. These objects can be links to specific PerceptMemories stored in the current working memory. Not all actions are object related (for example the action "say") which makes this part of the tuple optional.

- Do-until context: this part of the tuple specifies the percept and working memory context that is necessary to keep executing the action once it is activated after satisfying the trigger context.
- Intrinsic value: this part represents a value for the tuple. It can be used to bias the action selector to select the actions with highest values. These intrinsic values are usually learned from interactions through a rewarding mechanism (Isla et al. 2001). These intrinsic values work similar to Q-values in reinforcement learning (Sutton and Barto 1998).

To avoid the problems that may arise from parallel execution of contradictory actions that have their trigger context satisfied, action tuples are grouped into action groups. Each action group can have its own action selection mechanism but only a single action from each group is allowed to execute at any time. Several action groups can be added to the architecture. By default there are two action groups: the attention group which selects the perceptual focus of attention and the primary group which determines the large-scale motion of the robot (Sutton and Barto 1998). Navigation is handled outside the action system in a dedicated module to keep the action selection system focused on other tasks.

In C4, credit assignment occurs when one action tuple becomes active and another becomes inactive. During this process the value of an action tuple is adjusted to reflect the likely consequences of performing the actions of the action tuple in the context of the state of the world indicated by its trigger-context. This is done via a mechanism similar to temporal difference learning (Isla et al. 2001).

The system keeps statistics about the correlation between activation of an action tuple and the children (in the percept tree) of its trigger-context children. Whenever the intrinsic value of the action tuple goes above a system wide threshold, high such correlations mean that the child percept is reliably capable of activating this action tuple. In such cases, a new action tuple is created that clones the original one with its trigger-context set to this child percept. This results in a more specific activation of the action tuple that reflects a learned association.

A similar mechanism can be used to modify the percept tree. In this case, children percepts take the form of statistical models built from observations that capture those aspects of the state space that seem correlated with the reliability of a given action.

When applied to HRI, goal-directed action was needed to account for the understanding of the interaction in from the top down instead of the standard association based learning mechanism found in C4 (Breazeal et al. 2004). This can be achieved by adding high level cognitive capabilities on top of the standard C4 architecture. These include goal based decision making, hierarchical task representation, task learning, and task collaboration.

To represent goals, the action tuple was extended with the notion of a *goal* which can be added to the trigger-context or the do-unitl context. There are two types of goals. *State-change* goals that evaluate to true in the trigger-context only if the associated state is not perceived in the working memory (this allows for bypassing actions when their goals are already achieved). In the do-until context, *state-change* goals keep the robot trying until their state is achieved. *Just-do-it* goals on the other hand activate the action always if their state is not perceived when added to the trigger-context and do not cause repetition in the do-until context.

Goal hierarchies can be used to separate the overall task-goal from action goals. This allows the robot to bypass parts of a task when the overall task-goal is achieved even when action goals are not. Breazeal et al. (2004) applied this system successful to collaborative learning.

6.3.2 Situated Modules

One of the pioneering architectures that targeted social robots was the situated modules architecture (Ishiguro et al. 1999). This architecture was based on ideas from behavioral robotics and tried to give special attention to practical aspects of implementing robots that interact with changing environments and people. The architecture was the software base for Robovie (a robot designed specifically for HRI applications—See Fig. 1.4a) (Ishiguro et al. 2001).

The architecture was designed to achieve five main requirements:

1. The developer can easily add behavioral modules.
2. Behavioral modules have to be situated which means being able to deal with a specific situation in a local environment.
3. Execution order is adjusted automatically by the system to generate appropriate behavior in different environments.
4. The system can represent the relation between different behaviors in a task on a module network.
5. The developer can add, remove and update behavioral modules online.

The architecture first proposed by Ishiguro et al. (1999) as a general behavioral architecture without specific focus on social robotics. Later it was modified to accommodate natural HRI during daily interactions between robots and humans. Special care was given in this architecture to give the robot operator easy high level control of the robot as will be clear shortly.

The basic building block of this architecture is the situated module. A situated module is a computational unit that performs a specific behavior in a specific local environment (called the *situation*). Figure 6.6 shows the internal structure of situated modules designed for HRI applications.

The two main components of any situation module are the *preconditions* segment which evaluates the applicability of the module when scheduled for execution and a behavior segment which executes the actions represented by the module. For general application of the architecture (e.g. in navigation tasks as in Ishiguro et al. 1999), no further details are given about the internal structure of the situated module but for HRI applications the behavior segment gets more specific structure. The basic idea is taken from linguistic research which defines adjacency pairs (e.g. greeting/response, question/answer) as the building blocks of conversations. Assuming that HRI sessions are similar to conversations, a situated module is designed to represent an action/response pattern similar to an adjacency pair. The architecture mainly

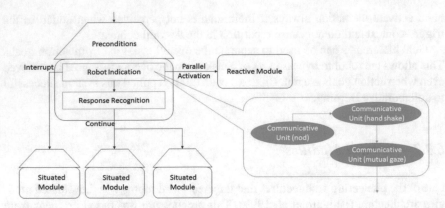

Fig. 6.6 A situated module as envisioned for HRI applications showing its preconditions and action parts and the three possible outcomes of it (continuation to next module, interruption or activation of reactive module). The robot indication is a program of communicative acts implemented as communicative units (See Kanda et al. 2004)

targets robotic applications in which the robot is actively seeking for interaction with humans leading to actions being from the robot and responses from the human. This defines the internal structure of the behavior segment given in Fig. 6.6 where the action part is called *indication* and the human response is measured using the *recognition* part of the behavior segment. Indications of HRI situated modules are built from basic blocks called communicative acts that represent basic simple unitary actions with well defined communicative roles like nodding, mutual gaze, mutual attention, approaching, turning to human, etc. We call the situated modules with this structure HRI-situated-modules (HRISW).

Reactive modules are simpler units that represent basic reactive behaviors needed in all contexts (e.g. obstacle avoidance). They can be activated and suppressed by the *higher level* situated modules similar to the case with the subsumption architecture (See Sect. 1.7.1).

Situated modules are always executed in a sequence which means that at most one situated module will be active at any time. There are three ways to leave an HRISW. Firstly, the recognition phase may result in a specific recognized human behavior and based on that behavior the next module in the current execution list is executed (more about execution order control later). Secondly, the human may not recognize robot's behavior, may not be interested in the interaction, or may be busy for any reason and no recognizable response is detected. In this case, an interruption happens and a different situated module is selected. The main difference between continuation and interruption is that the first represents the ideal unhindered execution of the interaction pattern being implemented while the later represents situations out of the normal and can handle exception. Finally, environmental conditions or human behavior may entail some reactive behavior to be executed. In this case, the appropriate reactive module is executed in parallel with the currently executing situated module. For

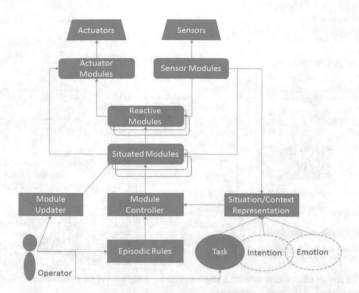

Fig. 6.7 The situated modules architecture, see (Kanda et al. 2004; Ishiguro et al. 1999, 2001)

example, when a human touches the robot, the reactive module *look-where-you-are-touched* may be executed in parallel with the current executing interaction module like talking.

Now that we know the anatomy of a situated module and understand the possible way to leave it once it starts executing, we show how these modules fit within the general situated modules architecture. Figure 6.7 shows the general form of the situated module architecture based on a consolidation of different articles detailing the architecture and its application to different HRI situations (Kanda et al. 2004; Ishiguro et al. 1999, 2001).

Situated modules are designed by hand and inserted into the system by the developer through the *module updater* which also allows the developer to removed unused modules and edit them.

Execution of situated modules is controlled centrally by the *module controller* which activates links between situated modules in the behavior network which selects next module to execute in the continuation and interruption cases. This behavior network is generated based on the task and a set of active episodic rules that are all under the control of the human operator.

The most important construct for behavior control are the *episodic rules*. An episodic rule is a specification of the condition(s) to select a specific situated module as the next module for execution either through interruption or continuation. Conditions of the episodic rules are continuously compared with the currently executing situated module and the history of execution and rules for which the conditions are satisfied are selected. If multiple rules are selected, the one with highest priority is allowed to execute and select the next situated module for execution. This generated

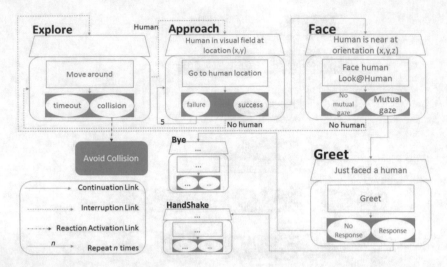

Fig. 6.8 Example of a social behavior implemented through a set of situated modules and their connections specified through episodic rules

a network of connection between situated modules that is adapted through activation and deactivation of episodic rules.

Kanda et al. (2004) proposed a simple declarative language for the specification of situated modules. Figure 6.8 shows an example social behavior implemented using this architecture. The words start by executing the *Explore* behavior which continues to execute until interrupted by a human appearing in the visual field leading to the execution of the *Approach* situated module. This module's behavior gets the robot to the location of the human subject. If the human disappeared from the visual field, the *Explore* module is activated through an interruption link. If *Approach* failed to arrive at the human's location (e.g. due to a timeout or an obstacle), it keeps trying for 5 times. Once *Approach* succeeds, the next module becomes *Face* through a normal continuation link. *Face* rotates the robot until it is facing the human in a socially acceptable manner and if successful activates the *Greet* module which executes a culturally appropriate greeting (e.g. a bow for a Japanese person and may be a simple nod for a western person with a short statement like "Hello") and then it waits for the recognition part of the situated module to report the human response. If an appropriate response was elicited from the human partner, then a hand shake is executed through the *handShake* module otherwise a bye gesture is executed through the *Bye* module.

This example represents adequately the steps involved in building a social robot using the situated modules approach. First of all, the designer implements as many situated modules as necessary. Redundancy is an asset in this stage. While implementing these modules, the designer reuses previous code that was stored as communicative units. Reactive modules are also added manually but these are usually very simple. Secondly, the designer starts to consider different situations and adds appropriate episodic rules to generate the required network structure for the modules to achieve a specific behavior in one task. The designer can then update these rules, edit them or delete some of them until satisfied with the robot's behavior. When a new task is considered, the designer can reuse the same situated modules but will be required to adjust the episodic rules to achieve the desired behavior. The structure of the system allows the robot operator to modify episodic rules online or to add new behaviors through the *module updater* which is a very effective way to achieve high level programming while the robot is operated by a human.

Learning does not seem to have any part in the architectural level. Every situated module may be learned (at least its parameters) through experiments with human participants but the architecture itself and connections between modules is fixed by the designer. That is not the only way to use the situated modules architecture. For example Ishiguro et al. (1999) used a visual map to learn the connections between different situated modules to achieve safe navigation in an indoors environment.

The situated modules architecture represents a large proportion of the current way to program social robots. Basic behaviors are designed and linked by the system developer. Learning has little to attribute to the success of the system and careful engineering is essential to achieve good behavior. Creators of the situated module architecture built several helpers including a visualization system and rule editing software that helps researchers implement their ideas using it. Researchers have achieved—using these tools—impressive natural behavior using this architecture allowing Robovie to work in schools and shopping malls, yet, the need to manually code most of the aspects of robot's behavior may limit the extend this approach can be extended. A nice illustration for the need for learning may be found in a recent publication from the same research group which proposed a learning from human–human interaction records similar in spirit to the approach advocated in this book (Liu et al. 2014). In the rest of this book we will be focusing on our proposal that tries to learn all of these aspects of behavior with minimal reliance on human intervention.

The situated modules architecture may be able to achieve sociality through careful engineering and may be of great value for robotic research and for robots designed explicitly for specific roles but it does not fulfill the requirements of autonomous sociality we are targeting in this book because of the lack of a clear methodology for learning situated modules and their interactions.

6.3.3   HAMMER

The situated modules architecture did not have any developmental aspect which means that there were no stages of development that the robot is supposed to follow. This section introduces another robotic architecture that targeted social robotics applications but with a developmental bend (Demiris and Hayes 2002). The name HAMMER stands for Hierarchical, Attentive, Multiple Models for Execution and Recognition. This architecture is based on the simulation theory of mind (Carruthers and Smith 1996). We will discuss the simulation theory of mind in more details in

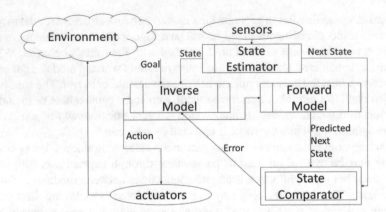

Fig. 6.9 The basic building block of the HAMMER architecture inspired by Demiris and Hayes (2002)

Chap. 7 but a brief introduction is in order here. The gist of the simulation theory is that humans use the same neural circuits used for action production in understanding actions of others as well as in understanding language about these actions. This means that the same neural mechanism can be used in the *forward* direction to generate action and in the *inverse* direction to understand it. This forward-inverse use of models is the hallmark of simulation based architectures in robotics.

The most basic building block in HAMMER is shown in Fig. 6.9. The computational system of the robot consists of several concurrently running blocks. Each one of these blocks consists of an inverse model (later called inverse control model) that can be used to control the robot to achieve some desired goal given an estimation of the current state (both of the robot and its environment) by sending actions/commands to the actuators and a forward model that can predict the next state of the system given the current state and the action (e.g. command) sent to the actuators. The difference between the predicted next state and the actual next state as sensed by the robot is calculated using the state comparator and sent back to the inverse model to modify its behavior closing the loop. This block is similar in spirit to the model predictive control approach in control theory.

The main insight from the simulation theory of mind employed by HAMMER is to reuse this basic building block for action understanding and imitation. Figure 6.10 shows the block when used for motion understanding. Firstly, sensed values are converted through a perspective taking process to the frame of reference of the agent which behavior is to be understood or imitated by the robot. Secondly, this state information is fed to the usual inverse-forward model arrangement but without sending the output of the inverse model to the actuators (we say that the actions are blocked). The error signal coming out form the state comparator are now fed to a confidence calculator which simply integrates this error information to form a scalar value that represents an estimation of the confidence on the hypothesis that the behavior represented by that block is active in the simulated agent. Confidence levels from different

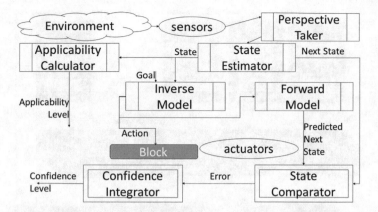

Fig. 6.10 The basic building block of the HAMMER architecture as used during simulation to understand the behavior of others

Fig. 6.11 Three HAMMER blocks with the inhibitory signals between them

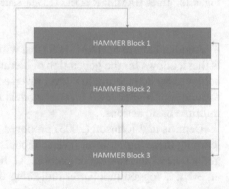

blocks are compared continuously and a winner-takes-all approach is utilized to esti-mate the active behavior in the simulated agent. Moreover, the input state to the block can be compared with the input states for which the action can in principle be activated and this results in another score for the applicability of the block. Only applicable blocks are allowed to compete during action understanding.

The winner-takes-all strategy is usually implemented in a distributed way by having inhibitory signals from each block based on its confidence level be connected to the other blocks reducing the target block's confidence (Fig. 6.11).

The discussion so far assumed that the state is available to the inverse model during action generation and understanding. This is not always achievable and one possible modification of the basic HAMMER architecture consists of adding an attention system that can be requested (by inverse models) to provide specific state information actively (Demiris and Khadhouri 2006). Figure 6.12 shows this arrangement of the architecture after adding the attention system.

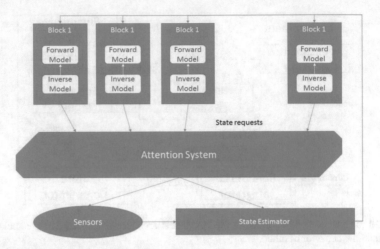

Fig. 6.12 Three HAMMER blocks sending state requests to the attention system

Another feature of HAMMER (that will be shared with our proposed architecture) is that it allows for the generation of hierarchy of HAMMER inverse models. This is achieved by connecting low level inverse models either in a sequence or in parallel to represent higher level inverse models corresponds to complex tasks employing multiple basic actions.

Demiris and Johnson (2003) proposed the following combinatorial approach for learning these higher level inverse models. While executing, several inverse models get activated and run to completion. At the completion of each such inverse model, an event is logged in a buffer indicating the start and end times of the execution of this model, its ID, and its confidence level. Now the buffer is tested for compatibility with any learned inverse model and if it was not compatible with any of them, a new higher level inverse model is created. For events in the buffer that are overlapping in time, the even with highest confidence is added to the new learned behavior if it shares and degrees of freedom with the other overlapping events otherwise both are added to the learned behavior to be executed in parallel.

How can we generate the models for the HAMMER blocks? One simple approach is to simply hardcode them. This was done for example by Demiris and Dearden (2005). A more interesting approach for our focus in this book is to autonomously learn them. This can be achieved through a process called motor babbling (Demiris and Dearden 2005). Motor babbling was first discussed in the context of infant development (Meltzoff and Moore 1997). The main idea is to generate random actions (probably generated from a Markov chain) and record the associated states after and before action execution. This information can be used to learn a forward model.

To achieve motor babbling, we need to fix the computational structure of the block. A convenient computational structure for HAMMER blocks is the Bayesian Belief Network (BN). A BN is directed acyclic graph where each node represents a random variable and links between nodes encode dependence. In the form used

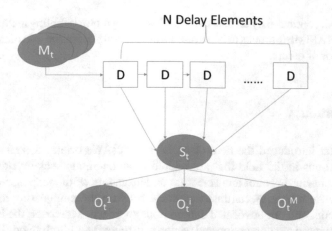

Fig. 6.13 The structure of the Bayesian Network used to realize the HAMMER block. From this structure we can have the forward and a rudimentary inverse model. The delay elements are used to represent the fact that action commands do not immediately affect the state

by HAMMER, links encode causal connections and the structure of the Graph is specially simple. Every state variable (e.g. an object in the environment or an internal degree of freedom) is considered as a random variable S. Every actuating action is considered as a random variable M and every perceived (sensed) value corresponding to every object or degree of freedom is an observed random variable O. For example, the state variable corresponding to an object in a scene can have several output variables like its position, velocity, color, size, etc. Now the Influence goes from actions or motor command M (observed) through state variables S (unobservable) to output variables O (observed) (Fig. 6.13).

During motor babbling, the collected M and O values are used to infer the structure and effect parameters defining the BN. An example due to Demiris and Dearden (2005) is learning the forward model representing opening and closing the gripper of a robot. A vision system for detection of the grippers in the camera image was developed and used to infer its location in the scene, its velocity and its size. A state variable for every gripper was added and these observable values were connected to it. The robot simply applied different motor actions to the grippers and recorded the resulting observables. The system could correctly infer that a delay of 11 steps is needed for a noticeable change in the robot state after issuing a motor command. Moreover, it could discover that only velocity is affected by the motor commands.

Now given a learned BN, it is easy to get the next state given a motor command by forward inference in the network ($p(S|M)$). An inverse model can also be found from the network using inverse inference $p(M|S, O)$. This inverse model can be used to decide the best command using for example MAP estimation (Demiris and Dearden 2005).

The robot continuously creates these BNNs during motor babbling leading to a rich library of HAMMER blocks to be used for both motion generation and understanding the behavior of others.

6.4   Summary

This chapter introduced the field of social robotics. We briefly described the two main directions in the field that are related to autonomous sociality: engineering of social behavior in robot and scientific understanding of the response of humans to these social robots. A contrast was made between the engineering approach to social robotics and our advocated autonomous approach. Moreover, the importance of expectations in our perception and response to robots was highlighted. The chapter also introduced three robotic architectures for social robots: C4, situated modules and HAMMER. The following two chapters will focus on learning from demonstration and imitation which will be the main learning mechanism in our proposed approach.

References

Bartneck C, Kanda T, Norihiro Hagita HI (2009) My robotic doppelganger: a critical look at the uncanny valley. In: RO-MAN'09: IEEE international symposium on robot and human interactive communication, IEEE, pp 269–276

Blumberg BM, Galyean TA (1995) Multi-level direction of autonomous creatures for real-time virtual environments. In: The 22nd annual conference on computer graphics and interactive techniques, ACM, pp 47–54

Breazeal C, Aryananda L (2002) Recognition of affective communicative intent in robot-directed speech. Auton Robots 12(1):83–104

Breazeal C, Berlin M, Brooks A, Gray J, Thomaz AL (2006) Using perspective taking to learn from ambiguous demonstrations. Robot Auton Syst 54(5):385–393

Breazeal C, Buchsbaum D, Gray J, Gatenby D, Blumberg B (2005a) Learning from and about others: towards using imitation to bootstrap the social understanding of others by robots. Artif Life 11(1/2):31–62

Breazeal C, Hoffman G, Lockerd A (2004) Teaching and working with robots as a collaboration. In: AAMAS'04: the 3rd international joint conference on autonomous agents and multiagent systems, vol 3. IEEE Computer Society, pp 1030–1037

Carruthers P, Smith PK (1996) Theories of Theories of mind. Cambridge University Press

Dautenhahn K, Billard A (1999) Bringing up robots or the psychology of socially intelligent robots: from theory to implementation. In: The third annual conference on autonomous agents, ACM, pp 366–367

Demiris Y, Dearden A (2005) From motor babbling to hierarchical learning by imitation: a robot developmental pathway. In: The 5th international workshop on epigenetic robotics: modeling cognitive development in robotic systems, Lund University Cognitive Studies, pp 31–37

Demiris Y, Hayes G (2002) Imitation as a dual-route process featuring predictive and learning components: a biologically plausible computational model. 13:327–361

Demiris Y, Johnson M (2003) Distributed, predictive perception of actions: a biologically inspired robotics architecture for imitation and learning. Connect Sci 15(4):231–243

Demiris Y, Khadhouri B (2006) Hierarchical attentive multiple models for execution and recognition of actions. Robot Auton Syst 54(5):361–369

Fong T, Nourbakhsh I, Dautenhahn K (2003) A survey of socially interactive robots. Robot Auton Syst 42(3):143–166

Ishiguro H, Kanda T, Kimoto K, Ishida T (1999) A robot architecture based on situated modules. In: IROS'99: IEEE/RSJ conference on intelligent robots and systems, vol 3. IEEE, pp 1617–1624

Ishiguro H, Ono T, Imai M, Maeda T, Kanda T, Nakatsu R (2001) Robovie: a robot generates episode chains in our daily life. In: ISR'01: the 32nd international symposium on robotics, vol 19, pp 1356–1361

Isla D, Burke R, Downie M, Blumberg B (2001) A layered brain architecture for synthetic creatures. In: IJCAI'01: international joint conference on artificial intelligence, vol 17. Citeseer, pp 1051–1058

Kanda T, Ishiguro H, Imai M, Ono T (2004) Development and evaluation of interactive humanoid robots. Proc IEEE 92(11):1839–1850

Kanda T, Kamasima M, Imai M, Ono T, Sakamoto D, Ishiguro H, Anzai Y (2007) A humanoid robot that pretends to listen to route guidance from a human. Auton Robots 22(1):87–100

Komatsu T, Yamada S (2007) How do robotic agents' appearances affect people's interpretations of the agents' attitudes? In: CHI'07: ACM conference on human factors in computing systems, ACM, pp 1935–1940

Kozima H, Yano H (2001) A robot that learns to communicate with human caregivers. In: The first international workshop on epigenetic robotics, pp 47–52

Liu P, Glas DF, Kanda T, Norihiro Hagita HI (2014) How to train your robot: teaching service robots to reproduce human social behavior. In: RO-MAN'14: the 23rd IEEE international symposium on robot and human interactive communication, IEEE, pp 961–968

Lockerd A, Breazeal C (2004) Tutelage and socially guided robot learning. In: IROS'04: IEEE/RSJ international conference on intelligent robots and systems, vol 4. IEEE, pp 3475–3480

MacDorman KF, Ishiguro H (2006) The uncanny advantage of using androids in cognitive and social science research. Interac Stud 7(3):297–337

Meltzoff AN, Moore MK (1997) Explaining facial imitation: a theoretical model. Early Dev Parent 6(34):179–192

Mori M, MacDorman KF, Kageki N (2012) The uncanny valley [from the field]. IEEE Robot Autom Mag 19(2):98–100

Ramey CH (2005) The uncanny valley of similarities concerning abortion, baldness, heaps of sand, and humanlike robots. In: Views of the uncanny valley workshop: IEEE/RAS international conference on humanoid robots, pp 8–13

Saygin AP, Chaminade T, Ishiguro H (2010) The perception of humans and robots: uncanny hills in parietal cortex. In: CogSci'10: the 32nd annual conference of the cognitive science society, pp 2716–2720

Sutton RS, Barto AG (1998) Reinforcement learning: an introduction, vol 1. MIT Press

Chapter 7
Imitation and Social Robotics

Imitation is one of the most utilized modes of learning in human infants and adults. We consciously imitate others when we want to learn new skills and unconsciously imitate them during interaction. Imitation can also be found in the animal kingdom in birds and higher apes. For these reasons, several fields of inquiry share an interest in imitation including developmental and social psychology, primatology, comparative and evolutionary psychology, ethology and robotics. This chapter provides a wide overview of the results of this research related to imitation learning in robotics and discusses the main aspects of imitation, and major challenges facing wide utilization of imitation in robotics.

7.1 What Is Imitation?

The word "imitation" is used by different researchers in different fields to mean different things. Nevertheless, there is a core of common aspects that can be used to make an operational definition of imitation for the purposes of this book. One of the earliest definitions in psychology was given by Thorndike (1898): "learning an act by seeing it done". This definition, while very simple, provides a good starting point for our discussions in this chapter.

The first word in this definition is "*learning*" which emphasizes the importance of acquiring new knowledge or skill during imitation. It is not clear though what is this knowledge or skill acquired. For example, seeing a person opening a stuck jar using a sharp edged knife, an imitator will be able to open another jar that (s)he failed to open previously which passes as imitation according to this definition. It is not obvious though what was learned in this case. Did the imitator learn how to use a knife? Did the imitator learn that jars can be opened by allowing air to pass under the lid? Did (s)he learn an exact trajectory of motion and pose orientations to open stuck jars? Or did (s)he learn that an appropriate time to use a knife in this manner— already known to her/him—is when wanting to open a stuck jar? Did the imitator

© Springer International Publishing Switzerland 2015
Y. Mohammad and T. Nishida, *Data Mining for Social Robotics*,
Advanced Information and Knowledge Processing,
DOI 10.1007/978-3-319-25232-2_7

open her/his jar because they just learned how to do so or simply because they are bored and the jar is now more salient because of the model's recent manipulation of a similar one? Depending on the answer to these questions, many different possibilities for explaining what happened arise and not all of them can be considered cases of imitation.

The second word in the definition is "*an act*". What is an act? The most obvious example of acts are body motions but words qualify as well. Even less substantial activities like learning strategies, and nonverbal behaviors accompanying speech can be considered as acts. An important feature of all of these acts is that they are different from their surrounding motions, words, or other activities. The ability to separate an act from its surroundings (i.e. action segmentation) is a very important step in imitation. If the would be imitator cannot find the boundaries of an act, it is very unlikely that it will be able to imitate it.

The final part of the definition reads "*by seeing it done*". It emphasizes the role of perception. The imitator has to perceive the act. Even though visual sensing seems to be implied if we take a literal view of the definition, most researchers will agree that other sensing modalities can qualify as well. We do not see the prosody of our partners during face to face interactions yet we imitate them. One thing that this simple definition implies though is that the model needs not be aware or even willing to produce the demonstration learned by the imitator (more on that in Chap. 12). Imitation is not always a form of tutelage. Another implication which is more technical but not less important is that the imitator must have a way to translate what it perceived into a form that it can use for reproduction. This becomes more problematic when the imitator and the model do not share the same body plan (which is usually the case in robotic imitation) or whenever the mapping of their actions to each other is not straightforward. Even when the imitator and the model have the same body plan and the translation between their actions seems trivial, the correspondence problem is still a complex one because what is perceived by the imitator is *not* the actions of the model as encoded in its motor control system but a projection of the external correlates of these actions on the model's body further transformed by the environment and sensing apparatus of the imitator.

In many cases it is not even clear what exactly was the action. For example when watching someone move her arm circularly while catching the lid of a jar, was the action to be imitated just the circular motion? was it the final effect of lid ease and removal? must the imitator use the same arm? must the imitator even use an arm or can she just use her mouth instead or a leg perhaps? does imitator need to keep her elbow in a similar relative pose as the one used by the model during the motion and if so relative to what exactly? should the imitator ignore the pose of the legs? should she ignore the grinding motion of the jaws that accompanied the action? There can be no clear answers to any of these questions as we can conceived situations in which each possible answer is a plausible one. This ambiguity is not an artifact of this definition but an inherent property in imitation that will occupy us much when dealing with imitation in social robots.

We use the term *imitation* to refer to behavior copying by the imitator of a feature of the model's behavior leading to learning when or how to produce that behavior, where this copying is *caused* by the observation of the model. Production imitation is what happens when the imitator learns how to do the—previously unknown to it—behavior of the model and context imitation is what happens when the imitator learns when to execute a behavior it already knows how to produce.

This definition of imitation, excludes behavior matching caused by chance as well as behavior matching caused by any other means other than the observation of model's behavior. This excludes, for example, stimulus facilitation (Zentall 2003) in which object manipulations associated with the model's behavior leads to increased object saliency which in turn leads to higher probability of interacting with it by the observer. Other behavior matching mechanisms that are excluded by this definition include contagion, social facilitation, affordance learning and emulation.

Contagion is a form of behavior copying that is outside the control of the observer and leads to no learning and no skill transfer. A common example is yawning which is contagious in humans. Notice that contagion is not excluded on the ground of being unconscious. The border between unconscious imitation and contagion is not always clear. We accept unconscious behavior copying as a form of imitation (e.g. entrainment of prosody during human–human interactions), yet contagion has the further feature that the copying is instantaneous and can lead to no persistent change in behavior. Prosody copying on the other hand may lead—with repeated exposure and at least in principle—to some persistent change in behavior.

Social facilitation happens when the mere existence of the model leads to increased probability of some behaviors in the observer that may appear as a form of behavior copying when preceded by a similar behavior by the model who may also be under the influence of social facilitation.

Affordance learning happens when the behavior of the model leads the observer to learn that some action can be done with an object (learned affordance). This new knowledge may increase the probability of interacting with the object executing the just learned action. Affordance learning is hard to distinguish from imitation on the grounds of external behavior. Pigeons (Columba livia) could learn how to gain access to food by just watching the relevant objects moved with a model causing the motion (Klein and Zentall 2003). Alpine parrots were also shown to be able to learn how to disentangle locks after observing them disentangled but, crucially, using actions other than the demonstrated ones (Huber et al. 2001).

Emulation is a related concept both to affordance learning and to imitation. Emulation happens when the observer's copying is for the change in object states rather than the actions of the model. The main difference between emulation and affordance learning is that the later implies that the observer did not have knowledge that the objects manipulated could be acted upon the way demonstrated by the model before the demonstration. Emulation is copying of final states which cannot be considered a form of goal copying. Final states are not goals because they do not carry the *intentional* connotations of goals. In order to call a form of copying "goal-oriented", it must be accompanied with some form of intentional attribution (i.e. the demonstrator intends the goal). Depending on the task and context, emulation may be even more complex than direct behavior copying, yet goal copying is always harder than direct motion copying and emulation. The uniqueness of human imitative skill lies on the complex interplay between goal copying and action copying adjusted to the context.

7.2 Imitation in Animals and Humans

The degree by which we ascribe imitative ability to a given species depends tremendously on how imitation is defined. Different animal species were found to possess a variety of copying behaviors. Biologists distinguish three forms of these behaviors:

product-oriented copying: The animal in this case tries to produce the same result of the actions demonstrated using any means at its disposal not necessarily the same means utilized by the model. This kind of copying is more related to emulation rather then imitation in the vocabulary of Sect. 7.1.
part-oriented copying: The animal tries to produce the same result using the same body parts used by the model but not necessarily using the same trajectories of motion.
process-oriented copying: The animal tries to faithfully copy the actions perceived from the model. This means that the same body parts (or corresponding ones) and roughly the same trajectories are followed during reproduction. This kind of copying is what we mean by imitation in this book.

One of the earliest studies that tried to find whether process-oriented copying existed in chimpanzees was an experiment by Hayes and Hayes (1952). In this study a home-raised chimpanzee called Viki was trained to reproduce actions on command. These kinds of studies (duped *Do–As–I–Do* studies) were repeated for many species including chimpanzees, orangutans, parrots, dolphins and dogs (Huber et al. 2009). Whiten et al. (2004) concludes after reviewing several studies of action copying in primates that:

> compared with children, who may show recognizable matching on all of the actions in the battery used, fidelity is typically low overall.

This low fidelity problem appears in most studies of the *Do–As–I–Do* variety. For example, Myowa-Yamakoshi and Matsuzawa (1999) showed that only 5.4 % of demonstrated actions were copied and these actions were invariably free motion actions involving two objects (e.g. putting a ball in a bowl). Actions involving no objects or object-to-self actions (e.g. touching the bottom of a bowl) were never copied.

Until the beginning of this century, the consensus was that even this low fidelity process-oriented copying existed only in higher apes. Some studies now challenged this positions showing forms of process oriented copying in birds, dolphins and even a dog (Topál et al. 2006). A general feature of most of these studies is that the animals were human-raised. For example, the dog Philip studied by Topál et al. (2006) was raised as a service dog helping his master opening doors and doing related actions and most chimpanzees studied in these experiments were human-raised.

The *Do–As–I–Do* experimental design does not distinguish clearly between process-oriented and product-oriented copying when objects are involved. Another design paradigm (that is used extensively with birds) is the *two–actions* design in which two groups of observers (subjects) are shown two kinds of demonstrations that differ only in the body part used for achieving identical environmental change. The two–actions design is so central to animal imitation studies and comparative psychology that it acts as an *operational* definition of imitation in this field. Imitation is treated as what the two–actions experiment measures and based on it several species were shown to have imitative abilities (sometimes called true-imitation for emphasis) (Byrne 2005).

One example of these studies showed that pigeons and Japanese quail show a preference of using the same body part used to press a lever (e.g. peak or foot) used by the model (Zentall 2004). These experiments fall more under the part-oriented copying category rather than the process-oriented category. Due to the simplicity of the actions involved in these studies, it is unlikely that the observers learned *how* to execute them. More likely, what was learned is *when* to execute them. This is sometimes called *context imitation* to distinguish it from *production* imitation in which the action itself is learned.

Part-oriented copying was also tested using the two–actions design in marmosets by Huber et al. (2009). Marmosets are more likely to open canisters using their hands if they did not observer a demonstration of canister opening or if they watched a demonstrator opening one with its hands. Nevertheless, after watching a demonstration of canister opening using the mouth, Marmosets become more likely to use their mouthes in opening canisters. This effect cannot be explained away by affordance learning because the learned affordance (openable by mouth) does not provide any advantage over the already known affordance (openable by hands) for the Marmoset. Moreover, it is not social facilitation as there is nothing about having the demonstrator in sight that would evoke opening-using-the-mouth behavior instead of opening-using-the-hands. It is not stimulus facilitation either as the marmosets opened the canisters with the same frequency only the body part employed was varied based on whether or not there was a demonstration. Emulation is not enough to explain the marmosets' behavior because there was no change in the outcome (canister opening) based on the demonstration. One final worry to lay to rest is whether the increase of opening-by-the-mouth behavior was just a result of some form of priming. Assuming that the brains of the marmosets have stored procedures for both kinds of opening behaviors (it is not clear that without a demonstration, marmosets *never* use their mouths to open canisters), just seeing the action done may prime them toward one of these stored procedures. Notice that this does not imply strong mirroring effects and there is no need to posit mirror neurons to justify that priming effect. The only way to handle this situation is to use delayed imitation so the observer is tested after some delay that is filled with other activities (better involving the same body parts employed by both models). This was done with Japanese quails with delays of up to 30 min and budgerigars with delays of up to 22 h (Byrne 2005). At least a weak form of process-oriented copying (e.g. part-oriented copying) is implied by these cases.

One critical difference between the *Do–As–I–Do* and *two–actions* paradigms is that the first is more appropriate to test imitation of complex tasks as the later usually employs simple actions from the imitator's repertoire. This is not always the case though. In two studies, four human-reared chimpanzees and six zoo gorillas were tested using the two–actions method but demonstrations involved three actions that must be executed serially to obtain an award. The chimpanzees successfully learned the complex task being demonstrated to them but only after several trials. The gorillas failed (Stoinski et al. 2001).

It is difficult to distinguish imitation from other copying behaviors mentioned above in the wild but some hints exist that it may probably play a role in acquiring population specific techniques for achieving the same action. Whenever an action can be done by two different methods where one of them is superior and there is a direct path of adjustment to convert the inferior action to the superior one, the persistence of the inferior technique in some exclusive population cannot be attributed to trial-and-error. Trial and error in these cases would act as a hill-climbing optimizer and would lead to the abandonment of the inferior technique in favor of the superior one as long as enough time is available. The persistence of the inferior technique exclusively in a population would then imply either explicit teaching or imitation. One such case can be found in how populations of chimpanzees catch Dorylus ants. Some populations exclusively use long sticks and bi-manual manipulations and others explosively use short ones and uni-manual manipulation. This difference though may be attributed to adaptation to slightly different environmental conditions (Byrne 2005).

The aforementioned discussion of animal imitation makes clear that process-oriented copying seems more difficult than product-oriented copying. Now consider the discussion of robotic learning from demonstration in Chap. 13. For robot's; one of the main challenges was goal-oriented imitation which in this case appears to be more difficult than just process-copying. This difference between robotic and animal imitation is not the only one and it may stem from a more general difference between animal and artificial cognition. Animals appear to have more built-in knowledge about the environment and other agents acting on it (engineered through ages of evolution) that makes them more apt at goal-inference—at least in familiar contexts relevant to survival—compared with artificial agents. Faithful process copying (e.g. production imitation)—on the other hand—is easier for artificial agents equipped with the appropriate (human-designed) correspondence mappings compared to animals that may not have any propensity to copy behaviors not affecting their survival.

It is clear—even from casual observation—that humans learn a lot of their skills through imitation. Infants have been shown to match tongue movements as early as 42 min after birth (Meltzoff and Moore 1997). Children also imitate at early ages and teenagers are known for their continual imitation of their peers and role models to the distress of their parents. As adults, we engage in both conscious and unconscious matching behaviors that may lead to genuine imitation in several cases.

There are two schools in developmental psychology about imitation in early life. The nativist school maintains that infants come to life equipped with basic abilities to engage in behavior matching and imitation while the developmental school assigns larger role for development and underplays the findings supporting behavior matching at early life as a form of imitation. Meltzoff's classical work on imitation in infants provides the main line of support for the nativist school (Meltzoff and Moore 1997; Meltzoff 2005). The developmental school employs a dynamical systems account of imitation taking the position that imitation develops during the first two years of the infant's life emerging out of infant's acquisition of different kinds of knowledge and motor, cognitive and social skills (e.g. Jones 2009; Ray and Heyes 2011).

Despite the differences between these two schools of thought, they both assign a role to developmental aspects in human imitation. The main difference is in the degree at which the mind comes to life equipped with enough structure to engage in imitation from the first day (Nishida et al. 2014).

A recent review by Oostenbroek et al. (2013) reported on three main alternatives for understanding neonatal imitation: (1) neonatal imitation is a genuine act of social communication mediated through an abstract representational system; (2) the phenomenon is actually an involuntary, inborn reflex limited to tongue protrusion; and (3) imitation in newborns is a product of arousal. stating that:

These views continue to be maintained without much promise of resolution.

Imitation is hypothesized to be at the root of several main characteristics of human cognition, interaction and life. For example, Nagy and Molnar (2004) showed that infants do not act passively just as imitators during natural interactions with caregivers but they provoke interaction by spontaneous execution of acts they learned through imitation earlier in the interaction (deferred imitation) which leads to a rudimentary form of turn-taking that may have bootstrapped the appearance of *dialog* in human communication.

Chartrand and Bargh (1999) experimentally showed that behavioral mimicry (the chameleon effect) has a significant effect on the interaction and increases empathy towards the interaction partner. HRI studies documented similar effects in human–robot interaction. Riek et al. (2010) used real-time head gesture mimicry to improve rapport between the human and a robot and Lee et al. (2004) programmed a robotic penguin to nod in return to detected nods in order to consolidate the back-channel communication in a natural way. Unconscious imitation is a stable feature of face to face interactions in humans and the degree of this unconscious imitation of nonverbal behavior predicts how agreeable a stranger partner appears to her interlocutors.

Imitation is also hypothesized to be a main mechanism for the transfer of skills and knowledge leading to the appearance of culture and its maintenance. For example, Prinz (2005) suggests that imitation plays a vital role in moral development. The concepts of "good" and "bad" cannot be defined by showing examples and moral behavior cannot be learned by just *mimicking* or matching the behavior of others. Imitation plays a role in moral development through a different route through unconscious emotional contagion. For example, when a child hurts someone, she perceives the undesirable emotional response of this hurt person and through emotional contagion (a rudimentary form of imitation), she feels a similar feeling. Without this simple mechanism, it is very difficult to distinguish good from bad. Huesmann and

Guerra (1997) adds to this unconscious emotional mechanism, the need to form normative moral beliefs by the child. These beliefs are generalized from the behaviors of others (specially care-givers) through imitation.

Imitation is also implied in enabling infants to *learn* a theory of mind (ToM) based on the simulation theory which will be described in more details in Sect. 8.2. The essence of the simulation theory is that we understand the actions and mental states of others by simulating them in our own brains. Imitation is used here as a technique to learn how to simulate other agents. For example, it was shown by Strack et al. (1988) that imitating facial expression elicits the associated emotional state. Thus, by imitating people, the infant may be able to associate their facial expressions with the emotions elicited due to the imitation which in turn forms an early form of emotional empathy that allows the infant to learn how to interpret the emotions of others.

Rao et al. (2004) proposed four stages of imitation development: (1) body babbling, (2) imitation of body movements, (3) imitation of actions on objects and (4) imitation based on inferring intentions of others.

Tomasello and Carpenter (2005) suggest different developmental stages starting with mimicking or emulation without understanding of intention in the first 9 months of life followed by imitative learning when the child starts to acquire an understanding of goal-directed action at around her first birthday. Goal emulation comes later at 14–15 months of age with an understanding of intentions followed by role reversal imitation at 18 months with an understanding of unfulfilled intentions, reciprocal intentions and communicative intentions. Prior intentions, symbolic intentions and hierarchies of goals are understood later during the first three years of life.

This brief discussion of imitation in animals and humans reveals important points about imitation that can inform designers of social robots: Firstly, imitation is not just means-ends copying or final state emulation. It is the complex interplay between copying the means and attaining the goals that defines imitation. Secondly, unconscious imitation is expected from social interlocutors. Social robots should be able to simulate this human trait to gain more acceptance by their human partners. Finally, behavior matching in all its forms provide a toolbox for learning new behaviors during social interactions. We will discuss several approaches to imitation learning in robotics in Chaps. 12 and 13. In the following section, we will focus on the social aspects of imitation in robotics research. A subject that is still to receive its due attention by researchers.

7.3   Social Aspects of Imitation in Robotics

Imitation is a social phenomenon. That seems clear and even tautological, yet most research in robotic imitation seems to ignore it. Imitation does not happen without interaction and this is why it is important to understand the social aspects of imitation related to the interaction context. This section provides two examples of studies targeting the social aspects of imitation. The first is an engineering approach to endow robots with social understanding based on psychologically inspired imitation. This exemplifies the engineering strand of social robotics in the terminology of Chap. 6.

The second is a study, that we conducted, on the effect of back imitation on perception of the robot's imitative skills. This is an example of the scientific strand in social robotics mentioned in Chap. 6.

7.3.1 Imitation for Bootstrapping Social Understanding

Breazeal et al. (2005) provided one of the early examples of studying the social aspects of imitation in the context of HRI. They proposed the use of imitative learning to bootstrap social understanding by a humanoid robot.

The starting point of this work was the Active Intramodal Mapping (AIM) model of Meltzoff and Moore (1997) which has an common (intra-modal) representation for both action production and perception.

Breazeal et al. (2005) notices the computational advantages of AIM's model for robotics despite its accuracy as a model for imitation in infants. The concept of intra-modal representation can be extended easily to other forms of imitation rather than being confined to facial expression.

The system uses the $C4$ cognitive architecture (See Sect. 6.3.1) and is implemented on an animal-like 64-DoF robot called Leonardo.

The percept tree consisted of movement percepts that detect motion of human facial features and contingency percepts that detect when these motions are due to movements of the robot. The system uses a single action group with two actions: motor babbling action, and imitation action.

The motor babbling action allows the robot to explore its pose space by selecting a random pose, going to it, then holding it for approximately four seconds. It is triggered by sensing an approaching human. The human is expected to imitate the robot during motor babbling by copying its pose during the four seconds it is fixed (simulating back imitation of care givers that will be discussed in more details in the next section). This form of back-imitation allows the robot to associate its actions (current facial pose), with their facial appearance (on the face of the human partner). The intra-modal representation used in this study is the pose space of the robot (Breazeal et al. 2005). This choice accords with our own experiments with different possible intra-modal representations for a navigation task (Mohammad et al. 2012) which showed better performance for motor-like representations.

The contingency percept is implemented as a simple heuristic that detects time-contingency between the start of robot and human motions. More complex contingency evaluations based on organ identification is not possible at the early states of learning. Two-layers feed-forward neural-networks can then be used to learn organ identification and improve the performance of contingency evaluation (similar to the second stage of the four stages proposed by Rao et al. (2004) and discussed in the previous section).

Once organ identification is completed, the imitation action can start to be executed. It is triggered by a set of percepts each detecting stable motion in one organ of the human face. The robot then tries to approximate this motion using its pose-graph incrementally using a hill-climbing algorithm (Breazeal et al. 2005).

Using this interplay between imitation and back-imitation, the robot is able to learn to imitate facial expressions and understand them in terms of its own motor system (using the intra-modal representation). This ability is now used to bootstrap social referencing.

Social referencing is the ability of one person to utilize another's interpretation of a situation in interpreting it and reacting to it. Infants start to show social referencing by the end of the first year in the form of secondary appraisal. This ability forms a simple yet important step toward social understanding of other people.

To achieve social referencing, the robot is equipped with an innate emotional system based on basic-emotions that is activated when its own face takes specific facial expressions (that are associated with these emotions in humans). When a new facial expression is activated, the current emotional state of the robot is attached to it. This allows the robot to learn affective appraisal.

The final stage involves the utilization of the attentional system (See Sect. 6.3.1) and emotional system to associate emotionally communicated appraisals from the human (based on its ability for intra-modal representation) to a novel object or toy (Breazeal et al. 2005).

In this work, interplay between imitation, attention and emotional modeling were utilized to achieve a socially meaningful milestone in social development, namely, social referencing.

7.3.2 Back Imitation for Improving Perceived Skill

Section 6.2 discussed some aspects of the human response to robots. In this section we focus on some of the imitation related responses to robots. As was the case in Sect. 6.2, this section is not trying to provide a comprehensive review of the subject but just focuses attention on some of the experimental results that may inspire new research directions in this under-researched area.

Meltzoff (1996) reports that:

human parents are prolific imitators of their young infants.

This highlights the ubiquity of back imitation (adult imitating infant) in natural interactions between infants and their care-givers. In a series of papers, we studied the effects of back imitation on human's perception of the robot in several evaluation dimensions (Mohammad and Nishida 2014a, b, 2015).

HRI studies documented a positive effect for having the robot imitate humans. For example, Riek et al. (2010) reported improvement in rapport between a human and a robot caused by head gesture mimicry from the robot. These studies have focused on the effect of robot's imitation on our response to it. To complete the picture, it is useful to consider the effect of us imitating the robot on our response to it. One motivation for conducting this study is the unconscious anthropomorphism known to color our response to robots and even TVs (Reeves and Nass 1996). This anthropomorphic tendency may be enhanced if we imitated the robot because this imitation can bias us toward perceiving its motion in a human-like manner (in order to successfully imitate it). This may in turn lead to higher acceptance and intention of future interaction (Mohammad and Nishida 2014b).

We conducted three experiments to evaluate this effect. The first experiment measured whether or not there is any effect for back imitation on the perception of the robot's imitative skill and whether this effect requires actual interaction with the robot or can be elicited in third persons *watching* the interaction. Twenty four subjects participated in this experiment. Half of them interacted with the robot in three conditions (online participants) while the other half watched these conditions (offline participants).

The three conditions were the same in our first two experiments and we will describe them briefly here: In the first condition called no-imitation (NI) the participant imitated a simulated agent that looks and moves just like the robot and the robot also imitated the same simulated agent. In the mutual-imitation (MI) condition, the participant imitated the robot (and the simulated agent also imitated the robot). The (BI) condition was a variation on the MI condition in which when the participants failed to imitate the robot, it imitates the participant for one pose then the session goes back the BI condition. Notice that the experimental design forced the participant to imitate something in all conditions and that the only difference between NI and BI conditions was the order by which commands are sent to the robot and the simulated agent. After finishing a session in one of these three conditions, the participant became the leader and the robot imitated it. It is about this second part (which was identical in all conditions), that the participant was reporting when judging the imitation skill, human-likeness, naturalness, and intention of future interaction with the robot.

We could show that the robots that were imitated in the BI and MI conditions were judged to have higher imitative skill and more human-likeness compared with the robots that were not imitated (the NI condition) but only by online participants not by offline participants. This showed that imitating the robot did have an effect on our judgment of its imitative skill and human-likeness (Mohammad and Nishida 2014b).

Conducting a larger experiment with the same design but with 36 online participants and no offline participants revealed the same pattern of results. Higher attribution of imitative skill, human-likeness, and intention of future interaction with the robots that we imitate (BI and MI conditions) (Mohammad and Nishida 2014a).

Finally, an even more direct comparison was performed. We had eighteen participants play a game with two NAO robots. One of the NAO robots was assigned the *leader*'s role and moved first. The participant and the other NAO robot imitated the pose of this leader robot. The leader here is the robot imitated by the human while the follower is not (similar to the difference between BI and NI conditions in previous experiments). In the second session, the participant became the leader and the two robots imitated him with the same algorithm (only a small random delay was inserted in the motions of the robots not to make the motion identical but without any bias toward any of the two robots). Even in this clear situation, in which the participant is asked to compare the performance of two robots that are interacting with her in the same time and with the same algorithm, more attribution of imitative skill and human likeness was reported for the leader robot (the one that the human previously imitated) (Mohammad and Nishida 2015).

Taken together, these results suggest that back imitation can be an effective way to familiarize humans with robots and improve their judgment of the robot's skill and human-likeness. If this finding is confirmed through other experiments with larger population and more varied participant sample (all our subjects were Japanese university students), it can be used to inform designers of learning from demonstration interactions. By providing a back-imitation game like *follow the leader* used in this work during the familiarization session that usually precede the demonstrations, the participant's subjective evaluation of the robot imitative skill leading to more demonstrations.

7.4   Summary

This chapter introduced the different shades of behavior copying related to imitation like contagion, emulation, stimulus facilitation, affordance learning, and social facilitation. We discussed various forms of behavior copying in animals reporting experiments using both the *do–as–I–do* and *two–actions* paradigms and differentiating between process-oriented, part-oriented, and product-oriented copying. This discussion revealed the need for a fourth category of goal-oriented copying to cover the cases involving intentional attribution of goals to the demonstrator. The chapter then discussed the two main schools in infant imitation (the nativist and developmental schools) commenting on the need of development for both approaches to understanding infant imitation and provided two specific developmental schedules proposed by researchers in developmental psychology. We also discussed the social aspects of imitation in social robotics focusing on the phenomenon of back imitation when humans imitate robots and reported two studies related to it. The first was an engineering study that tried to utilize back imitation to bootstrap social referencing and the later was a scientific study trying to understand the effect of back imitation on our perception of the robot's imitative skill and human likeness.

References

Breazeal C, Buchsbaum D, Gray J, Gatenby D, Blumberg B (2005) Learning from and about others: towards using imitation to bootstrap the social understanding of others by robots. Artif Life 11(1/2):31–62

Byrne RW (2005) Detecting, understanding, and explaining animal imitation. MIT Press, pp 255–282 (Chap 9)

Chartrand T, Bargh J (1999) The chameleon effect: the perception behavior link and social interaction. J Pers Soc Psychol 76(6):893–910

Hayes KJ, Hayes C (1952) Imitation in a home-raised chimpanzee. J Comp Physiol Psychol 45(5):450–459

Huber L, Rechberger S, Taborsky M (2001) Social learning affects object exploration and manipulation in keas, nestor notabilis. Anim Behav 62(5):945–954

Huber L, Range F, Voelkl B, Szucsich A, Virányi Z, Miklosi A (2009) The evolution of imitation: what do the capacities of non-human animals tell us about the mechanisms of imitation? Philos Trans R Soc B: Biol Sci 364(1528):2299–2309

Huesmann L, Guerra N (1997) Children's normative beliefs about aggression and aggressive behavior. J Pers Soc Psychol 72:408–419

Jones SS (2009) The development of imitation in infancy. Philos Trans R Soc B: Biol Sci 364(1528):2325–2335

Klein ED, Zentall TR (2003) Imitation and affordance learning by pigeons (columba livia). J Comp Psychol 117(4):414–419

Lee C, Lesh N, Sidner CL, Morency LP, Kapoor A, Darrell T (2004) Nodding in conversations with a robot. In: extended abstracts on human factors in computing systems, ACM, pp 785–786

Meltzoff AN (1996) The human infant as imitative generalist: a 20-year progress report on infant imitation with implications for comparative psychology. Social learning in animals: the roots of culture, pp 347–370

Meltzoff AN (2005) Imitation and other minds: the "like me" hypothesis. Perspect Imitation: Neurosci Soc Sci 2:55–77

Meltzoff AN, Moore MK (1997) Explaining facial imitation: a theoretical model. Early Dev Parenting 6(34):179–192

Mohammad Y, Nishida T (2014a) Human-like motion of a humanoid in a shadowing task. In: CTS'14: international conference on collaboration technologies and systems, IEEE, pp 123–130

Mohammad Y, Nishida T (2014b) Why should we imitate robots? In: Extended abstracts of the international conference on autonomous agents and multi-agent systems. International foundation for autonomous agents and multiagent systems, pp 1499–1500

Mohammad Y, Nishida T (2015) Why should we imitate robots? Effect of back imitation on judgment of imitative skill. Int J Soc Robot 7(4):497–512

Mohammad Y, Ohmoto Y, Nishida T (2012) Common sensorimotor representation for self-initiated imitation learning. In: Advanced research in applied artificial intelligence, Springer, pp 381–390

Myowa-Yamakoshi M, Matsuzawa T (1999) Factors influencing imitation of manipulatory actions in chimpanzees (pan troglodytes). J Comp Psychol 113(2):128–136

Nagy E, Molnar P (2004) Homo imitans or homo provocans? Human imprinting model of neonatal imitation. Infant Behav Dev 27(1):54–63

Nishida T, Nakazawa O, Mohammad Y (2014) From observation to interaction. Springer, pp 199–232 (Chap 9)

Oostenbroek J, Slaughter V, Nielsen M, Suddendorf T (2013) Why the confusion around neonatal imitation? A review. J Reprod Infant Psychol 31(4):328–341. doi:10.1080/02646838.2013.832180

Prinz JJ (2005) Imitation and Moral Development, vol 2. MIT Press, pp 267–282

Rao RP, Shon AP, Meltzoff AN (2004) A Bayesian model of imitation in infants and robots. Imitation and social learning in robots, humans, and animals, pp 217–247

Ray E, Heyes C (2011) Imitation in infancy: the wealth of the stimulus. Dev Sci 14(1):92–105. doi:10.1111/j.1467-7687.2010.00961.x

Reeves B, Nass C (1996) How people treat computers, television, and new media like real people and places. CSLI Publications and Cambridge University Press

Riek LD, Paul PC, Robinson P (2010) When my robot smiles at me: enabling human-robot rapport via real-time head gesture mimicry. J Multimodal User Interfaces 3(1–2):99–108

Stoinski TS, Wrate JL, Ure N, Whiten A (2001) Imitative learning by captive western lowland
    gorillas (gorilla gorilla gorilla) in a simulated food-processing task. J Comp Psychol 115(3):272–
    281
Strack F, Martin LL, Stepper S (1988) Inhibiting and facilitating conditions of the human smile: a
    nonobtrusive test of the facial feedback hypothesis. J Pers Soc Psychol 54(5):768–777
Thorndike EL (1898) Animal intelligence: an experimental study of the associative processes in
    animals. Psychol Rev: Monogr Suppl 2(4):1–109
Tomasello M, Carpenter M (2005) Intention reading and imitative learning, vol 2. MIT Press, pp
    133–148
Topál J, Byrne RW, Miklósi A, Csányi V (2006) Reproducing human actions and action sequences:
    "do as i do!" in a dog. Animl Cognit 9(4):355–367
Whiten A, Horner V, Litchfield CA, Marshall-Pescini S (2004) How do apes ape? Learn Behav
    32(1):36–52
Zentall TR (2003) Imitation by animals how do they do it? Curr Dir Psychol Sci 12(3):91–95
Zentall TR (2004) Action imitation in birds. Anim Learn Behav 32(1):15–23

Chapter 8
Theoretical Foundations

Research in HRI focuses on the social aspect of the robot while intelligent robotics research strives to achieve the smartest possible autonomous robot. We argue in this chapter that realizing sociability and realizing autonomy are not two independent directions of research but are interrelated aspects of the robot design.

The robotic architecture used for any social robot can limit or enhance its prospects of achieving appropriate autonomous and social behavior in the presence of humans. The following chapters of the book will provide two broad architectures for approaching autonomous sociality that stemmed from our research in the past decade. These two architectures are the Embodied Interactive Control Architecture (*EICA*) discussed in this chapter and the following three chapters and the Fluid Imitation Engine (FIE) which will be discussed in Chap. 12. Both of these architectures rely on imitation in some form to learn the needed behaviors. The main difference between the two approaches is in the relation between autonomy and sociality. While EICA relies on autonomous *imitation* to realize natural social interaction, FIE relies on interaction to realize effective imitation. This interplay between interaction, imitation and autonomy is the common thread that unifies these two architectures and our approach to social robotics. This chapter discusses some of the theoretical concepts that guide our approach and their implications for the design of robotic architectures for HRI.

8.1 Autonomy, Sociality and Embodiment

Throughout this book, we are using the concept of autonomy, yet we did not provide a clear definition of this term. Researchers in psychology and social sciences use the term in a variety of ways. Self-determination theory argues that there are three essential psychological needs: autonomy, competence and relatedness (Ryan 1991). Psychologists differentiate between two perceived sources of one's behavior: some actions are said to have internal perceived locus of causality (IPLC) while others have external perceived locus of causality (EPLC). The former are called autonomous

© Springer International Publishing Switzerland 2015
Y. Mohammad and T. Nishida, *Data Mining for Social Robotics*,
Advanced Information and Knowledge Processing,
DOI 10.1007/978-3-319-25232-2_8

actions while the later are sometimes called *heteronomous* actions. It is important to notice here that autonomy is not independent but is perceived volition (Ryan and Deci 2000). De Charms (2013) distinguishes between autonomy and independence in humans. The former highlighting the internal focus of behavior control while the later highlights the ability—in principle—to enact this behavior. For example, a person can be *autonomously dependent* on another for guidance or support (Chirkov et al. 2003).

In AI and robotics, the terms autonomy and independence are more or less used as synonyms. In robotics and AI literature, autonomy is usually equated with the ability of the agent to decide *for itself* how to behave when confronted with unforeseen situations in its environment. This does not though capture the different possibilities entailed by the term. For example, consider two robots: the first robot has an implementation of a navigation algorithm that was developed for it to navigate inside a building the map of which is stored in its memory. The second robot runs a SLAM algorithm that builds the map through its own interaction with the environment. The two robots can navigate their environment effectively and achieve the same performance, yet we do not attribute the same kind of autonomy to both. The reason is that the first robot is autonomous in the sense that it does not need a human to give it commands during its navigation (e.g. it is autonomous *from* human operators *toward* navigation of the environment). The second robot has this same level of autonomy but it also has another kind of autonomy. It needs no map. This means that the second robot is both autonomous *from* human operators and map builders *toward* navigation of the environment. This highlights the *relational* aspect of autonomy. It is always a relation between the agent and other entities on one hand and its goals on another. Autonomy presupposes goals in this account.

To summarize: Autonomy is a relational concept that cannot be defined without reference to the agent goals and environment. *An agent X is said to be autonomous from an entity Y toward a goal G if and only if, X has the power to achieve G without needing help from Y* (Mohammad and Nishida 2009). We would like our robots to be as autonomous as possible from real-time operators, but also from designer choices (to simplify the design process), and from environmental factors (to allow them to behave appropriately in many situations).

The relational aspect of autonomy means that it is always from something and toward something. Being from something implies being *embedded* in an environment, a context, from which the agent can be autonomous in some aspect or another of its behavior. This presupposes a weak form of embodiment (defined later). Being toward something, presupposes the existence of goals or desires. This is why being autonomous presupposes being an agent but being an agent presupposes the *possibility in principle* to be autonomous by just having goals. In that sense, autonomy, agency and embodiment are related concepts.

Sociability is the ability of an agent to elicit social response from humans by being able to follow human-style social interaction with them (Breazeal 2004). The ability of following *human-style social interaction* is an essential aspect of sociability as just eliciting a social response may tell more about people than about robots due to our tendency to anthropomorphize objects and media (Reeves and Nass 1996).

Dautenhahn (1998) suggests that socially intelligent agents should be in a sense *like us* and provides guidelines for the design of these agents based on roles of humans. Humans are embodied agents, humans are active agents, humans are individuals, humans are social beings, humans are storytellers, humans are animators, humans are autobiographic agents, humans are observers. These roles of humans are take by Breazeal (2004) to imply that sociable robots should be embodied, have life-like quality, human-aware, understandable and understanding of social norms, and capable of socially situated learning.

In this work, sociality is defined as *the ability to use normal modes of interaction, and stable social norms during close encounters with humans*. This definition is focusing on short term fast interaction compared with the aforementioned definitions of sociability and socially intelligent agents that focus more on long-term relation to humans. Sociality with this definition is a pre-condition for implementing sociable robots but is not enough to make the robot sociable. Sociability and social intelligence need explicit reasoning about social relations and expectations of others. Sociality in the short term sense given here is a real time quality which resists traditional deliberative reasoning and suggests a reactive implementation.

Autonomous sociality—our long term goal—can then be defined as *the autonomy of an agent from human operators or designers toward achieving sociality*, where autonomy and sociality are understood in terms of the definitions given earlier in this section. This translates computationally to reliance exclusively on unsupervised learning techniques to reduce the need for human intervention in robot learning.

Autonomy and sociality are related. Castelfranchi and Falcone (2004) argues convincingly that autonomy and sociality (the ability to live and interact in a society) are inter-related and as much as the society limits the autonomy of its members, it enhances it at the same time through different processes.

Early AI research focused on implementing intelligence using logical manipulation implying a Cartesian transcendental understanding of the mind. The body was assumed to have nothing to do with intelligence. This view was challenged by several newer approaches to cognition that stressed the importance of studying intelligence in the context of an environmental embodiment (Brooks 1991) and focused on the dynamical coupling of the agent with its environment instead of exclusively considering internal symbol manipulations of representations of this environment (Vogt 2002).

Ziemke (2003) identified five different notions of embodiment. The widest notion is called *structural embodiment* and it means that the agent is structurally coupled with its environment. This means that perturbation channels exist between the two. In other words, a structurally embodied agent is one that is affected by and can affect its environment. These mutual effects need not be physical in any sense. A software agent that is embedded in the computer's memory can be considered as a structurally embodied agent in some cases. This notion of embodiment is not very useful as any open-system in the physical sense is structurally embodied. Even a rock on a river's floor is structurally embodied.

Another notion of embodiment discussed by Ziemke (2003) is the notion of *historical embodiment* which implies a history of structural coupling with the environment. Riegler (2002) defines embodiment as follows: *A system is embodied if it has gained competence within the environment in which it has developed* (Riegler 2002). This notion of embodiment is less inclusive than structural embodiment but it is still wide enough to include agents with no physical bodies and, once more, may allow a rock in a river to be included if we consider that the shape of this rock is affected by the river while its position, orientation and shape in turn affects the flow of water in the river and that this shape and configuration is a direct result of a *history* of dynamical coupling with the river.

A third notion of embodiment that is used by Pfeifer and others is the notion of *physical embodiment* which requires the agent to have a physical body (Pfeifer et al. 2001). Self organization plays an important role in this notion and the ability to sense the environment through sensors and affect it through actuators are taken as essential features of any embodied agent (Brooks et al. 1998).

Other notions of embodiment identified by Ziemke (2003) include organismic, oranismoid embodiment that are much more related to biological entities (e.g. animals and plants) than to artificial agents. We do not consider these notions further in this book.

Both humans and robots can easily achieve structural, historical, and physical embodiment. In that sense, they do not provide much guidance in designing autonomously social robots. A more important notion of embodiment for our work is *social embodiment* (Barsalou et al. 2003; Ziemke 2003; Duffy 2004). Social psychology was always concerned by the effects of others on one's cognition and effects of cognition on action which made this field extremely receptive of the notion of embodiment (Meier et al. 2012). Here, embodiment refers to the notion that feelings, thoughts, and behaviors are all grounded in sensory experiences and bodily states. Four types of social embodiment effects were identified by Barsalou et al. (2003):

> First, perceived social stimuli do not just produce cognitive states, they produce bodily states as well. Second, perceiving bodily states in others produces bodily mimicry in the self. Third, bodily states in the self produce affective states. Fourth, the compatibility of bodily states and cognitive states modulates performance effectiveness (Barsalou et al. 2003).

The importance of bodily states and its relation to social interaction in social embodiment provides a guiding principle and a challenge to research in social robotics. The guiding principle is that natural nonverbal behaviors of the robot are essential for effective social interaction with people. The challenge is that this kind of behavior should match the expected human behavior in similar situations. This is because this matching would produce the required *compatibility* between bodily and cognitive states required by social embodiment and expected by humans from life-long interactions with other humans. This does not entail that social robots have to be humanoids, it merely requires them to have human-understandable relation between their internal state and bodily state.

This notion of *social embodiment* may be compared with the notion of *structural embodiment* but assuming that the environment is a social environment. In this work, we are more interested in socially embodied agents that develop their social interaction abilities through interaction with their social environments (Mohammad and Nishida 2009). This is how the challenge discussed in the previous paragraph is to be met in our work. We posit that in order to build human-understandable relation between the internal and bodily state of the robot in the social sense, the robot needs to ground this relation (i.e. its nonverbal behavior) in historical interaction within its social world. This suggests a notion of *historical social embodiment* analogous to historical embodiment (Ziemke 2003). To achieve this form of social embodiment, the robot will need to understand the behavior of humans using the intentional stance. In other words, the robot needs to have a theory of mind.

8.2 Theory of Mind

Humans understand the behavior of other humans (and some animals) by assigning mental states to them and *reasoning* about how these states *affect or generate* their perceived behavior. This ability to reason about the mental state of other agents (whether people or otherwise) is called by various names by different researchers including *mentalizing*, *mind reading*, and *theory of mind* (ToM). It should be noted though that, while the term *theory* of mind, implies some form of representational structure with interconnected concepts and law-like relations; this is not by any means the only possibility for ToMs. As we will see in this section, ToM—at least in some views—requires no explicit theory. Despite this bias, the term ToM is the most widely used of the three mentioned above in robotics research and will be used throughout the book (Scassellati 2002).

There are several competing theories for how ToM works and how is it acquired by humans. We will briefly describe the main theoretical clusters then discuss the implications for social robotics.

Historically, the debate regarding ToM was mainly between theory theory (TT) theorists on one side and simulation theory (ST) theorists on the other side. Variants of TT included modular ToM (MT) and child-scientist theories (CT). A new comer to the debate is what can be called the enactive theory (ET). We will discuss these alternatives here with an eye on applicability and lessons for social robotics. Only a broad brush will be used to paint the view taken by each one of these five approaches to ToM and we will skip many of the details that sometimes define the debate between them. Our goal is not to provide an exhaustive review of different approaches to ToM in psychology and philosophy but to give he reader a general birds-eye view of the research landscape that will be used to support specific design decisions of our architecture for social robots later in the book.

The first approach to ToM that we consider here is the theory theory (TT). The term "theory theory" may have first been used by Morton (1980) to define the view that people rely on ...

a fairly extensive commonplace psychological theory, concerned both with dispositional traits such as those of character and mood and with the intentions that produce action (Morton 1980).

The main distinguishing feature of TT is that it assumes that a form of representation exists for the mental states of other people and their relations to behavior. We understand other minds in this view by *inference* from this representation. This representation may contain rules like "thirsty people want to drink", "people who see a tap, will believe it can give water", and "people who want to drink and can find water, will most likely drink". The first of these rules represents a relation between two internal states of mind, the second between a perception and a state of mind and the third between a state of mind and a behavior. These rules may be deterministic but are most likely only probabilistic. Using these kinds of rules, TT theorists suggest that people *understand* the behaviors of others in mental terms.

The most commonly used supportive evidence for TT comes from studies of children at ages of 3 and 4 years. It was shown by Wimmer and Perner (1983) that three years old children cannot attribute a false believe to another while four years old children can. A common scenario involves a child seeing two containers. A puppet puts a chocolate bar in the left container and leaves the scene. Another puppet enters the scene and moves the chocolate bar to the right container. The first puppet then returns to the room and the child is asked, which container will the puppet open? The correct answer is the left container as the puppet did not see the change of location and that is what 4 years old children usually give as an answer showing that they can attribute a false believe to the first puppet. Three years old children in most cases fail this test asserting that the first puppet will look in the right container. Another scenario is to show a child a container that usually holds some kind of objects (e.g. candies) and then ask her what does she think will be inside the container. She usually answers with the common object. The container is then opened and shown to the child to contain something else (e.g. a pen) and the child is asked what did she think was in the container *before it was opened*? Again 3 years old usually fail this test.

The idea that children fail at some point to pass the false belief test then succeed later was taken as a proof that they have some form of internal representation (theory) that changes over time supposedly according to some predetermined maturing or developmental schedule. This needs not be the case though. For example, if the simulation theory (to be discussed later) is true and we understand others by simulating their internal state as if we were in their positions, it would be perfectly possible that children fail the test before their fourth birthday because their simulation system is not fully developed. It may be the case that they simulate both persons in the false belief test and cannot distinguish their internal states which makes them attribute the knowledge of the change in the location of the chocolate bar not only the second puppet but also to the first puppet that was not in the room. It may also be the case that 3 years old children fail the test not because of some problem in their ToM but because of another reason. For example, it may be immaturity of their inhibitory control system which does not allow them to inhibit the direct answer that they *know* is correct about the chocolate bar location or the content of the candy container

(Carlson and Moses 2001). It may also be the cognitive difficulty of the task and the language it is framed in. For example, Johnson (2005) showed, that simplifying the task and using a nonverbal paradigm allows 15 months old children to pass the false-belief test.

There are several ways in which the internal representation or theory can be structured and acquired and we can find several flavors of TT in the literature depending on various views of this structure and process.

One extreme for the structure and acquisition of the internal representation in TT, takes the position that it has the same structure, functional composition, and dynamics of a scientific theory (Gopnik et al. 1997). This is usually called the child-scientist theory (CT). Not only ToM is assumed to be acquired this way, but also all kinds of theories about how the world work. In some cases, the specific Bayesian formalism (popular in philosophy of science studies) is used to describe this internal theory (Gopnik et al. 2004). CT does not try only to inform us about how children learn their ToM but proposes that it can be of use for understanding how scientists go about conducting science.

Gopnik and Wellman (1992) argue for a specific progression of *theories* that are driven by the child's interaction with the world and others. Around second birthday, the child is assumed to have only a non-representational ToM that consists solely of two constructs: *desires* and *perceptions*. Desires are mind-to-world states in the language of Searle that drive the person toward specific objects or states and perceptions are world-to-mind states that are awarenesses of objects. This simplistic theory is then changed at around the age of three to contain other constructs like beliefs but again in a non-representational forms. Beliefs in this case are understood similarly to perceptions. This non-representational nature makes it hard to keep false-beliefs as it is not possible to keep false-perceptions. At around the age of four, this theory morphs into a representational theory that corresponds to our folk psychology. All of these are related to what is called the action theory which covers nearly all of folk psychology. CT is not offered only as an approach to explain the acquisition of ToM but cover two more areas of cognition: appearances theory (how object appearances change and relate to their states) and kinds theory. This is far beyond our interests in this section and we will focus only on action theory.

There are three kinds of features for theories in CT: structural, functional and dynamic features. Structural features include abstractedness, causality, coherence, and ontological commitment. Functional features include prediction, interpretation and explanation. Dynamical features include defeasibility (being falsifiable by evidence), changeability in light of counter evidence, and characteristic ways of revision in light of such evidence (Gopnik et al. 1997). CT maintains that children's ToM shares with scientific practice all of these features and goes to the extend of suggesting that this is the basis upon which the scientific revolution of the sixteenth and seventeenth century was based. It is not only that children are small scientists in this theory, scientists are children as well. Even though not everyone agrees with this feature of CT (see for example Bishop and Downes 2002), it provides a plausible computational model for building robots that can learn social interaction in a similar incremental theory-building way.

The problem with CT for social robotics, is best understood by understanding how does CT supporters explain the delay in appearance of modern science until the sixteenth or seventeenth century if it is based on such capacities that all children have. Gopnik et al. (1997) justified the appearance of modern science at that time by the combination of several factors: Firstly, leisure—at least for some—enabled them to put more time in falsifying existing theories about how the world work. Secondly, a large body of counter-evidence to existing theories accumulated at that time and, finally, communication advances allowed for better inter-fertilization between different disciplines.

In social robotics we seem to be in the situation of the world before the sixteenth century in this regard. The robot cannot expect rich and continuous interactions similar to those available for children specially when it cannot already respond appropriately due to its lack of a ToM. This scarcity of interactions would reduce the amount of counter-evidence needed to falsify and modify any existing theory it may be equipped with initially. Moreover, the robot does not have much interaction with other learning robots to allow for transfer of information in most cases. Finally, it is not clear how can the robot bridge the gap between non-representational theories based on desires and perceptions and full folk psychology theories that can be used for effective reasoning—beyond the level of a 3 years old child—about other people essential for effective interaction.

CT was the inspiration for the learning mechanism of HAMMER (See Sect. 6.3.3 and Fig. 6.13) which used the Bayesian formalism for realizing HAMMER blocks that acted as mini-theories usable both for forward and backward reasoning. In the words of HAMMER creators:

> The inspiration for solving these problem here is taken from developmental psychology. Gopnik and Schulz (2004) compares the mechanism an infant uses to learn to that of a scientist: scientists form theories, make observations, perform experiments, and revise existing theories. Applying this idea here, forward models can be seen as a theory of the robots own world and how it can interact with it. The process a robot uses to obtain its data and infer a forward model from it is its experiment. To perform an experiment, a robot decides what motor commands to use, gathers a corresponding set of evidence and uses this evidence to learn the BBN structure (Demiris and Dearden 2005).

Baker et al. (2011) proposed the Bayesian ToM (BToM) as a computational model that formalizes the concepts in CT and extends it but focuses completely on the theory of action. Figure 8.1 shows the relation between different constructs in this approach. The principle of rational belief is used as the basis for constructing the probabilistic relation between perceptions of the robot's internal state and the environmental state on one hand and the beliefs on the other. The principle of rational action is used to relate actions to beliefs and desires. Notice that this model does not distinguish desires from intentions which means that it has no attentional mechanism. Evolution of beliefs, desires, and actions over time is governed in BToM by a dynamic Bayesian Network (DBN) in which current state is affected by previous state and action, current environmental state is affected only be previous environmental state, current observations are affected by current internal and environmental states, current beliefs are affected by previous beliefs, current observations and previous actions,

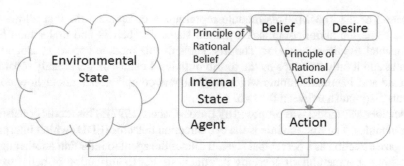

Fig. 8.1 Bayesian ToM showing the relation between beliefs, actions, and desires inspired by (Baker et al. 2011)

current actions are affected by current beliefs, and desires. BToM does not model the evolution of desire explicitly.

BToM can be thought of as an implementation of the full *action theory* assumed to emerge after four years of age in CT. Computational models of the earlier perception-desire theories assumed by CT can also be found in literature. Richardson et al. (2012) investigated the development of these theories in children aged from 3 to 6 years by assessing their internal state inferences after watching a short video showing a bunny and two fruits in three different conditions that affected the bunny's beliefs. The results suggested a developmental shift between three models. The first model attributes to the bunny the desire of whatever outcome it gets. The more advanced model used both perception (world states) and desires but not beliefs. This was called the copy-theorist after Gergely et al. (1995). The final model was the full BToM (Baker et al. 2011) which employed beliefs as well as desires and perceptions. This progression corresponds fairly well with the CT progression described earlier (Gopnik et al. 1997).

Another flavor of TT is the modular theory of mind approach (MT). While CT employs a general construct (theory generation and revision) as the basis for learning a theory of mind which is also employed for learning other theories (e.g. appearances and kinds theories), MT theorists posit the existence of an evolutionary selected *module* for reasoning about other people's minds.

Leslie (1984) provided one of the early accounts of MT. Three modules are implied by this model. The first model (ToBY) represents a theory about physical objects and causal relations between their change of states in the spatiotemporal domain. It was assumed by Leslie (1984) that this module is innate, yet some evidence suggests that it develops in the first six months of age (Cohen and Amsel 1998). This module allows the child to take the *physical stance* in Dennett's terminology (Dennett 1989). The next module to develop is system-1 of the ToMM. This system is responsible for modeling goals/desires and actions and begins to appear at the age of 6 months allowing the infant to understand behaviors in goal-directed manner. This module corresponds roughly to the copy-theorist (Gergely et al. 1995) in CT (Gopnik et al. 1997). It does not have any conception of beliefs. The final module to develop is system-2 of the ToMM. This module represents a complete ToM that allows the infant to reason about complicated mental states like beliefs and understand how can mental life drive behavior. The first signs of this module appear at around 18 months and it fully develops by the fourth birthday. Leslie's model cleanly separates animate and inanimate inputs which raises some complications about how could infants distinguish between the two.

Another MT model was proposed by Baron-Cohen (1997). This model consists of four modules. The first module is the eye direction detector (EDD) which interprets eye-gaze direction as a perceptual state. It allows the agent to infer that another agent is looking at something or is seeing it. This is similar to attribution of belief to the agent but is less symbolic and is tied to a specific perceptual mechanism (vision). The EDD is assumed to be the oldest evolutionarily of the four building blocks of ToMM as most vertebrates can use it to deduce the attention of predators (or prays for predators). The second module is called he intentionality detector (ID) and is responsible for detecting goals and desires in self-propelled motions. This module is assumed be ancient and present in many primates. It allows an agent to understand goal-directed behavior from other agents in its environment. These two modules together can be used to attribute desires, perceptions and beliefs to other agents (the building blocks of a ToM). The third module is called the shared attention module (SAM) and it takes dyadic representations from the ID and EDD and generates triadic representations involving the self, the other, and objects of mental states. For example, it can take the representation of "He sees me" from the EDD and "I see the toy" and constructs a representation of "He sees that I see the toy". Both shared attention and attribution of internal states can be constructed by SAM. The final module of Baron-Cohen's model is the ToMM which encapsulates a full ToM using what is called M-representations that are capable of representing arbitrary complex mental states like "he sees that I know that he beliefs that I desire that he wants that she sees the toy".

One supportive evidence that is usually marshaled in support of MT is the specific impairment of mindreading ability (ToM) in autistic children as compared with normal children or even children with Down syndrome. An early study that supported this hypothesis was conducted by Baron-Cohen et al. (1985) who compared the performance of normal pre-school normal children, Down syndrome children, and autistic children on a false-belief task. All children had a mental age of above 4 years, although the chronological age of the second two groups was higher. Eighty-five percent of the normal children, 86 % of the Down syndrome children, but only 20 % of the autistic children passed the test. This specific impairment (the argument goes) suggests that ToM is a specific module that is affected in autistic children (Nishida et al. 2014). Critics of the theory point out that ToM cannot be though of as a standard module either because it does not have a limited domain of stimulus to use and be activated through (Goldman 2006) or because it is affected by higher cognitive functions (Gerrans 2002).

ToMM provided inspiration for several studies in social robotics thanks to the clear computational path it provides for implementing ToM. Scassellati investigated a system combining aspects from Leslie's and Baron-Cohen's approaches to implement a theory of mind in a humanoid robot (Scassellati 2002, 2003). Ono et al. (2000) added to Baron-Cohen's model an utterance understanding module to understand the utterances of others by interacting with the ToMM module.

A simple approach to incorporate something like TT in social robotics is to endow the robot with a *complete* folk-psychology theory and use deliberative reasoning—in the style of good old-fashioned Artificial Intelligence (GOFAI)—to reason about the internal mental states of people. This approach would be a return to the engineering approach to social robotics that we argued against in Chap. 1. Moreover, it is a challenging approach given that—despite several efforts—we do not have a satisfactory common-sense knowledge base upon which to base this kind of theory. Like most other incarnations of GOFAI, this approach will also suffer from grounding and poor response time problems.

Nevertheless, all is not lost. We have shown already that CT and MT (variations of TT) could be used successfully for social robotics. They provided guidance and inspiration for specific implementations. The idea that *rationality* is assumed by the theory (found specially in CT and BToM) informs us that social robots should show rational behavior in order to activate our intentional stance attribution. From the modular-nativist approach (as well as studies in early neonatal imitation) we salvage the idea of the existence of a-priori structures that allow agents to develop their ToMs during natural interaction with people. These structures need not be encapsulated in Fodorian modules because of the need to interact heavily with other components of the cognitive structure of the agent, yet it is still possible to achieve some limited form of separation—at least in the computational resources—that allows us to focus on the development of ToM components without having to implement the complete cognitive structure of the agent.

A different approach to ToM that differs dramatically from TT in the forms of representations and underlying computational mechanisms is the simulation theory (ST). The main idea of the simulation theory is that we understand others by putting ourselves in their position and trying to see what will happen inside our heads if we were them (i.e. by simulating them). A major difference between the simulation theory and the theory theory is that there is no need for generalized science-like laws that we generate by observing other's behavior (or that we have innately). The understanding is then driven by simulation rather than propositional inference or any kind of inference.

Goldman (2006) tracked the history of simulation theory to Hume, Adam Smith and Kant. In modern times, the theory in its current form and the use of the term *simulation* can be traced back to Robert Gordon when he suggested that we predict others' behavior by answering the question: *What would I do in that person's situation?* (Gordon 1986).

If the simulation theory is true, then we run internal simulations of other agents using the *same* brain regions and activities that is used for generating our own beliefs, emotions and motions. This immediately suggests two modes of failure. The simulation may leak into our own beliefs and emotions (contagion) and our own beliefs and emotions may leak into the simulation (attributing to the other some of our own beliefs, desires and emotions). There is no clear reason why these two kinds of failure should exist in the theory theories. As the reader may have already noticed, these two types of failures are common enough to give—at least some weak—support to the simulation theory. The first type of failure mentioned may be—at least partially—behind the very known forms of contagion in human behavior ranging from yawning to emotional contagion and adult unconscious imitation. The second type of failure is very common in everyday life and may be one of the contributing factors behind the failure of 3-years old children to pass the false belief task.

Several strands of evidence are now available to support ST. For example, Kanakogi and Itakura (2011) found that the onset of infants' ability to predict the goal of others' action was synchronized with the onset of their own ability to perform that action. Moreover, action prediction ability and motor ability for some action were found to be correlated. The evidence of a correspondence in the impairment of mindreading (ToM) and motor abilities in autistic children can also be marshaled as a supporting evidence for ST (Biscaldi et al. 2014). Nevertheless, this does not explain the delay in attributing beliefs to other people compared with desires and perceptions that is well explained by CT (Gopnik et al. 2004).

Another major criticism of ST is that it is not adequate for explaining self-attribution of mental states (at least in its pure form) as discussed in details by Goodlman who proposes a hybrid ST-TT approach that emphasizes the simulation aspect in much of the same spirit as the proposal we put forward computationally in this book (Goldman 2006).

ST has an obvious computational advantage for AI and robotics. It allows for reuse of computational resources designed for actuation in understanding of others. Reuse is a favorite for engineers at all times. Several studies in social robotics utilized one form or another of ST. For example, a series of studies used an embodied cognitive architecture called Adaptive Character of Thought-Rational/Embodied (ACTR/E) to implement a form of a ToMM in the architectural level (Trafton et al. 2013). A "Like-me" hypothesis is used to derive a simulation within this module to understand the mental states of others (Kennedy et al. 2009). Nishida et al. (2014) relied heavily on ST to construct a computational architecture for conversational agents.

In this book, we employ elements of ST, TT, and MT to allow robot learning of a ToM suitable for understanding the behavior of people in social interactions. Our approach is sub-symbolic as will be explained in more details in Chap. 9. A theory of mind is useful for predicting the dynamics of others' intentions, beliefs and desires. It is important to fully utilize it to have a way to represent these constructs. We turn our attention now to intention modeling in AI and how it can be adapted to social robotics.

8.3   Intention Modeling

Intention recognition is one of the major challenges facing realization of autonomously social robot (Burke et al. 2004). Natural interaction between people relies heavily on our ability to understand the intentions of others—even at an implicit level—and use this understanding in guiding our behavior both in cooperative and non-cooperative domains. A large body of research is dedicated to this problem since the early days of AI and most of this work relies on a specific view of intention as some hidden state of the person that needs to be *recognized* the same way that we recognize people's identity given a picture of them in face recognition applications. We propose that—despite its widespread acceptance—this view of intention driven from folk psychology is not the best way of thinking about that concept at least in social robotics.

The traditional view of intention does not only define it as a hidden-state but it attributes more structure to intention that was paramount in the way HRI researchers thought about it. Intention in this view is a *describable* hidden state or a symbol. By that we mean that—at least in principle—it is possible to verbally describe the intentions of people. It is the kind of mental structure that can be put to words. Some other aspects of our mental life may not be describable as most of subconscious cognition. This assumption about the nature of intention as a describable entity among related mental entities like beliefs and desires led directly to symbolic and deliberative algorithms and techniques for both describing and discovering it under the general framework of Belief–Desire–Intention (BDI) architectures.

The English word *Intention* has several meanings:

- WordNet: A *volition* that you intend to carry out
- Encarta:

  - Something that somebody *plans* to do.
  - The quality or state of having a *purpose* in mind.

- Webster dictionary: An anticipated *outcome* that is intended or that guides your *planned actions*

The first definition focuses on the *goal*, the second on the *plan* and the third combines both in some sense. The traditional view of intention seems to focus more on the *goal* aspect of the concept which is stable and describable.

8.3.1 Traditional Intention Modeling

What does we mean when we say that a person *intends* something? Despite the intricacies of intention definition, the question itself implies that intention is something that a person can do. A specific intention seems to be a mental state that is related to some content (i.e. whatever is intended).

Intentionality, in philosophy, is usually understood as *the power of the mind to be* about *something*. Intentionality in that sense is not a feature of intention alone but of other mental constructs like belief and hope. The *aboutness* is—nevertheless—not the main feature of intention that we are trying to model in robots as it is implied by the computational manipulations of internal representations.

Intentionality according to the traditional view seems to be something that we have in our minds in relation to other cognitive constructs. Something describable, with specific identity and either is there or is not there. We either intend something, intend its negation or we do not intend anything. We may oscillate between these states but we always are in one of them. This is the view of intention assumed by folk psychology and is underlying most of the research in intention modeling in HRI and AI. In a court of law, a crime is either committed *with intention* or an accident happened *unintentionally*. Either one of these is true and it seems that AI and HRI borrowed this discrete nature of the concept.

It is worthwhile to note that intention in this view (as well as beliefs, desires, and goals) has all the qualifications of a *symbol* in the traditional AI sense. Evolution of the internal state is coded in this framework as continuous adoption and revocation of different intentions.

As an example of employing this symbolic view of intention, consider the work of Wang et al. (2013) which uses a statistical approach for inferring human intention from motion. Two examples where given for intention modeling. The first is modeling the *intended* location of the ball in table-tennis and the second is the intended gesture during HRI sessions. In both cases, there is a clear domain for intentions(i.e. a rectangle in $\mathbb{R}^2$ representing the table in the first case and the list of predefined gestures in the second case). The human is assumed to have a single intention belonging to this domain at every point in time and the goal of the robot is to recover this hidden intention from the behavior of the human. We will use this example through out this section to contrast it with our dynamic view of intention.

8.3.1.1   Intention Modeling in AI and HRI

One of the most influential models of intention in AI in general is the *planning theory of intention* proposed by Bratman (1987) which considers intention as the main attitude that directs future planning. There is a close coupling between intention and goals in this view. The practical usefulness of intention in this framework is to guide practical reasoning to the most promising areas related to the current situation to increase the efficiency of the agent. In a sense its main use is as an attention mechanism.

The Belief–Desire–Intention (BDI) framework is a representative example of this approach to intention modeling. Here intentions are representations of commitments to specific plans (Braubach et al. 2004). This commitment has an attentional aspects by focusing the reasoning power of the agent or robot (Morreale et al. 2006). This framework represents intentions by symbols and uses deliberative manipulations intensively and is one of the main frameworks for intention modeling not only in AI but in HRI research as well (see for example Stoytchev and Arkin 2004; Parsons et al. 2000).

Even when the traditional BDI approach is not explicitly used in modeling intention, this view of intention as a goal for a plan that directs behavior is usually assumed implicitly. Consider the work of Wang et al. (2013) mentioned above. Even though the framework used ts probabilistic where Bayesian inference is utilized instead of logical manipulations used usually in BDI models; still the system assumes that intention is a predictor of a specific ball location in the table-tennis example or an action in the action recognition task.

8.3.1.2 Intention Communication in AI and HRI

In social situations, intentions of interacting partners get joined. What does the traditional view of intention use to model this interaction between partners' intentions? In general, the intention-as-a-plan view naturally divides this intention interaction into two separate processes: intention expression and intention recognition.

Intention expression is a behavioral process that generates external motions and behaviors representing the hidden intentional state.

Intention recognition is a perceptual process that recognizes the hidden internal intentional state from sensed external behaviors. In most cases the intention to be recognized is assumed to be one specific yet unknown member of a set of possible intentions.

The importance of intention expression and recognition for social robots led Alami et al. (2005) to propose integrating HRI within the cognitive architecture of the robot based on the Joint Intention Theory (JIT) which relies on the aforementioned view of intention but expands it to handle intention communication to form coherence between the goals of interacting agents.

The JIT itself is designed after the planning view of intention discussed in this section. Successful teams in this theory are ones in which all members commit to a joint persistent goal that synchronizes their own personal intentions (Cohen and Levesque 1990). JIT uses symbolic manipulation and logic to manage the intention/belief of team members. This approach proved successful in dialog management, and simple collaborative scenarios.

8.3.2 Intention in Psychology

This view of intention as something we posses from some domain of possible intentions has the advantage of simplicity as well as providing a natural decomposition of intention communication into expression and recognition stages. Nevertheless, this view has several problems that we will investigate in this section from the points of view of experimental psychology and neuroscience. These difficulties do not mean that this simple model is not useful for modeling intention in social robotics. In many simple cases, this static view of intention is sufficient and even desirable because of its simplicity.

Two main types of intention can be distinguished in psychological research (Gollwitzer 1999): goal and implementation intentions.

Goal intentions simply specify a goal or an answer to the question: "what should happen?". Implementation intentions couple perceived situations with specific behaviors to achieve some goal. In another word, implementation intentions are policies not just symbolic goals. They are the precursors of actual intentional behavior.

Experimental psychology suggests that only a small portion of our day-to-day activities and behaviors are generated based on prior goal intentions (from 20 % to 30 % in some accounts) (Sheeran 2002). Why is that the case? In most cases the reason is that we fail to create effective plans from goal intentions. One resolution to this problem is to construct implementation intentions instead of goal intentions. Being a policy, implementation intentions come ready with the plan and this was shown to increase goal attainability (e.g. Gollwitzer and Sheeran 2006; Paschal Sheeran and Gollwitzer 2005).

It is believed that the existence of an implementation intention increases the probability of detecting the associated situation, and converts the effort-full plan initiation action needed to attain goal intentions into effortless unconscious behavior launching based on environmental cues (Gollwitzer 1999).

Goal intentions are the least effective according to the aforementioned discussion, yet they are the kind of intentions directly representable in symbolic BDI frameworks. Robots do not have the problem of distraction or failure to construct a plan (as long as the planner can find one) but, as with the human case, representation of intentions as policies (implementation intentions) may improve adaptability to environmental changes and reduce the computational power required to execute them.

Intention causes action. Nothing seems more common-sense from the viewpoint of the static view of intention. This is, after all, the basis of the planning theory of intention. Searle (1983) distinguish between two kinds of intention in relation to action execution: prior intentions (similar to goal intentions) and intention-in-action. The later is the *direct* cause of action while the former may be a contributing factor in creating the intention-in-action (for more details, see Searle 1983).

This simple causal relation between intention-in-action and action, despite being common-sensical, was challenged in some neuroscience research suggesting that at least conscious intention may be the result of action preparation rather than the cause of it (Haggard 2005).

8.3.3 Challenges for the Theory of Intention

Any theory of intention must meet two challenges. The first challenge is modeling intention in a way that can represent all aspects of intention mentioned in the discussion above including being a goal and a plan, while being able to represent implementation intentions.

The second challenge is to model intention communication in a way that respects its continuous synchronization nature which is not captured by the recognition/expression duality of the traditional view. This means recasting the problem of intention communication from two separate problems to a single problem we call *Mutual Intention* formation and maintenance.

Mutual intention is defined in this work as a dynamically coherent first and second order view toward the interaction focus shared by interaction partners.

First order view of the interaction focus, is the agent's own cognitive state toward the interaction focus.

Second order view of the interaction focus, is the agent's view of the other agent's first order view.

Two cognitive states are said to be in *dynamical coherence* if and only if the two states co-evolve according to a fixed dynamical law.

8.3.4   The Proposed Model of Intention

To solve the problems with the traditional view, we propose a new vision of intention that tries to meet the two challenges presented in the last section based on two main ideas:

1. Dynamic Intention, which tries to meet the representational issue (Challenge 1).
2. Intention through Interaction, which meets the communication modeling issue (Challenge 2).

8.3.4.1   Dynamic Intention

This hypothesis can be stated as: *Every atomic intention is best modeled as a Dynamic Plan*.

A *Dynamic Plan*—in this context—is defined as an executing process—with specific set of final states (goals) and/or a specific behavior—that reacts to sensory information directly and generates a continuous stream of actions to attain the desired final states (goals) or realize the required behavior.

This definition of atomic intentions has a representational capability that exceeds the traditional view of the intention point as a goal or a passive plan because it can account for implementation intentions that was proven very effective in affecting the human behavior.

This definition of atomic intentions is in accordance with the aforementioned neuroscience results, as the conscious recognition of the intention-in-action can happen any time during the reactive plan execution (it may even be a specific step in this plan) which explains the finding that preparation for the action (the very early steps of the reactive plan) can proceed the conscious intention.

A single intention cannot then be modeled by a single value at any time simply because the plan is still in execution and its future generated actions will depend on the inputs from the environment (and other agents). This leads to the next part of the new view of intention as explained in the following section.

8.3.4.2   Intention Through Interaction

This hypothesis can be stated as:

1. Intention can be best modeled not as a fixed unknown value, but as a dynamically evolving function over all possible outcomes.
2. Interaction between two agents couples their intention functions creating a single system that co-evolves as the interaction goes. This co-evolution can converge to a mutual intention state.

The intention of the agent at any point of time—in this view—is not a single point in the intention domain but is a function over all possible outcomes of currently active plans. From this point of view every possible behavior (modeled by a dynamic plan) has a specific probability to be achieved through the execution of the reactive plan (intention) which is called its *intentionality*.

The view presented in this hypothesis is best described by a simple example. consider the evolution over time while a human is drawing a "S" character. In the beginning when the reactive drawing plan is initiated, its internal processing algorithm limits the possible output behaviors to a specific set of possible drawings (that can be called the intention at this point). When the human starts to draw and with each stroke, the possible final characters are reduced and the probability of each drawing change. If the reactive plan is executing correctly, the intention function will tend to sharpen out with time. By the time of the disengagement of the plan, the human will have a clear view of his/her own intention which corresponds to a very sharp intention function. It should be clear that the mode of the final form of the intention function need not correspond to the final drawing except if the coupling law was carefully designed (Mohammad and Nishida 2007b).

8.4   Guiding Principles

Our discussions of autonomy and historical social embodiment (Sect. 8.1) suggest that the robot is better equipped with appropriate learning mechanisms and left to discover for itself how to achieve socially acceptable interaction protocols. On the other hand, earlier discussions of imitation in Chap. 7 and its role in animal and human cognition suggest that it could provide a valuable technique for learning these protocols. This is the approach that will be taken in the following chapters to achieve autonomous sociality.

Two principles guided our design of the embodied interactive robot architecture (EICA) described in Chaps. 9–11. The first principle is what we call the *historical social embodiment principle* which stresses the need for life-long interaction with the social environment.

The second principle is the *intention through interaction principle* which states that intention should be modeled as a dynamically evolving function that changes through interaction rather than a fixed hidden variable of unknown value. Interaction between two agents couples their intention functions creating a single system that co-evolves as the interaction goes. This co-evolution can converge to a mutual intention state of either cooperative or conflicting nature if the dynamical coupling law was well designed (Mohammad and Nishida 2007a, 2009, 2010).

A common concept in both principles is that of interaction as a kind of *co-evolution* between the agent and its social environment. This co-evolution can be modeled by a set of *interaction protocols* and these protocols are what the social robot should be designed to learn.

8.5 Summary

This chapter provided the theoretical underpinnings of the embodied interactive control architecture (EICA) that will be proposed in Chaps. 9–11 to tackle the autonomous sociality challenge. We started by defining more carefully what is meant by autonomy and sociality in this book and related them to the ideas of embodiment and agency. This led us to the importance of endowing social robots with a theory of mind that can help them understand human's internal mental life and its relation to behavior. Discussion of different theories for the structure and development of this theory of mind in humans led us to explore the theory theory, simulation theory, modular theory, and child scientist theory among others and show how many of them are already being utilized for social robotics research. Having a theory of mind implies having the ability to represent intentions of humans. Analysis of intention modeling in AI and HRI as well as research about intention in psychology, philosophy and neuroscience led to the concept of dynamic intention as a function that evolves with interaction with the environment and get coupled to the intention functions of other agents to achieve mutual intention in natural interactions. Finally, we summarized the discussions in this and the previous chapter in two principles that guided our design of the proposed architecture: The *historical social embodiment principle* which stresses the need for life-long interaction with the social environment and the *intention through interaction principle* which stresses the dynamic evolution and coupling of intention functions.

References

Alami R, Clodic A, Montreuil V, Sisbot EA, Chatila R (2005) Task planning for human-robot interaction. SOC-EUSAI'05: Joint conference on smart objects and ambient intelligence: innovative context-aware services: usages and technologies. NY, USA, New York, pp 81–85

Baker CL, Saxe RR, Tenenbaum JB (2011) Bayesian theory of mind: Modeling joint belief-desire attribution. In: CogSci'11: the 32nd annual conference of the cognitive science society, pp 2469–2474

Baron-Cohen S (1997) Mindblindness: an essay on autism and theory of mind. MIT Press

Baron-Cohen S, Leslie AM, Frith U (1985) Does the autistic child have a 'theory of mind'? Cognition 21(1):37–46

Barsalou LW, Niedenthal PM, Barbey AK, Ruppert JA (2003) Social embodiment. Psychol Learn Motiv 43:43–92

Biscaldi M, Rauh R, Irion L, Jung NH, Mall V, Fleischhaker C, Klein C (2014) Deficits in motor abilities and developmental fractionation of imitation performance in high-functioning autism spectrum disorders. Eur Child Adolesc Psychiatry 23(7):599–610

Bishop MA, Downes SM (2002) The theory theory thrice over: the child as scientist, superscientist or social institution? Stud Hist Philos Sci Part A 33(1):117–132

Bratman ME (1987) Intentions, plans, and practical reason. Harvard University Press, Cambridge

Breazeal CL (2004) Designing sociable robots. MIT Press

Brooks RA (1991) Challenges for complete creature architectures. pp 434–443

Brooks RA, Breazeal C, Irie R, Kemp CC, Marjanovic M, Scassellati B, Williamson MM (1998) Alternative essences of intelligence. In: AAAI'98: the national conference on artificial intelligence, Wiley, pp 961–968

Burke J, Murphy B, Rogers E, Lumelsky V, Scholtz J (2004) Final report for the DARPA? NSF interdisciplinary study on human-robot interaction. IEEE Trans Syst Man Cybern Part C 34(2):103–112

Carlson SM, Moses LJ (2001) Individual differences in inhibitory control and children's theory of mind. Child Dev 72:1032–1053

Castelfranchi C, Falcone R (2004) Founding autonomy: the dialectics between (social) environment and agent's architecture and powers. Lect Notes Artif Intell 2969:40–54

Chirkov V, Ryan RM, Kim Y, Kaplan U (2003) Differentiating autonomy from individualism and independence: a self-determination theory perspective on internalization of cultural orientations and well-being. J Pers Soc Psychol 84(1):97–109

Cohen LB, Amsel G (1998) Precursors to infants' perception of the causality of a simple event. Infant Behav Dev 21(4):713–731

Cohen PR, Levesque HJ (1990) Intention is choice with commitment. Artif Intell 42(2–3):213–261. doi:10.1016/0004-3702(90)90055-5

Dautenhahn K (1998) The art of designing socially intelligent agents: science, fiction, and the human in the loop. Appl Artif Intell 12(7–8):573–617

De Charms R (2013) Personal causation: the internal affective determinants of behavior. Routledge

Demiris Y, Dearden A (2005) From motor babbling to hierarchical learning by imitation: a robot developmental pathway. In: The 5th international workshop on epigenetic robotics: modeling cognitive development in robotic systems, Lund University Cognitive Studies, pp 31–37

Dennett DC (1989) The intentional stance. MIT Press

Duffy B (2004) Robots social embodiment in autonomous mobile robotics. Int J Adv Rob Syst 1(3):155–170

Gergely G, Nádasdy Z, Csibra G, Biro S (1995) Taking the intentional stance at 12 months of age. Cognition 56(2):165–193

Gerrans P (2002) The theory of mind module in evolutionary psychology. Biol Philos 17(3):305–321

Goldman AI (2006) Simulating minds: the philosophy, psychology, and neuroscience of mindreading. Oxford University Press

Gollwitzer PM (1999) Implementation intentions: Strong effects of simple plans. Am Psychol 54:493–503

Gollwitzer PM, Sheeran P (2006) Implementation intentions and goal achievement: a meta-analysis of effects and processes. Adv Exp Soc Psychol 38:249–268

Gopnik A, Meltzoff AN, Bryant P (1997) Words, thoughts, and theories, vol 1. MIT Press, Cambridge

Gopnik A, Schulz L (2004) Mechanisms of theory formation in young children. Trends Cogn Sci 8(8):371–377

Gopnik A, Glymour C, Sobel DM, Schulz LE, Kushnir T, Danks D (2004) A theory of causal learning in children: causal maps and Bayes nets. Psychol Rev 111(1):3–32

Gopnik A, Wellman HM (1992) Why the child's theory of mind really is a theory. Mind Lang 7(12):145–171

Gordon RM (1986) Folk psychology as simulation. Mind Lang 1(2):158–171

Haggard P (2005) Conscious intention and motor cognition. Trends Cogn Sci 9(6):290–295

Johnson S (2005) Reasoning about intentionality in preverbal infants, vol 1. Oxford University Press

Kanakogi Y, Itakura S (2011) Developmental correspondence between action prediction and motor ability in early infancy. Nat Commun 2:341–350

Kennedy WG, Bugajska MD, Harrison AM, Trafton JG (2009) "like-me" simulation as an effective and cognitively plausible basis for social robotics. Int J Soc Robot 1(2):181–194

L Braubach DM A Pokahr, Lamersdorf W (2004) Goal representation for bdi agent systems. In: PROMAS'04: the AAMAS workshop on programming in multi-agent systems, pp 44–65

Leslie AM (1984) Spatiotemporal continuity and the perception of causality in infants. Perception 13(3):287–305

Meier BP, Schnall S, Schwarz N, Bargh JA (2012) Embodiment in social psychology. Top Cogn Sci 4(4):705–716

Mohammad Y, Nishida T (2007a) Intention thorugh interaction: Towards mutual intention in human–robot interactions. In: IEA/AIE'07: the international conference on industrial, engineering, and other applications of applied intelligence, pp 115–125

Mohammad Y, Nishida T (2007b) NaturalDraw: interactive perception based drawing for everyone. In: IUI'07: the 12th international conference on intelligent user interfaces, ACM Press, New York, NY, USA, pp 251–260. http://doi.acm.org/10.1145/1216295.1216340

Mohammad Y, Nishida T (2009) Toward combining autonomy and interactivity for social robots. AI Soc 24:35–49. doi:10.1007/s00146-009-0196-3

Mohammad Y, Nishida T (2010) Controlling gaze with an embodied interactive control architecture. Appl Intell 32:148–163

Morreale V, Bonura S, Francaviglia G, Centineo F, Cossentino M, Gaglio S (2006) Reasoning about goals in bdi agents: the practionist framework. In: Joint workshop from objects to agents. http://citeseerx.ist.psu.edu/viewdoc/download?doi=10.1.1.62.203&rep=rep1&type=pdf

Morton A (1980) Frames of mind: constraints on the common-sense conception of the mental. Clarendon Press, Oxford

Nishida T, Nakazawa A, Ohmoto Y, Mohammad Y (2014) Conversational informatics: a data intensive approach with emphasis on nonverbal communication. Springer

Ono T, Imai M, Nakatsu R (2000) Reading a robot's mind: a model of utterance understanding based on the theory of mind mechanism. Adv Robot 14(4):311–326

Parsons S, Pettersson O, Saffiotti A, Wooldridge M (2000) Intention reconsideration in theory and practice. In: Horn W (ed) ECAI'00: the 14th european conference on artificial intelligence, Wiley, pp 378–382

Paschal Sheeran TLW, Gollwitzer PM (2005) The interplay between goal intentions and implementation intentions. Pers Soc Psychol Bull 31:87–98

Pfeifer R, Scheier C, Illustrator-Follath I (2001) Understanding intelligence. MIT Press

Reeves B, Nass C (1996) How people treat computers, television, and new media like real people and places. CSLI Publications and Cambridge University Press

Richardson H, Bake C, Tenenbaum J, Saxe R (2012) The development of joint belief-desire inferences. In: CogSci'12: the 34th annual meeting of the cognitive science society, vol 1, pp 923–928

Riegler A (2002) When is a cognitive system embodied? Cogn Syst Res 3(3):339–348

Ryan RM (1991) A motivational approach to self. Perspect Motiv 237:38–46

Ryan RM, Deci EL (2000) Self-determination theory and the facilitation of intrinsic motivation, social development, and well-being. Am Psychol 55(1):68–76

Scassellati B (2002) Theory of mind for a humanoid robot. Auton Robots 12(1):13–24

Scassellati B (2003) Investigating models of social development using a humanoid robot. Int Joint Conf. Neural Netw IEEE 4:2704–2709

Searle JR (1983) Intentionality: an essay in the philosophy of mind. Cambridge University Press

Sheeran P (2002) Intention-behavior relations: a conceptual and empirical review. In: Hewstone M, Stroebe W (eds) European review of social psychology, pp 1–36. Wiley

Stoytchev A, Arkin RC (2004) Incorporating motivation in a hybrid robot architecture. J Adv Comput Intell Intell Inform 8(3):269–274

Trafton G, Hiatt L, Harrison A, Tamborello F, Khemlani S, Schultz A (2013) Act-R/E: an embodied cognitive architecture for human-robot interaction. J Hum-Robot Interact 2(1):30–55

Vogt P (2002) The physical symbol grounding problem. Cogn Syst Res 3:429–457

Wang Z, Mülling K, Deisenroth MP, Amor HB, Vogt D, Schölkopf B, Peters J (2013) Probabilistic movement modeling for intention inference in human-robot interaction. Int J Robot Res 32(7):841–858

Wimmer H, Perner J (1983) Beliefs about beliefs: representation and constraining function of wrong beliefs in young children's understanding of deception. Cognition 13(1):103–128

Ziemke T (2003) What's that thing called embodiment. In: CogSci'03: the 25th annual meeting of the cognitive science society, Mahwah. Lawrence Erlbaum, NJ, pp 1305–1310

Chapter 9
The Embodied Interactive Control Architecture

In the previous chapter, we pointed out the main theoretical foundations of EICA and its guiding principles: intention through interaction and historical social embodiment. EICA has two components: a general behavioral robotic architecture with a flexible action integration mechanism upon which interaction protocol learning to support autonomous sociality is implemented. In this chapter, we discuss the design of the behavioral platform of EICA focusing on its action integration mechanism and provide example applications of social robots that can be implemented directly using the constructs of this platform without the need for interaction protocol learning. Chapter 10 will provide the details of our autonomous sociality supporting architecture that is built on top of the platform discussed here.

9.1 Motivation

The main construct for achieving autonomous sociality are interaction protocols as discussed in Sect. 8.4. These protocols can be either implicit or explicit protocols but they correspond to ways of synchronizing and organizing the behaviors of different partners in a social interaction according to their roles in it. Our goal is to develop an architecture that can support learning of these protocols autonomously. Because these protocols represent behaviors executed at different time-scales, they will require a variety of implementation techniques. This chapter introduces a generic robotic architecture that was designed to handle this situation by allowing different processes implementing these protocols to co-exist and propose actions for the robot while providing a general action integration mechanism that can generate consistent behavior by combining these actions.

Robotic architectures have a long history in autonomous robotics literature. The earliest of these architectures were deliberative in nature within the sense–think–act paradigm where logical manipulations of a world representation are used extensively. One problem common to most of these architectures was poor real-

© Springer International Publishing Switzerland 2015
Y. Mohammad and T. Nishida, *Data Mining for Social Robotics*,
Advanced Information and Knowledge Processing,
DOI 10.1007/978-3-319-25232-2_9

time performance and inability to adapt to changes in the environment. The second wave of architectures focused on solving these problems by being reactive and encapsulating the robot's computational functions into a set of *behaviors* designed to reduce the latency between perception and action. These architectures, nevertheless, lacked the internal representations needed for scalability to complex situations including social contexts. The third type of architectures were hybrid architectures that tried to marry the advantages of these two diametrically different approaches.

Hybrid architectures employed in most cases fixed pre-determined relations between deliberation and reaction (see for example, Arkin et al. 2003, Karim et al. 2006). The platform discussed here is a hybrid architecture that does not limit the relation between deliberation and reaction. Deliberative and reactive processes occupy the same space and interact with each other using the same communication mechanisms. Only reactive processes are allowed to directly access the action integration mechanism under the assumption that they represent dynamic intentions as described in Sect. 8.3.

9.2 The Platform

EICA consists of a set of active components connected by communication channels and their actions are integrated through an action integration mechanism.

Figure 9.1 shows the building blocks of this architecture. Two types of components exist in the architecture: active components and passive channels. Active components are the executing components of the architecture and they encapsulate the processing building blocks of the robot. Channels are used to transfer data and control signals between active components. Active components in EICA correspond to behaviors in behavioral systems or agents in multi-agent systems.

There are three types of active components in EICA. The simplest kind are *reflexes* that encapsulate fast processing direct couplings between perception and

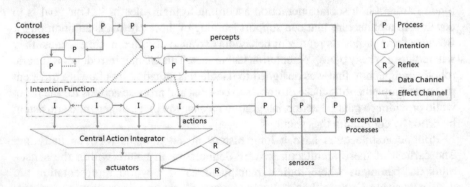

Fig. 9.1 Embodied interactive control architecture's (EICA's) underlying hybrid robotic architecture

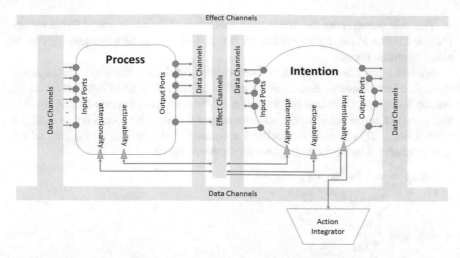

Fig. 9.2 Schematics of EICA components

action. These components are connected directly to the sensors and actuators of the robot and their actions override actions from higher components. This is in direct contrast with subsumed lower processes in the subsumption architecture (Brooks et al. 1998). These reflexes should be used *only* to implement reactive responses to dangerous situations in the environment that require immediate reactions like collision avoidance.

Other than reflexes, there are two active components in EICA: processes and intentions. Figure 9.2 shows a schematic of a process and an intention. The figure shows that they share most of their internal structure. Both have data ports for interchanging data through data channels (to be explained later). Three special ports exist for intentions and only two of them are available for processes. The special ports available for both intentions and processes are activation-level and attentionality.

Activation-level is a value from zero to one that determines the level of importance of an active entity when exchanging activation and deactivation signals with other active entities. If activation-level is zero, the entity does not run. It is simply kept in secondary storage. If activation-level is higher than zero, its value is used to calculate the effect of its component (either a process or intention) on other components as will be described when discussing effect channels.

Attentionality is used to implement a form of attention focusing. Active entities are allowed to use processing power relative to their attentionality value. This means that a process with low attentionality will get less processor cycles than another one with higher attentionality.

Intentions have a third special port called intentionality. This value is used to adjust the effect of intentions on the action integration mechanism as will be explained in Sect. 9.4.

This specification for the *Active* type was inspired by the Abstract Behaviors of (Nicolescu and Matarić 2002) although it was designed to be more general and to allow attention focusing.

Special ports (attentionality, activation-level, and intentionality) are adjusted through effect channels. Each special port is connected to an effect channel as its output. Every *effect channel* has a set of n inputs connected to ports (including special ports) of active components and one output that may be connected to multiple special ports. This output is calculated according to the *operation* attribute of the *effect channel*. The currently implemented operations are:

- Max:$y = \max_{i=1:n} (x_i \,|a_i > \varepsilon)$
- Min:$y = \min_{i=1:n} (x_i \,|a_i > \varepsilon)$

- Avg: $y = \dfrac{\sum\limits_{i=1}^{n} (a_i x_i)}{\sum\limits_{i=1}^{n} (a_i)}$

  where x_i is an input port, a_i is the activation-level of the object connected to port i and y is the output of the effect channel.
- DLT: y is calculated from a lookup table after discretizing the inputs. This effect channel is useful for implementing discrete joint probability distributions and makes it possible to translate any Bayesian Network directly into an EICA implementation.

By convention, processes connected directly to sensors are called *perceptual processes* and their outputs are called *percepts*. This terminology is borrowed from the C4 architecture (Sect. 6.3.1).

Active components exchange data through *data channels*. These channels implement the publisher-subscriber communication pattern. They have multiple modes of operation the details of which are not necessary for understanding the rest of this chapter (for details, see Mohammad and Nishida 2009).

Other than the aforementioned types of active entities, EICA employs a central *Action Integrator* that receives actions from intentions and uses the source's intentionality level as well as an assigned priority and mutuality for every DoF of the robot in the action to decide how to integrate it with actions from other intentions using simple weighted averaging subject to mutuality constraints. The details of action integration in EICA are discussed in Sect. 9.4.

The current implementation of EICA is done using standard C++, and is platform independent. The system is also suitable for implementation onboard and in a host server. It can be easily extended to support distributed implementation on multiple computers connected via a network. The EICA implementation is based on Object-Oriented design principles so every component in the system is implemented in a class. Implementing software EICA Applications is very simple: Inherit the appropriate classes from the EICA core system and override the abstract functions.

9.3 Key Features of EICA

It should be noted here that EICA was designed with natural interaction in mind and so a lot of infrastructure was implemented to facilitate intention communication, but—nevertheless—EICA has enough expressive power to implement many other robotic application like autonomous navigation and accurate manipulator controllers. Some of the key features that separate EICA from available robotic architectures will be summarized in the following points.

1. Intention Modeling: EICA is designed to implement the dynamic view of intention advocated in Sect. 8.3. *Intentions* provide the basic building blocks of computation and they are the only active processes allowed to access the action integrator. Internal interaction between control processes determines the *intentionality* attributed to each of these intentions. Taken together, active intentions provide the intention function described in Sect. 8.3. Interactions between different robots implemented using EICA couples these intention functions through the higher control processes realizing the *intention through interaction* principle described in Sect. 8.4.

2. Action Selection/Integration: Action integration in EICA is done in two stages: a distributed stage implemented by adjusting the activation-level attributes of control processes and intentions and a central stage implemented through the action integrator process. Through this two-stages process, EICA is able to achieve a continuous range of integration policies ranging from pure action selection to action averaging common to behavioral systems. The details of this mechanism are given in Sect. 9.4.

3. Low level continuous attention focusing. By separating the activation-level from the attentionality and allowing activation-level to have a continuous range, EICA enables a distributed form of attention focusing. This separation allows the robot to select the active processes depending on the general context (by setting the activation-level value) while still being able to assign the computation power according to the exact environmental and internal condition (by setting the attentionality value). The fact that the activation-level is variable allows the system to use it to change the possible influence of various processes (through the operators of the effect channels) based on the current situation.

4. Relation between Deliberation and Reaction: In most hybrid architectures available the relation between deliberation and reaction is fixed by design. An example of that is the deliberation as configuration paradigm. Some architecture designers supported the scheduling of deliberation process along with reactive processes and forced both to compete for processing time (Arkin and Balch 1997), but EICA took this trend further by imposing no difference on the architectural level between deliberative and reactive processes. For example, deliberative processes can take control and directly activate or deactivate intentions. This is useful for implementing some interaction related social behaviors like turn taking that is usually implemented as a deliberative process.

9.4   Action Integration

Action integration is a major concern for any distributed architecture like EICA in which multiple computational components (intentions in this case) can issue actuation commands. This is of special importance for behavioral architectures because computation in these architectures are decomposed into vertical stacks that are run concurrently instead of the horizontal decomposition common to earlier deliberative architectures (Mohammad and Nishida 2008). There are two general approaches to action integration: selective and combinative approaches. Selective approaches select one action from the set of proposed actions at any point of time and executes it. Combinative approaches generate a final command by combining proposed actions using a simple mathematical averaging operation in most cases. The action integration mechanism proposed in this chapter can be classified as a hybrid two layered architecture with distributive attention focusing behavior level selection followed by a fast central combinative integrator and was first proposed by Mohammad and Nishida (2008).

Mohammad and Nishida (2008) proposed six requirements for an HRI friendly action integration mechanism based on analysis of selective, combinative and hybrid action integration mechanisms usually used in robotics. Only, the most important of them will be discussed here.

The first requirement is allowing a continuous range from purely selective to purely combinative action integration strategies. This is essential for HRI because of the intuitive anthropomorphism that characterizes human's understanding to robot's behavior (Breazeal et al. 2005). This means that the minute details of the robot's external behavior may be taken as intentional signals. To make sure that these details are not affected by the existence of multiple active behaviors in the robot, most HRI architectures utilize some form of selective action integration. For example, only a single action from each action group is allowed to execute in C4 (Sect. 6.3.1) and a single situated module is active at any point in Situated Modules architecture (Sect. 6.3.2.). Nevertheless, this may lead to jerky behavior as the robot jumps between behaviors (the usual *robotic* motion). The ability to handle combinative strategies can be of value for reducing such jerkiness and achieving human-like smooth motions that can easily be interpreted as social signals by the robot's interlocutors. Moreover, research in nonverbal communication in humans reveals a different picture in which multiple different processes do collaborate to realize the natural action. For example, human spatial behavior in close encounters can be modeled with two interacting processes (Argyle 2001). It is possible in the selective framework to implement these two processes as a single behavior, but this goes against the spirit of behavioral architectures that emphasizes modularity of behavior (Perez 2003).

Given the controllability of selectivity in action integration, a mechanism is needed to adjust it according to timely perceptual information (both from external sensors and internal state). This leads to the second requirement of enabling control of the selectivity level in the action integration mechanism based on perceptual and internal sensory information.

EICA is a parallel architecture with tens and may be hundreds of running active components. It is extremely difficult to integrate the actions suggested by all of these active components by hand. A first simplification of this problem was the design decision of allowing *only* intentions to issue action commands to the central action integrator. Still, if action integration strategy depended on global factors like the specifics of the effect channel connections between running processes and intentions, adjustment of the strategy will be extremely challenging. The third requirement is then to make the action integration mechanism local in the sense that it does not depend on the global relationships between behaviors. For a negative example consider the Hybrid Coordination approach presented in Perez (2003). In this system every two behaviors are combined using a *Hierarchical Hybrid Coordination Node* that has two inputs. The output of the HHCN is calculated as a nonlinear combination of its two inputs controlled by the activation levels of the source behaviors and an integer parameter k that determines how combinative the HHCN is, where larger values of k makes the node more selective. The HHCNs are then arranged in a hierarchical structure to generate the final command for every DoF of the robot (Perez 2003). The structure of this hierarchy is fixed by the designer. That means that when adding a new behavior, it is not only to decide how to implement it but where to fit it in this hierarchy. That is why we need to avoid global factors affecting action integration.

The number of behaviors needed in interactive robots usually is very high compared with autonomously navigating robots if the complexity of each behavior is kept acceptably low, but most of these behaviors are usually passive in any specific moment based on the interaction situation. This property leads to the requirement that the system should have a built-in attention focusing mechanism.

Fig. 9.3 The proposed action integration mechanism

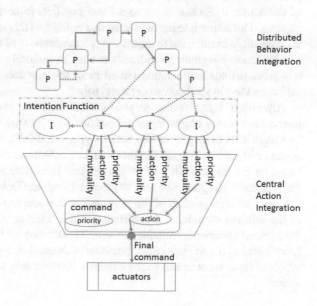

Figure 9.3 shows the block diagram of this hybrid action integration system used in EICA.

9.4.1   Behavior Level Integration

Control Processes implement the behavior-level distributed integration mechanism in EICA. This is achieved by adjusting the three special ports of intentions (activation-level, attentionality and intentionality). Activation-level and attentionality will determine the computational resources allocated to each intention as discussed earlier in this chapter while intentionality will determine how much each intention affects the central action integration mechanism to be discussed later.

Control processes are connected to each other and to intentions through *effect channels*. Effect channels can be arranged into hierarchical structures like the HHCNs in Perez (2003), but rather than carrying action and influence information; the effect channels carry only the influence information between different processes of the system.

9.4.2   Action Level Integration

This layer consists of a central fast action integrator that fuses the actions generated from various intentions based on the intentionality of their sources, the mutuality and priority assigned to them, and general parameters of the robot. The *activation-level* of the action integrator is set to a fixed positive value to ensure that it is always running. The action integrator is an *active* entity and this means that its *attentionality* is changeable at run time to adjust the responsiveness of the robot.

The action integrator periodically checks its internal master *command* object and whenever it finds some action stored in it, the *executer* is invoked to execute this action on the physical actuators of the robot.

Algorithm 1 shows the *register-action* function responsible of updating the internal *command* object based on actions sent by various intentions of the robot. The algorithm first ignores any actions generated from intentions below a specific system wide threshold τ_{act}. The function then calculates the *total priority* of the action based on the intentionality of its source, and its source assigned priority. Based on the mutuality assigned to every degree of freedom (DoF) of the action, difference between the total priority of the proposed action and the currently assigned priority of the internal command, the system decides whether to combine the action with the internal command, override the stored command, or ignore the proposed action. Intentions can use the *immediate* attribute of the *action* object to force the action integrator to issue a command to the executer immediately after combining the current action.

Algorithm 1 Register Action Algorithm (Mohammad and Nishida 2008)

---

function REGISTER- ACTION(Action a, Intention s)
  if $s.activation - level < \tau_{act} \lor s.intentionality < \tau_{int}$ **then**
    exit
  end if
  $c \leftarrow$ current combined command
  $p \leftarrow a.priority + max\_priority \times s.intentionality$
  for every DoF i in the a **do**
    $combined \leftarrow true$
    if $p < c.priority \land s \neq c.source$ **then**
      $c.source \leftarrow s$
      $c.dof(i) \leftarrow a.dof(i)$
      $c.priority(i) \leftarrow p$
      $c.hasAction(i) \leftarrow true$
    end if
    if $c.source \neq null$ **then**
      $c.source \leftarrow null$
      $c.priority(i) \leftarrow \max(p, c.priority(i))$
    else
      $c.source \leftarrow s$
      $c.priority(i) \leftarrow p$
    end if
    if $a.mutual = true \lor c.source = s$ **then**
      $c.source \leftarrow s$
      $c.dof(i) \leftarrow a.dof(i)$
      $c.priority(i) \leftarrow p$
    else
      $c.dof(i) \leftarrow \frac{p \times a.dof(i) + c.priority \times c.dof(i)}{p + c.priority}$
    end if
    if $combined = true$ **then**
      return false
    end if
    $c.actionable \leftarrow true$
    if $a.notCombinableWithLower$ **then**
      $c.stopCombiningLower \leftarrow true$
    end if
    if $a.immediate$ **then**
      execute c
    end if
  end for
end function

9.5 Designing for EICA

One of the main purposes of having robotic architectures is to make it easier for the programmer to divide the required task into smaller computational components that can be implemented directly. EICA is designed to help achieving a *natural* division of the problem by the following simple procedure. First the task is analyzed to find the basic competencies that the robot must possess in order to achieve this task.

Those competencies are not complex behaviors like *attend-to-human* but finer behaviors like *look-right*, *follow-face*, etc. Those competencies are then mapped to the *intentions* of the system. Each one of these intentions should be carefully engineered and tested before adding any more components to the system.

The next step in the design process is to design the control processes that control action integration of these intentions. To do that, the task is analyzed to find the underlying processes that control the required behavior. Those processes are then implemented. The most difficult part of the whole process is to find the correct parameters of those processes to achieve the required external behavior. Once all the required behaviors are implemented, any adequate learning algorithm can be used to learn the optimal values of their parameters. In this chapter we describe e a Floating Point Genetic Algorithm (FPGA) to achieve this goal. The details of this algorithm are given in Sect. 9.6 (Mohammad and Nishida 2010). This simple design procedure is made possible because of the separation between the basic behavioral components (intentions) and the behavior level integration layer (processes).

The following two chapters will report another design procedure that replaces careful design for the distributed action integration layer (control processes) with a developmental learning system based on data mining algorithms. Even the low level intentions can be learned using this approach leaving only the choice of perceptual processes and available actuation commands to be the only responsibility of robot designers.

9.6   Learning Using FPGA

The final behavior of the robot depends crucially on the parameters of each action component as well as those controlling the timing of activation and deactivation of these components. Manual adjustment can be used to set these parameters but at great cost (Mohammad and Nishida 2007). In this section we introduce a floating point genetic algorithm (FPGA) that was first proposed by Mohammad and Nishida (2010) for learning any set of parameters in the EICA architecture.

The algorithm is first initialized with n individuals. Each individual consists of m floating point numbers representing the parameters to be selected ($P_1^i : P_m^i$ for individual i). The algorithm iterates for G_m generations doing the following:

1. Calculate the fitness of all the individuals f_i and if the highest fitness ($max\ (f_i)$) is higher than the current elite individual fitness (f_e) then make the associated individual the current elite individual and remember its fitness value ($f_e = max\ (f_e, \{f_i\})$).
2. Select top N individuals by calculating the fitness function of every individual in the pool. Call this set the regenerating pool. Every individual r_i in the regenerating pool is associated with its fitness f_i.

3. Randomly select $n/2$ pairs from the regenerating pool using normalized f_i as the probability distribution for selection. This will generate the mating set of the form $\{m_i\}$ where m_i is a set of two individuals $\langle r_k, r_j \rangle$.
4. Apply the fitness guided cross-over operator explained later between the two members of every mating pair generating two new individuals. Call the resulting set the un-mutated offspring set. This set will contain exactly n individuals.
5. Apply the generation guided mutation operator explained later to every one of the un-mutated offspring with probability p_{mut} to generate the new generation.

The fitness guided cross-over operator is defined as a mapping between two input individuals (I_1^i and I_2^i) and two output individuals (I_1^o and I_2^o) as follows, assuming that $f_1 > f_2$:

1. Find four candidate individuals $I_{1:4}{}^c$ are defined using:

$$I_1^c = I_1^i, I_2^c = I_2^i,$$
$$I_3^c = (1 + \alpha_1) I_1^i - \alpha_1 \times \left(I_2^i\right),$$
$$I_4^c = (1 + \alpha_2) I_1^i - \alpha_2 \times \left(I_2^i\right),$$
(9.1)

where α_1 and α_2 are selected randomly satisfying the constraints $\alpha_{min} < \alpha_1 < 0.5$ and $0.5 < \alpha_2 < \alpha_{max}$. If I_3^c or I_4^c is not feasible (one or more of the parameters are outside the feasible range) then randomly select another value for the corresponding α_i mixing factor and repeat this process until all of the four individuals are feasible or a predetermined number of iterations (c_m) have passed.
2. Calculate the fitness of these candidate individuals (f_i for $i = 1 : 4$) and normalize it using:

$$p_i = \frac{\left(f_i^{\frac{G_m-g}{G_m-1}}\right)^{r_c}}{\sum \left(f_i^{\frac{G_m-g}{G_m-1}}\right)^{r_c}},$$
(9.2)

where g is the generation number, G_m is the number of generations, and r_c determines how much the generation number affects the algorithm.
3. Using p_i as a probability distribution select two output individuals.

The operator first generates four hypotheses including different mixing ratios of its two inputs and then uses the fitness of these four hypotheses to select the final two offspring. Since the final selection is done probabilistically, lower fitness individual can make it to the offspring to provide better diversity in the next generation. This dependence on the fitness is low in the first generations and becomes higher with advanced generation to explore more space in the beginning while focusing the search at the end. This operator is similar to other operator suggested for floating point genetic algorithms in general (Mahanti et al. 2005; Devaraj and Yegnanarayana 2005).

The generation guided mutation operator is defined as a mapping between one input individual (I^i) and one output individual (I^o) as follows:

1. Calculate two probability distributions over the integers from 1 to m (the number of parameters to be optimized) as follows:

$$P^+(k) = \frac{\sigma_k}{\sum\limits_{i=1}^{m} \sigma_i}, \quad P^-(k) = 1 - P^+(k), \tag{9.3}$$

where $\sigma_i = \frac{1}{n-1} \sum\limits_{k=1}^{n} (I_i\langle k \rangle - \mu \langle k \rangle)^2$.

2. Calculate the probability distribution of mutation over the parameters (P^μ) using:

$$\widehat{P}^\mu(k) = \frac{g}{G_m} P^-(k) + \left(1 - \frac{g}{G_m}\right) P^+(k),$$
$$P^\mu(k) = \frac{\widehat{P}^\mu(k)}{\sum\limits_{i=1}^{m} \widehat{P}^\mu(i)}. \tag{9.4}$$

3. Select an integer (β) from 1 to m using the distribution P^μ.
4. Mutate parameter β of the input individual using:

$$I^o(\beta) = \begin{cases} I^i(\beta) + \eta \left(\frac{G_m-g}{G_m}\right)^{r_m} \left(I_{\max}(\beta) - I^i(\beta)\right) & \text{if } \gamma > 0.5, \\ I^i(\beta) - \eta \left(\frac{G_m-g}{G_m}\right)^{r_m} \left(I^i(\beta) - I_{\min}(\beta)\right) & \text{if } \gamma \leq 0.5, \end{cases} \tag{9.5}$$

where η and γ are random numbers between 0 and 1. η is used as the mutation strength and γ is used to select either to increase or decrease the value of the parameter. $I(i)$ is the value of parameter number i of individual I, $I_{max}(i)$, $I_{min}(i)$ are the maximum and minimum acceptable values of the parameter i, and r_m determines how much the mutation operator is affected by the generation number.

This mutation operator first selects the mutation site to be (with high probability) near the point of low variance in the beginning of the evolution in order to generate more diversity in the next generation by increasing the variance of this parameter, then the operator gradually shifts to selecting the parameter of maximum variation in advanced generations to explore more areas of the parameters space in by changing these dimensions that are not yet settled while not affecting the settled dimensions.

9.7   Application to Explanation Scenario

Event though this chapter introduced only the low-level platform of EICA without the HRI specific components to be described in Chap. 10, this platform can be used (and was used) to implement social robotics applications directly due to the flexibility

of action integration and attention focusing it provides and the simple design procedure (Sect. 9.5) combined with the learning algorithm outlined in Sect. 9.6. We will
report here two implementations of gaze-control in the explanation scenario outlined
in Sect. 1.4.

9.7.1 Fixed Structure Gaze Controller

The gaze controller implemented in this section uses the design procedure mentioned
in Sect. 9.5 and is inspired by research on human nonverbal behavior in close
encounters.

Four intentions were designed that encapsulate the possible interaction actions that
the robot can generate, namely, looking around, following the human face, following
the salient object in the environment, and looking at the same place the human is
looking at. These intentions where implemented as simple state machines (for details
of the internal structure of these intentions, see Mohammad and Nishida 2010).

A common mechanism for the control of spatial behavior in human–human
face-to-face interactions was the approach-avoidance mechanism (Argyle 2001).
The mechanism consists of two opposing process. The first process (approach) tries
to minimize the *distance* between the person and their interaction partners (subject to
limitations based on power distribution structure). The second process (avoidance)
tries to avoid invading the personal space of the partner. The interaction between
these two processes generates the final spatial distribution during interaction.

Mohammad and Nishida (2010) borrowed this idea to implement the behavioral
integration layer of the gaze controller. These two processes were not enough for
our scenario because of the existence of objects in the explanation scenario and the
need of the robot to attend to these objects. A third process (mutual attention) was
specially designed to achieve this goal.

The behavioral level integration layer of this fixed structure controller thus uses
three processes as follows (Mohammad and Nishida 2010; Mohammad et al. 2010):

1. *Look-At-Instructor*: This process is responsible of generating an attractive virtual
   force that pulls the robot's head direction to the location of the human face.
2. *Be-Polite*: This process works against the *Look-At-Instructor* process to provide
   the aforementioned *Approach-Avoidance* mechanism.
3. *Mutual-Attention*: This process aims to make the robot look at the most salient
   object in the environment when the instructor is looking at it.

Five perception processes were needed to implement the aforementioned
behavioral processes and intentions (Mohammad and Nishida 2010; Mohammad
et al. 2010):

1. *Instructor-Head*: Continuously updates a list containing the position and direction of the human head during the last 30 s sampled 50 times per second.

2. *My-Head*: Continuously updates a list containing the position and direction of the robot head during the last 30 s sampled 50 times per second.
3. *Instructor-Gaze*: Calculates the intersection between the direction of gaze of the instructor and the plan on which the objects used in the explanation are located.
4. *Gaze-Map*: Continuously updates a representation of the distribution of the human gaze both in the spatial and temporal dimensions.
5. *Speaking*: Uses the power of the sound signal to detect the existence of human speech.

Mohammad and Nishida (2010) analyzed the performance of this simple gaze controller and showed that it can achieve human-like gaze behavior in the explanation scenario even without access to the verbal content of instructor's utterances. For example, the robot achieved mutual gaze 26.72 % of the interaction time compared 28.15 % of the time for a human in the same situation. It engaged in mutual attention (looking at the same object as the instructor) 60.62 % of the interaction time compared with 57.23 % for a human subject.

For more detailed evaluation of the performance of the two aforementioned gaze controllers six new sessions in which an instructor is explaining the operation and connections of a medical device (Polymate physiological signal acquisition system from TEAC) to a human listener were conducted. The two gaze controllers described in this section and the previous section were compared with three simpler controllers that either moved the head randomly (Random), looked continuously at the most salient object (Fixed) or at the instructor's face (Stare).

9.8   Application to Collaborative Navigation

The first robot that was controlled using the EICA architecture is a miniature e-puck robot designed to study nonverbal communication between subjects and the robot in a collaborative navigation situation. The goal of this robot was to balance its internal drive to avoid various kinds of obstacles and objects during navigation with its other internal drive to follow the instructions of its operator who cannot see these obstacles and to give understandable feedback to help its operator correct her navigational commands.

The main feature of the control software in this experiment was the use of Mode Mediated Mapping which means that the behavioral subsystem controlling the actuators is only connected to the perceptual subsystem representing sensory information through a set of processes called *modes*. Each mode continuously uses the sensory information to update a real value representing its controlled variable. The behavioral subsystem then uses these modes for decision making rather than the raw sensory information. Mohammad et al. (2008) showed that this simple implementation allows the robot to give nonverbal feedback to the user that is more effective than using verbal feedback.

9.9 Summary

This chapter introduced the behavioral platform upon which the Embodied Interactive Control Architecture (EICA) was built. Computational processes in this platform are either intentions or processes with the intentions representing the intention function discussed in Chap. 8 and the processes providing distributed behavior integration. The second stage of action integration is provided by a central action integrator process. This action integration system was designed to allow the full range of selective and combinative action integration strategies. We discussed the bottom-up approach for designing social robots using just this platform and discussed the role of parameter learning with an example implementation of a floating point genetic algorithm. Experimental evaluations of the architecture in both the explanation and guided navigation scenarios are also reported.

References

Argyle M (2001) Bodily communication. Routledge; New Ed edition

Arkin RC, Balch TR (1997) AuRA: principles and practice in review. J Exp Theor Artif Intell (JETAI) 9(2/3):175–188

Arkin RC, Fujita M, Takagi T, Hasegawa R (2003) An ethological and emotional basis for human–robot interaction, robotics and autonomous systems. Robot Auton Syst 42(3/4):191–201

Breazeal C, Kidd C, Thomaz A, Hoffman G, Berlin M (2005) Effects of nonverbal communication on efficiency and robustness in human–robot teamwork. In: IROS'05: IEEE/RSJ international conference on intelligent robots and systems, IEEE, pp 708–713

Brooks RA, Breazeal C, Irie R, Kemp CC, Marjanovic M, Scassellati B, Williamson MM (1998) Alternative essences of intelligence. In: AAAI'98: the national conference on artificial intelligence, Wiley, pp 961–968

Devaraj D, Yegnanarayana B (2005) Genetic-algorithm-based optimal power flow for security enhancement. IEE Proc: Gener, Transm Distrib 152(6):899–905. doi:10.1049/ip-gtd:20045234

Karim S, Sonenberg L, Tan AH (2006) A hybrid architecture combining reactive plan execution and reactive learning. In: PRICAI'06: 9th biennial Pacific Rim international conference on artificial intelligence, pp 200–211

Mahanti G, Chakraborty A, Das S (2005) Floating-point genetic algorithm for design of a reconfigurable antenna arrays by phase-only control. APMC'05: Asia-Pacific microwave conference, vol 5, pp 3–11. doi:10.1109/APMC.2005.1606987

Mohammad Y, Nishida T (2007) Intention thorough interaction: towards mutual intention in human–robot interactions. In: IEA/AIE'07: the international conference on industrial, engineering, and other applications of applied intelligence, pp 115–125

Mohammad Y, Nishida T (2008) Two layers action integration for HRI—action integration with attention focusing for interactive robots. In: ICINCO'08: the 5th international conference on informatics in control, automation and robotics, pp 41–48. doi:10.5220/0001482600410048

Mohammad Y, Nishida T (2009) Toward combining autonomy and interactivity for social robots. AI Soc 24:35–49. doi:10.1007/s00146-009-0196-3

Mohammad Y, Nishida T (2010) Controlling gaze with an embodied interactive control architecture. Appl Intell 32:148–163

Mohammad Y, Nishida T, Okada S (2010) Autonomous development of gaze control for natural human–robot interaction. In: The IUI 2010 workshop on eye gaze in intelligent human machine interaction, IUI'10, pp 63–70

Mohammad Y, Xu Y, Matsumura K, Nishida T (2008) The $H^3 R$ explanation corpus: human-human and base human–robot interaction dataset. In: ISSNIP'08: the 4th international conference on intelligent sensors, sensor networks and information processing, pp 201–206

Nicolescu MN, Matarić MJ (2002) A hierarchical architecture for behavior-based robots. In: AAMAS'02: the first international joint conference on autonomous agents and multiagent systems, ACM, New York, USA, pp 227–233. http://doi.acm.org/10.1145/544741.544798

Perez MC (2003) A proposal of a behavior-based control architecture with reinforcement learning for an autonomous underwater robot. PhD thesis, University of Girona

Chapter 10
Interacting Naturally

The previous chapter introduced the behavioral platform of EICA upon which we built our approach to social robotics. This chapter provides the details of this architecture in light of the theoretical foundations presented in Chap. 8 and shows how different parts of the architecture fit together to provide human-like interaction capabilities for the robot. The following chapter will discuss the learning algorithms used to develop this system and learn all of these processes from watching human–human and other human–robot interactions and how these processes can be adapted through actual interaction with people to provide better social performance.

10.1 Main Insights

Chapter 8 discussed the theoretical underpinning of our approach to social robotics. Two main principles were proposed based on this analysis: historical social embodiment and intention through interaction principles. These two principles will be utilized here to provide design guidelines for the architecture of the social robot.

Based on the intention through interaction principle, social interactions should be arranged as *interaction protocols* that computationally define the coupling functions between the intention functions of interacting partners. An interaction protocol is a computational structure that encapsulates the dynamics of this coupling in a specific interaction context. Different partners in the interaction are to be represented by *roles* that define their behavior during the interaction. This chapter will detail how these protocols are represented and activated in our proposed architecture.

Based on the historical social embodiment principle, these interaction protocol should be learned and adapted throughout the lifetime of the robot instead of being hard-coded or engineered within it from the start. It should be clear that this does not mean that the protocols *cannot* be engineered to some degree. For example, it is unavoidable for the designed to select the set of percepts available to the control processes and the set of actuators that the robot can utilize for feedback. Nevertheless,

© Springer International Publishing Switzerland 2015
Y. Mohammad and T. Nishida, *Data Mining for Social Robotics*,
Advanced Information and Knowledge Processing,
DOI 10.1007/978-3-319-25232-2_10

watching interactions of other people (which is a social act) and engaging in such interactions should leave their marks on the robot's computational processes in the form of adaptation or learning. Other than this theoretical motivation, engineering interaction protocols is a challenging task given the complexity of human social behavior. Relying on learning techniques then has both theoretical and practical justifications. Chapter 11 will detail the process that is utilized by our architecture for learning and adapting these protocols.

How are these protocols to be structured? The simulation theory (Sect. 8.2) suggests that we understand the minds of others by simulating them. Interaction protocols are structured as a set of roles with dynamic interactions between them. Given these two premises, we can propose that every role in the interaction protocol should be represented by a set of control processes in EICA that are then connected together through effect channels to provide the dynamic interaction required. This has the effect of representing other partners in the interaction as specific processes within the computational architecture of the robot as suggested by the simulation theory. Moreover, these same processes can be used to actually interact using the same protocol but in different role in the future.

This ability to interact in all roles of the interaction protocol is clearly an advantage from an engineering point of view. For example in the explanation scenario (Sect. 1.4), the same computational structure will support interacting both as the explainer (instructor) and as the listener. Moreover, as will be clear in the following chapter, this arrangement makes it possible to learn both roles simultaneously.

Face-to-face interactions are governed in human–human encounters by a variety of *synchronization* protocols at different time-scales. Compare for example the fast spontaneous gestural dance with turn-taking during verbal dialog. This suggests that the interaction protocols governing these behaviors are hierarchical in nature with different horizontal protocols coupling the two interaction partners at different time-scales.

Processing in the human brain usually combines top-down and bottom-up information flows. For example, the adaptive resonance theory (ART) posits such two-directions flow for the location memorization mechanism implemented by the hippocampus (Carpenter and Grossberg 2010). This suggests that information flow in the processes representing interaction protocols should employ both top-down and bottom-up processing.

This discussion highlights four main insights behind the design of EICA are:

1. Interaction protocols are to be implemented as interacting processes representing all roles of the interaction.
2. Interaction protocols are hierarchical in nature while allowing dynamic exchange of influence and data between processes of each role at each level of the hierarchy.
3. Influence and data flows within these protocols should combine top-down and bottom-up directions.
4. Information processing within interaction protocols should not distinguish the cases when the robot is in any of the roles of the interaction protocol. This indistinctness between different roles is a direct consequence of employing a common structure for acting and understanding of others' actions.

10.2   EICA Components

Figure 10.1 shows the general structure of the proposed architecture. The main goal of
EICA is to discover and execute an interaction coupling function that converges into
a mutual intention state. Social researchers discovered various levels of synchrony
during natural interactions ranging from role switching during free conversations and
slow turn taking during verbal interaction to the hypothesized gestural dance (Kendon
1970). To achieve natural interaction with humans, the agent needs to synchronize its
behavior with the behavior of the human at different time scale using different kinds
of processes ranging from deliberative role switching to reactive body alignment.

The architecture is a layered control architecture consisting of multiple interaction control layers. Within each layer a set of interactive processes provide the
competencies needed to synchronize the behavior of the agent with the behavior of
its partner (s) based on a global role variable that specifies the role of the robot in
the interaction. The proposed system provides a simulation theoretic mechanism for
recognizing interaction acts of the partner in a goal directed manner. EICA combines aspects of the simulation theory (during social interaction with people) and the
theory-theory during learning of the interaction protocols.

As with any EICA based system (See Sect. 9.5), we need to identify three sets
of components to specify the computation structure of the robot: intentions, perceptual processes and control processes. Perceptual processes are limited by the
sensing capabilities of the robot and we will not consider them any further in this
general discussion. The two remaining parts of the system are intentions and control
processes.

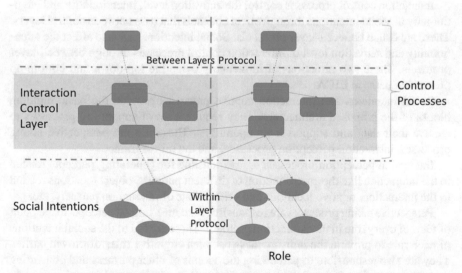

Fig. 10.1   Interaction Protocols as implemented in EICA. Basic social behaviors are implemented
as intentions that are controlled using a hierarchy of interaction control processes representing
different roles in the protocol. Both within layer and between layer protocols are shown

Each social intention is implemented as a pair of intentions in EICA. One of them is called the forward intention and it drives actuators and the other (inverse intention) is connected to perceptual processes. The inverse intention detects the actions of interaction partners and uses them to find the probability that the corresponding forward intention is activated in the partner *assuming that (s)he/it is using the same protocol as the robot*. This arrangement is inspired from mirror neuron research in neuroscience (Gallese and Goldman 1998). These interaction intentions are shown in the bottom row of Fig. 10.1. Notice that social intentions are organized in sets representing the roles of different agents in the interaction protocol. The interactions between partners in the real world at the fastest level of synchrony are represented in the architecture by within-layer protocols which are effect channels running *between* social intentions of different roles in the protocol.

Social intentions provide the lowest level of interaction capacities represented by the architecture. The activation of these capacities is controlled through higher level interaction control processes that are organized, again, in roles but also organized vertically in a hierarchy to represent the interaction protocol at increasing abstraction and slower synchrony levels.

Every interaction control process consists of two EICA control processes similar to social intentions: a forward and an inverse process. This reliance on forward and inverse processes is similar to the HAMMER block (Sect. 6.3.3). The main differences are that all the processes in EICA are learned through watching interactions, that these processes are organized in roles corresponding to the interaction protocol and that the system does not limit the implementation technology of these processes to Bayesian Networks. In our implementations, we used both BNs and NNs for these processes.

Interaction control processes control the activation level, intentionality and attentionality of social intentions through effect channels that pass between control layers. These are called between-layer protocols. Social intentions can also affect the attentionality and activation level of interaction control processes through between-layer protocols. These two directions of influence provide the top-down and bottom-up processing paths in EICA.

Social intentions and interaction control processes provide the basic building blocks of the proposed architecture. They require a set of supporting processes to receive their data and support their operations. These are the perspective taking processes, interaction perception processes, and the mirror trainer.

Interaction perception processes are responsible for generating percepts related to the interaction like the gaze direction of different partners, object locations related to the interaction, relative locations and orientations of interaction partners, etc.

Perspective taking processes are responsible for converting all percepts to the point of view of every role in the interaction protocol. These are fed to the social intentions of each role to provide the indistinctness between cognitive roles discussed earlier. They are also responsible of perceiving the actions of other partners and converting them to appropriate forms for the inverse intentions within social intention pairs to predict the activation level of different intentions of the partner under the "Like-me" hypothesis (Meltzoff 2005).

The mirror trainer is responsible for keeping forward and inverse intentions and processes in sync when either of them is changed through adaptation or added through learning (See Chap. 11).

10.3 Down–Up–Down Behavior Generation (DUD)

The previous section presented the basic constituents of our proposed EICA architecture for modeling interaction protocols for social robots. This section addresses the behavior generation mechanism that is used to generate the final behavior of the social robot. Top-down and Bottom-up behavior generation mechanisms are combined at every interaction control layer and simulation is used as the primary mechanism for representing the minds of other agents involved in the interaction.

Algorithm 2 provides an overview of the proposed behavior generation mechanism. As it is always the case with EICA, all processes at all interaction control layers are running in parallel. The algorithm provides only a shorthand serialized explanation of the interaction between these processes.

Algorithm 2 DUD Behavior Generation Mechanism (Mohammad and Nishida 2015)

1. Set *role* variable to the current role in the interaction.
2. Use perspective taking processes to generate input to all inverse intentions and propagate the calculation of all inverse interaction control processes of all layers.
3. Asynchronously calculate intentionality of all forward intentions using within-layer, up-going, and down-going factors.
4. Asynchronously calculate activation-level of all forward interaction processes using within-layer, up-going, and down-going factors.
5. Asynchronously update inputs and calculate outputs of forward intentions and propagate the calculations to all forward interaction control processes.
6. Repeat these steps until *role* changes or all activation levels become zero (end of interaction).

We will use the explanation scenario (Sect. 1.4) for illustrating the operation of this system. In this scenario we have two roles: instructor and listener. We assume that the robot is interacting in the listener role hereafter but it is easy to extend the discussion to cases in which the robot takes the instructor's role or even to cases with more than two roles in the interaction.

In this case, the intentions represented in the basic interaction acts (Fig. 10.2) can be processes like *look at partner*, *nod*, *look at object*, *face partner*, *approach partner*, *avoid partner*, *point to object*, etc. Section 11.1 shows how can these intentions be learned by imitating human behavior in natural human–human interaction. In this chapter, we assume that such intentions are given and that the mirror trainer successfully synchronized the forward and inverse intention in each pair.

An interaction control process (interaction process for short) in the first layer may be something like *mutual attention* which will adjust the activation levels of

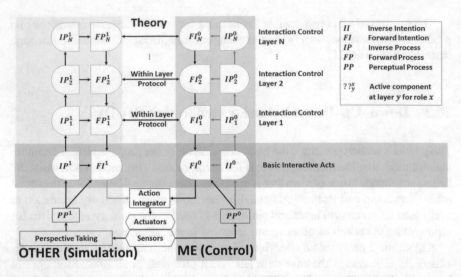

Fig. 10.2 Representation of interaction protocols in the EICA architecture. The robot is represented by the control stack running only in the forward direction (*right*), while other agents in the interaction are represented by simulation stacks running in both directions (*left*). See the text for details

the intentions *look at partner* and *look at object* to achieve mutual attention to the same object attended to by the instructor. A higher yet control process may be *listen* which adjusts the activation level of *mutual attention*, *mutual gaze*, etc. that exist in the first control layer to achieve a natural listening behavior.

EICA utilizes a hierarchy of dynamical systems corresponding to different interaction control layers that represent the interaction protocol at different speeds and abstraction levels. There are three influence paths that are activated during behavior generation called the *theory*, *simulation* and *control* paths (Fig. 10.2). These paths represent effect-channels in EICA not data channels (See Chap. 9). Data channels can connect any processes in the same layer or processes in the adjacent control layers of the same role.

The *control* path runs top-down and models the internal control of the robot itself based on the role assigned to it in the interaction. It contains only forward intentions and will implement the kind of nonverbal behavior described above for the *listen* case. For this path to work effectively, information is needed about the partner's intentional behavior and the interaction protocol itself and these are represented by the simulation and theory paths in order. Notice that the inverse processes for the role taken by the robot are disabled (grayed in Fig. 10.2) because the robot already knows its own intentions through the control path. This disabled direction of activation though may be useful for representing self-reflection or for implementing some form of feedback control between the layers which we do not pursue further here.

The *simulation* path represents a simulation of partner's behavior assuming its current perceived situation as encoded through perspective taking processes. The lowest level of this path are forward and inverse intentions corresponding to a simulation of the partner (e.g. the instructor in our running example). These intentions are copies of the intentions running in the control path that represent the internal state of the partner(s) assuming that she is controlling her actions using the same architecture (the "Like-me" hypothesis). The forward intentions in this path need information about the situation as seen through the eyes of the partner represented and this information is made available by the perspective taking processes. Perspective taking in the explanation scenario for example can be implemented by changing the frame of reference of all objects in the scene by projecting them to a frame of reference attached to the location and direction of attention of the partner. Both forward and inverse processes and intentions are active in the simulation path to represent both top-down and bottom-up processing directions in the model of the simulated agent.

The final path is the *theory* path which is not represented by any active processes but by the effect channels between processes representing the self and processes representing other partners in the interaction. This theory represents the levels of specification of the interaction protocol and is responsible of synchronizing the behavior of the robot with its partners in the interactions at all speeds and levels of abstraction. It is the need for representing interaction protocols that made it necessary to have this path which departs from the pure simulation theoretic approach of ST as discussed in Sect. 8.2.

The three paths are interacting to give the final behavior of the robot. It is possible to have multiple interaction protocols implemented in the robot in the form of different stacks of interaction control processes to represent different interaction contexts. This is similar to the case of situated modules in the Situated Modules architecture of Sect. 6.3.2.

As shown in Fig. 10.2, the activation level of each forward control process is controlled through three factors:

1. The activation level of the corresponding inverse process. This is called the up-going factor.
2. The activation level of other forward corresponding to other roles in the interaction (from the control or simulation paths). This is called the within-layer factor.
3. The output of higher level interaction processes corresponding to the self for the control path and to the same role in the simulation path. This is called the inter-layer protocol factor. This factor represents the top-down activation representing the relation between the interaction protocols at different time scales and is called the down-going factor.

The architecture itself does not specify which type of effect channels to be used for combining these factors. In our implementations though we used either DLT or AVG effect channel types (See Sect. 9.2 for details).

After understanding how each process and intention is activated we turn our attention to the meaning of the inputs and outputs of them. The input-output mapping implemented by intentions and processes is quite different and for this reason we will treat each of them separately. Moreover inverse and forward versions are also treated differently.

The input for the forward intentions representing the current role (in the control path) is taken directly from the sensors while its output goes to the actuators. The inputs for the corresponding forward intentions representing all other roles in the interaction (the simulation path) are taken from perspective taking processes and the output is simply discarded. In actual implementation the forward intentions corresponding to other agents are disabled by having their attentionality set to zero to save their computation time.

The inputs to reverse intentions in the theory path are the last n_z values of all the sensors sensing the behavior of the partner. The output is the probability that the corresponding forward intentions are active in the partner control architecture assuming that the partner has the same control architecture (the central hypothesis of the theory of simulation). This output is the up-going factor affecting the activation level of the corresponding forward intention. It is also used for adapting the parameters of this forward intentions as will be shown in Sect. 11.3.

Reverse processes are doing exactly the same job with relation to forward processes. The only difference is that because of the within-layer factor reverse processes in general have a harder job detecting the activation of their corresponding forward processes in the partner. To reduce this effect, the weight assigned to the reverse processes is in general less than that assigned to reverse intentions in the effect channels connected to their respective forward processes.

10.4   Mirror Training (MT)

The mirror trainer is not invoked during behavior generation but is invoked during learning to make sure that the inverse interaction control processes are in synchrony with forward interaction control processes. This synchrony means that the inverse version can *detect* the activity of the *forward* version and can estimate its activation level (activation-level for processes and intentionality for intentions).

The algorithm used to do mirror training is very simple. The mirror trainer instantiates a copy of the forward intention or process that is to be learned, gives it random inputs, and gets its output. These are then used as positive examples. The trainer then instantiates other forward processes and reads their output for the same inputs and these are the negative examples. All inverse processes in current implementation are implemented as radial basis functions neural networks (RBFNNs). RBFNNs are proved to be global function modelers and can represent most functions of interest. The mirror trainer then uses well known RBFNN training algorithms with the collected sets of positive and negative examples to learn the mapping.

Because of the difficulties in learning the reverse processes due to the existence of within-layer up-going factors, the mirror trainer trains the system with various values of these factors and at run time the reverse process selects the neural network that is trained with the nearest values of these factors to the actual values. This reduces the effect of these factors without increasing the computational time much.

A second possibility is to use the time-series representing the motion from the forward intention or processes as training examples for a HMM as explained in Sect. 2.4.3. We use this approach for learning inverse intentions and use RBFNNs for inverse processes.

10.5   Summary

This chapter provided an overview of the EICA architecture as being used for social robotics application. We discussed the motivation behind the general design of the architecture and how it related to the underlying behavioral architecture discussed in Chap. 9.

Processes of this architecture are organized in different interaction control layers and each process has two twin computational components namely the forward and inverse versions of it. The forward version is used to generate the required behavior while the inverse version is used to detect this behavior in the perceived data streams. Mirror training is used to keep these two processes in sync. The proposed behavior generation mechanism (called Down–Up–Down algorithm) was also discussed and its inspiration from cognitive models presented. In the following chapter we detail the algorithms used in the three stages presented in Chap. 1 to create all intentions and processes in EICA by mining human–human interaction records and then adapting them with actual human–robot interactions.

References

Carpenter GA, Grossberg S (2010) Adaptive resonance theory. Springer
Gallese V, Goldman A (1998) Mirror neurons and the simulation theory of mind-reading. Trends Cogn Sci 2(12):493–501
Kendon A (1970) Movement coordination in social interaction: some examples considered. Acta Pyschologica 32:101–125
Meltzoff AN (2005) Imitation and other minds: the "like me" hypothesis. Perspect Imitation: Neurosci Social Sci 2:55–77
Mohammad Y, Nishida T (2015) Learning interaction protocols by mimicking: understanding and reproducing human interactive behavior. Pattern Recognit Lett 66(15):62–70

Chapter 11
Interaction Learning Through Imitation

In Chap. 10 we presented an overview of proposed architecture and detailed how can it generate behavior given that the intentions and processes involved are already available. This chapter describes a three–stages developmental approach to learn intentions and processes that can be used to generate interaction protocols to be run by the EICA architecture. The first two stages require logs of human–human interactions that are processed by the robot to learn the intentions (first stage) and interaction control processes (second stage) involved. After these two stages the robot is able to behave, more or less, in accordance to the interaction protocol it learned and is ready to engage in real human–robot interactions. The third stage involves adaptation of th interaction protocol learned in the first two stages while interacting with people.

11.1 Stage 1: Interaction Babbling

This section focuses on the first developmental stage of the proposed architecture called Interaction Babbling. Before this stage the only available parts of the agent are its sensors, actuators, perceptual processes and perspective taking processes (See Fig. 10.2). This entails that the agent is not a "blank slate" in the beginning as the act of selecting these components by the designer constrains the possibilities of the agent. For example, without a camera or a similar sensor; it is impossible for the robot to learn from visual stimuli. Even when a camera is available, if the perceptual processes do not provide information about color, color perception would be impossible because the intentions and control processes of the robot are connected to the sensors only through perceptual processes as shown in Fig. 10.2.

The goal of this stage is to learn the basic interactive acts layer which consists of intentions representing basic actions related to the interaction. Only forward intentions need to be learned as inverse intentions can then be created using the mirror

© Springer International Publishing Switzerland 2015
Y. Mohammad and T. Nishida, *Data Mining for Social Robotics*,
Advanced Information and Knowledge Processing,
DOI 10.1007/978-3-319-25232-2_11

trainer (Sect. 10.4). Two steps are involved: firstly, the perceptual/actuation features of the intention are learned from the output of perceptual processes. Secondly, these motions are used to learn controllers that can generate the same motions using the robot's actuators.

11.1.1 Learning Intentions

Algorithm 3 Interaction Babbling

1. Filter input stream to generate interaction related time series (e.g. convert all motion capture data to ego-centric coordinates).
2. Apply change point discovery (e.g. RSST from Sect. 3.5) to input stream to discover probable locations of recurrent patterns (the constraints).
3. Apply a motif discovery algorithm (e.g. MCFull from Sect. 4.6) using the constraints calculated in the previous step to discover intention representation in the input space as recurring motifs.
4. Find the mean pattern representing each motif and use it to generate a controller able of moving the robot to generate the same behavior (this is the controller generation process of Sect. 11.1.2).

The problem of learning interactive intentions can be casted as an unsupervised discovery of recurrent patterns (the basic actions) from a stream of unmarked input data (behavior). This is the standard motif discovery algorithm discussed in Chap. 4. In our case the input is a multidimensional time series corresponding to the outputs of different *perceptual processes* (See Sect. 10.2). The general algorithm we used for solving this problem is shown in Algorithm 3.

Firstly, robot's sensors and perspective taking processes are used to capture the behavior of both the robot and its partner as a multidimensional time series. In some cases the raw output of the hardware sensors needs to be filtered or processed at this step. For example, our motion capture system (PhaseSpace) generates the locations of predefined points in the partner body, but these locations are in a global coordinate system attached to the cameras of the motion capture system. During interaction it does not matter whether the partner is at point $\langle 1, 1, 1 \rangle$ or $\langle 34.343, 64.243, -34545.45 \rangle$, but what matter is his/her *relative* location to the partner. This can easily be found by simple coordinate conversion (affine transformation) knowing the current location of the robot. This kind of preprocessing is necessary to maximize the possibility of discovering relevant patterns.

Secondly, a change point discovery algorithm from Chap. 3 is used to find interesting locations in the outputs of these perception processes (for example RSST from in Sect. 3.5). This step is done to convert the problem form unconstrained motif discovery to a constrained motif discovery problem (Sect. 4.6) that can be solved in linear (or sub-linear) time. This speedup step is necessary as the lengths of time series we need to process can easily be in millions of points and traditional motif discovery algorithms that work in superlinear time cannot handle these long time series with acceptable speed.

Thirdly, A motif discovery algorithm (Chap. 4) is applied to discover recurrent patterns in the processed input streams. Each discovered pattern is then attached to a forward intention.

Fourthly, the mean of the discovered pattern (from its occurrences) is used to build a controller able of generating the required behavior as will be shown in Sect. 11.1.2.

Finally, the occurrences of each pattern are used to induce a Hidden Markov Model that represents the discovered pattern. This HMM is then used to detect the required behavior during activation of the robot (inverse intention). This is done through the mirror trainer (Sect. 10.4).

11.1.2 Controller Generation

The final step in Algorithm 3 is to generate a controller that can make the robot behave as dictated by the mean of each discovered motif (pattern). If we have direct access to the joint angles of the participants used for generating this motion, it is possible to directly use algorithms from Chap. 13. In our experimental situation, we did not have direct access to this data which means that we need to solve an inverse kinematics or inverse dynamics problem (depending on the kind of data we have) before applying any standard learning from demonstration system from Chap. 13. In this section, we explain a different approach that allows the robot to learn the inverse problem (kinematic or dynamic depending on the input) while learning the controller represented by the forward intention associated with the motif discovered in the previous step in perceptual processes' outputs. This is achieved in two step. In the first step (motor babbling), the robot uses random exploration of its actuators to discover a set of functions that can either increase or decrease a single dimension from the output of its perceptual processes keeping all other dimensions fixed. The second step (piece-wise linear controller generation) then uses these functions to learn a linear approximation of the motif mean learned in the intention learning phase (Sect. 11.1.1).

During motor babbling, the robot builds a repertoire of motor primitives related to the action dimensions that will be used in controller generation. Two functions are to be learned for each dimension of the outputs from the perceptual processes (Fig. 10.2). The first function F_i^+ receives a constant rate δ_i and increases that dimension by that output by applying commands to the actuators of the robot while keeping other dimensions constant and the second function F_i^- decreases that dimensions with the same rate δ_i. Effectively these functions convert the problem of multivariable control to single input single output control problem.

A simulator can be used to reduce he risks to the robot during this stage. The robot starts in a predefined initial state and tries random motion actions (commands to its motors C) keeping track to the changes happening to each of its action dimensions. This training data is then used to learn F_i^+, F_i^-. Thus F_i^+ and F_i^- both solve the following constrained optimization (minimization) problem with positive and negative δ_is:

Objective:

$$\sum_{\substack{j = 1 : n_a, \\ i \neq j,}} \left(a_j \left(n + 1\right) - a_j \left(n\right)\right)^2, \tag{11.1}$$

Constraints:

$$\begin{aligned} &\text{for} 1 \leq i < n_a, \\ &|a_i \left(n + 1\right) - a_i \left(n\right)| \geq \delta_i^-, \\ &|a_i \left(n + 1\right) - a_i \left(n\right)| < \delta_i^+. \end{aligned} \tag{11.2}$$

Control Variables:

$$C \equiv \left[\Delta m^+ \left(n\right), \Delta m^- \left(n\right)\right], \tag{11.3}$$

where $a_i \left(n\right)$ is the nth sample of the ith action dimension, Δm^+ is the *sum* of the two commands given to the motors and is proportional to the speed, Δm^- is the *difference* between these two commands and is proportional to the rotation angle (differential drive arrangement), δ_i^+ is the upper limit on the required rate of change, and δ_i^- is the lower limit. The pair Δm^+ and Δm^- constitute the command sent to the motors C.

The problem is formulated as a Markov Decision Process (MDP) and is solved using standard Q-Learning with the aid of a simulator that can produce $A \left(n + 1\right)$ given $A \left(n\right)$ and $C \left(n\right)$.

The resulting F_i^+ and F_i^- can now represent a straight line in the action dimension i with an approximate slope of $\delta_i = \left(\delta_i^- + \delta_i^+\right)/2$. These two functions thus serve to linearize the relation between the action dimensions and motor commands and in the same time decouples different action dimensions.

A particular property of the specific action stream dimensions we selected for our applications (Sect. 11.4) is that a different slope $\delta_2 = a \times \delta_1$ can be achieved simply by multiplying the final command by the constant a. This means that F_i^+ and F_i^- need to be learned for a single value for δ_i and their output can then be scaled to achieve any required slope in the corresponding action dimension.

Algorithm 4 Piecewise-Linear Controller Generation Algorithm (Mohammad and Nishida 2010b, 2015)

---

1. Calculate mean pattern using:

$$m(t) = \sum_i m_i(t),$$ (11.4)

   where m_i is an occurrence of the motif and m is its mean.
2. Generate a piecewise linear approximation ($\hat{m}$) of the mean using SWAB algorithm (Keogh et al. 2001).
3. Calculate the slope v of every line l in $\hat{m}$ and call F_i^+ or F_i^- depending on the sign of the slope to generate the required linear motion.

Algorithm 4 summarizes the steps of our controller generator. Firstly, the mean of each motif is calculated using all of its discovered occurrences. Secondly, the mean is approximated as a series of piecewise linear segments. For this approximation, we use the SWAB algorithm described in (Keogh et al. 2001). Finally, the slop of every line segment is calculated and passed to the corresponding controller at run time.

The piecewise-linear controller generation algorithm discussed in this section provides only one possible implementation of the controller generation phase. Its main advantage is the ability to learn control functions based on motor babbling. Nevertheless, this assumes that the perceived channels can be controlled independently which is not always true. Other algorithms for learning a controller from perceived channel signals can utilize learning from demonstration technology to be introduced in Chap. 13.

11.2 Stage 2: Interaction Structure Learning

This section introduces the second stage of the developmental process aiming at learning the interaction control processes of all interaction control layers (one layer at a time). This stage is called interaction structure learning and during which the robot (agent) builds a hierarchy of interaction control processes representing the interaction structure at increasing levels of abstractions using the basic interactive acts represented by the intentions already learned in the first stage (See Sect. 11.1.1). This stage is applied offline without actual interaction between the robot and human partners.

11.2.1 Single-Layer Interaction Structure Learner

Interaction protocols are not created equal. In some cases, the interaction protocol is simple enough to be represented by a single interaction control layer. This is true of most explicit interaction protocols. Consider the guided navigation example of

Sect. 1.4. In this case, the operator (human) is controlling the actor (robot) using free hand gesture. The interaction protocol in this case is very simple and can be encoded in a directed graph that connects gestures of the operator (G) to actions of the actor (A). In such cases, a single layer interaction structure learner can be used to learn single interaction control process involved. In our example, each gesture and each action will be encoded by an intention that was learned in the first stage of development (Interaction babbling as explained in Sect. 11.1). The main goal of the interaction structure learner in this case is to use the intentionality assignments from forward intentions and activation probabilities from reverse intentions to learn a single interaction control process for each agent that represents the behavior of that agent in the form of a probabilistic graphical model. The two processes representing the operator and actor (in our example) constitute the single interaction control layer in that case. This algorithm is called single-layer interaction structure learner (SISL).

The main idea behind SISL is to model the interaction of different partners as a single interaction control process implementing a directed acyclic graph encoding the causal connections between the different intentions. For example, an intention may represent the gesture of opening the palm while raising the hand to face level with the back of the palm facing the face of the operator. The robot may notice from watching several guided navigation examples, that shortly after this intention is activated, another intention which represents zero motion of the operator is executed. Leaning this relation allows the robot to discover that the gesture "stop" activates the intention "stop" in the actor (without the linguistic baggage of course). This simple example shows the importance of causality analysis for learning the interaction protocol.

We start by constructing for each intention i of role r (M_i^r), an associated occurrence list L_i^r that specifies the timesteps at which each occurrence of M_i^r starts.

The intuition from the previous example can be formalized as follows: if intention i for role r_1 causes the agent in role r_2 to execute the action primitive j then L_j^{r2} will *most of the time* has values that are τ_{ij} time-steps after L_i^{r1} where τ_{ij} is a constant representing the delay of the causation.

In the real world, many factors will affect the actual delay. Using the law of large numbers, it is admissible to assume the delays between these action onset points can be approximated with a Gaussian distribution with a mean of $\hat{\tau}_{ij}$ where $|\hat{\tau}_{ij} - \tau_{ij}| < \varepsilon$ for some small value ε. The main idea of our algorithm is to check for the normality of the delays and to use the normality statistic as a measure of causality between the two processes. This is simply the change causation algorithm described in Sect. 5.5.

Prior knowledge about the system can be incorporated at this phase. For example if we know that there can be no self-loops in the causality graph (a change in a process does no directly cause another change later) we can restrict the pairs of L_i, L_j by having $i \neq j$. If we know that processes with higher index can never cause process with lower index, then we can restrict i to be less than or equal to j. Extension to more complex constraints is fairly straightforward.

After completing this operation for all L_i^{r1} and L_j^{r2} pairs (where $r1 \neq r2$), we have a directed graph that represents the causal relationships between the series involved in the form of a Bayesian Network with the edges augmented with information about explicit delays. We call this graph the Augmented Bayesian Network (ABN) hereafter. The problem now is that this graph may have some redundancies. This ambiguity can be resolved by simply removing the causal relation with the longest time delay. A better approach would be keep multiple possible graphs and then use a causal hypothesis testing system like Hoover's technique (Hoover 1990) for selecting one of them.

11.2.2 Interaction Rule Induction

The method described in Sect. 11.2.1 is effective for learning a single interaction control layer that has a single interaction control process which amounts to a simple interaction protocol. In some cases, we prefer to model the interaction by a set of probabilistic *rules* that can be used for more complicated interaction protocols like the implicit protocol of the natural listening scenario in Sect. 1.4. This can be achieved by the interaction rule induction (IRI) algorithm.

The system architecture assumed by the IRI algorithm is shown in Fig. 11.1. The main differences between this special form of the architecture and the standard form presented in Fig. 10.2 are the following: Firstly, only two interaction control layers are allowed: the interaction rules layer and the session protocol layer. This is in contrast

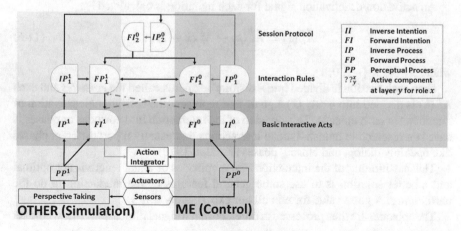

Fig. 11.1 The interaction protocol representation for the interaction rule induction protocol. The main differences from Fig. 10.2 is that only two interaction control layers are allowed and the implicit information passing between rules in the first layer

with the *deep* architecture assumed in Fig. 10.2. Secondly information pass in the first control layer not only from intentions in the same role but from other intentions as well. Finally, the within-layer protocol in the second layer is implemented in a single process that controls the activation of rules in the interaction rules layer.

The idea behind Fig. 11.1 is to model the interaction as a set of reactive rules that relate the behavior of different agents (roles) at one time step with their partners' behavior at the previous time-step which implements an implicit turn taking exchange. The session protocol in the second layer then adjusts the activation level of each rule from the second layer to achieve the overall session control of the interaction. This model is more appropriate to implicit protocols like the one implied by the explanation scenario.

The main idea behind the algorithm is to first divide each interaction in the training pool into a disjoint set of *episodes* that represent different *stages* of the interaction. A rule discovery algorithm is used to *explain* that activation and deactivation of intentions that happen at each episode as a function of other running intentions. Once all of these rules (IRs) are discovered, their activation with the episodes is used to build the Session Protocol (SP) that will be used in realtime to activate/deactivate the learned IRs.

Let's define a set of signals $A_i^j(k)$ that represent the activation level of intentions i in session j at time step k. We discretized these signals to get a binary form $\hat{A}$ that represents either active or inactive intentions using:

$$\hat{A}_i^j(k) = \begin{cases} 1 & A_i^j(k) > \tau \\ 0 & otherwise \end{cases}. \tag{11.5}$$

An activation/deactivation signal for each intention is calculated as:

$$\tilde{A}_i^j(k) = \hat{A}_i^j(k) - \hat{A}_i^j(k-1), \tag{11.6}$$

where $\tilde{A}_i^j(k) \in \{-1, 0, 1\}$.

Now each session is divided into n disjoint segments called the *episodes* and each $\tilde{A}_i^j(k)$ is divided into n $^s\tilde{A}_i^j(k)$ signals that represent the activation/deactivation of intentions in each episode. This division of the interaction into episodes or phases is based on research in human–human interactions that usually report different phases like opening, dialog, and closing phases (Kendon 1970).

This partitioning of the interaction into n episodes of equal length is not optimal and a better solution is to use some general features of the interaction to do the partitioning. A good value for n in all our experiments was 3.

The episodes are then processed in order. For each signal dimension i, the whole set $\left\{^s\tilde{A}_i^j\right\}$ for the current episode is collected in memory. There are two challenges here: Firstly, the activation of an intention may depend on the past of other intentions not only on their current values (e.g. the natural delay between instructor's and listener's movements in route guidance reported by Kanda et al. 2007). This prevents us from using algorithms like DBMiner (Han et al. 1996) that can find rules of the form:

$$X_i = x_i \wedge X_j = x_j \wedge$$
$$..... \wedge X_k = x_k \Rightarrow X_l = x_l. \tag{11.7}$$

Secondly, the activation of a single intention may depend on the values of multiple other intentions which prevents us from using simple association rule discovery algorithms that account for time delays like the algorithm proposed by Das et al. (1998) which finds rules in the form:

$$A \xrightarrow{T} B, \tag{11.8}$$

which means that when A happens at time step t, B happens at time step $t + T$.

The rules, we are trying to find, have the general form:

$$X_i (t - \tau_i) = x_i \wedge X_k (t - \tau_k) = x_k,$$
$$..... \wedge X_z (t - \tau_z) = x_z \rightarrow \tilde{A}_i (t) = \pm 1, \tag{11.9}$$

where X_i is a feature extracted from $\tilde{A}_i$.

We use three features:

$$S_i^j (t) = \max \left(0, \sum_{k=1}^{t} \left(\hat{A}_i^j (k) - 0.5 \right) \right), \tag{11.10}$$

where S_i^j features represent the discrete integration of the binary approximation of the intentionality of different intentions, while $\tilde{A}_i^j$ represent their differentiation. In case of S_i^j, $\hat{A}_i^j$ is first shifted down by 0.5 to make S forget about old activations. S_i^j and $\hat{A}_i^j$ are both segmented into the episode sets $\left\{ ^s S_i^j \right\}$ and $\left\{ ^s \hat{A}_i^j \right\}$ in the same way as $\left\{ ^s \tilde{A}_i^j \right\}$.

To find these rules for some activation sequence $\left\{ ^s \tilde{A}_i^j \right\}$, we first find the optimal delays τ_k for all other features in this episode and then shift each feature back by its corresponding delay. After that, the problem becomes a classification problem that can be easily solved by a CART classifier. The features used to train the classifier are all the values of $^s X_k^j (t - \tau_k)$ and the target labels are $^s \tilde{A}_i^j (t)$. We train two trees: one to detect activation events ($^s \tilde{A}_i^j (t) = +1$) and the other to detect deactivation events ($^s \tilde{A}_i^j (t) = -1$).

The novel step in this approach is the technique used to discover the delays τ_k. Again we use an algorithm that depends on causality analysis. Simply we find the delay that maximizes the Granger causality between the two streams.

Once each rule is learned, the whole interaction corpus is scanned for all cases of activation/dectivation that can be explained by it and these cases are removed by setting the associated $\tilde{A}_i^j$ to zero. This is done in all episodes not only the current one. The activation pattern of this rule is then saved in the Session Protocol.

11.2.3 Deep Interaction Structure Learner

The most complex case for learning interaction control layers involves learning deep architecture represented by the full system in Fig. 10.2. The IRI algorithm presented in Sect. 11.2.2 cannot be extended to more than two interaction control layers while the SISL (Sect. 11.2.1) can only learn a single control layer. Recent findings from the deep learning community shows that several machine learning tasks are best learned using deep architectures (Hinton et al. 2006).

We employ here the deep interaction structure learner (DISL). Successful algorithms in Deep Learning use an iterative process in which each layer is learned using mostly unsupervised techniques then a small set of training sets can be used to learn connections between layers and provide final classification for classification problems. DISL works in a similar fashion by learning each layer based on the activation levels (or intentionalities) of the lower layer and the complete system is then integrated.

The Interaction Structure Learner is invoked by a set of training examples and a set of verification examples. Every example is a record of the sensory information ($_i^c s$) and body motion information ($_i^c m$) of every agent i during a specific interaction c of the type to be learned.

The outputs of the interaction structure learner are:

1. The number of layers needed to represent a specific interaction protocol (l_{max}).
2. The number of interaction control processes in each layer (n_l).
3. An estimate of the parameters of every forward interaction control process ($_r P_j^l$) in every layer l for every role in the interaction r. The corresponding reverse process can be learned using the mirror trainer ($_r^{rev} P_j^l$).

DISL needs to know the expected set of frequencies at which synchrony between different agents (roles) represented by the interaction protocol exist. This can be found from known human–human interaction data. If this set is not available the system must try many frequencies until a good synchronization frequency is found.

The activation level of every intention or process in layer l is connected to the outputs of the process in layer $l + 1$ which inputs are in turn connected through a set of k delay elements to the activation levels of all the processes in layer l. Formally:

$$a\left(_i P_j^l, n\right) = \sum_{g=1}^{n_{agents}} \sum_{m=0}^{n_{l+1}} O^j\left(_g P_m^{l+1}, n\right) \times a\left(_g P_m^{l+1}, n\right), \qquad (11.11)$$

$$O^j \left({}_i P_m^{l+1}, n \right) = f_m^j \left(\{ {}_a P_b^l (k) \} \right),$$
$$k = 1 : K, a \in \{agents\}, b = 0 : n_l - 1. \tag{11.12}$$

Currently we use Radial Basis Functions Neural Networks (RBFNNs) to learn the functions f_m^j for all processes in all layers:

$$f_m^j (\mathbf{x}) = w_0 + \sum_{u=1}^{p} w_u e^{\left(\frac{d(\mu_u, \mathbf{x})^2}{2\sigma_u^2} \right)}. \tag{11.13}$$

The steps of DISL can be summarized as:

1. Convert the sensory inputs of all interactions into activation levels of the intentions using reverse intentions.
2. Learn the interaction control processes starting from layer 1 incrementally one layer at a time until one layer contains only one process for every agent as follows:

   a. Find the best number of clusters for every partner in layer $l - 1$ and call it ${}_i n^l$.
   b. For layer l, use the activation levels of layer $l - 1$ to train ${}_i n^l$ RBFNNs representing the processes of every role i in this layer using:
   Input:$<_{*-i}^k P_*^{l-1} (n - z(f))_i^k P_*^{l-1} (n) >$
   Output:$<_i^k P_*^{l-1} (n + 1) >$,
   where $k_i P_j^l$ is k times down-sampled version of the activation level of process j in layer l for role i, f is the current synchrony frequency, and $z(f)$ is the delay corresponding to this frequency.
   c. Apply this learned process to the validation set and find the difference between its outputs and the action activation levels at layer $l - 1$.
   d. Accept the process if this error level is acceptable.
   e. Repeat the previous steps for different values of k.
   f. Invoke the mirror trainer to learn the reverse process.

The performance of this algorithm is greatly affected by two factors:

- The cluster counting method used to find the best number of processes in each layer. Currently we use a modified version of the algorithm suggested in Sugar and James (2003).
- The *acceptable* () function used to reject or accept trained RBFNNs. Currently a process is accepted if the error level is less than τ for at least n_{min} continuous steps for ζ of the interactions (currently fixed at 0.1, 10, and 0.7 respectively).

This algorithm is not guaranteed to converge, but—when converging—there are two different situations depending on the number of processes for every role in the highest layer(${}_i n^{l_{max}}$):

1. $_in^{l_{max}} = 1$: This means that the behavior of this role in this kind of interactions is passive and can be decided solely depending on the behavior of other roles (e.g. in the following example the listener gaze behavior in explanation scenarios was found to be of this kind in the example scenarios used).
2. $_in^{l_{max}} = 0$: This means that no single process can represent the high level behavior of this role of the interaction. This is usually associated with proactive roles (e.g. the instructor in the explanation scenarios used in the following example).

11.3   Stage 3: Adaptation During Interaction

We now turn to the final stage of development for the robot. Now the robot has a complete interaction protocol learning in first two stages. Using the DUD algorithm (Sect. 10.3), the robot can start to engage in social HRI interactions. These interactions can be used to improve the protocol learned in the second stage. The algorithm used for this stage will depend on whether we have a shallow architecture (e.g. the single-layer architecture of Sect. 11.2.1) or a deep architecture (e.g. the deep architecture described in Sect. 11.2.3). This section details the algorithms used to adapt the interaction protocol based on HRI sessions in these two cases.

11.3.1   Single-Layer Interaction Adaptation Algorithm

Section 11.2.1 explained how to learn an augmented Bayesian Network (ABN) to represent the interaction protocol by a single control layer. When the robot interacts with other people, it can use the logs of these interactions to generate new ABNs. The adaptation algorithm is then reduced to the need to incrementally generate ABNs that take into account these new ABNs in order to improve the performance of the robot in future trials.

To combine two different ABNs we need to discover the correspondences between nodes/intentions in the two ABNs. We can simply compare the perceptual process representation between the intentions represented by the two nodes but this approach is local in the sense that it matches nodes without looking into the global structure of the networks. Consider for example two ABNs learned from two sessions in the guided navigation scenario. One node of each network represents a "stop" gesture by the operator and is connected to another node representing the "stopping" action by the actor. Now the two "stop" nodes may represent very different motions (gestures) and this will not allow the simple algorithm described above to see their similarity. Looking at the network as a whole, on the other hand, allows us to see that these two nodes are leading to the same action ("stopping") which will have the same representation in the two networks (zeros sent to both motors). This example shows the need to look not only for similarity in the pattern represented by the intention or processes under consideration but also to the global connection structure of the two ABNs. Mohammad and Nishida (2010b) proposed such a global approach that will be described briefly here.

The algorithm starts by compiling two lists of action nodes for every role from the two ABNs (namely AN^1 and AN^2). For every member of AN^1 (called an_i^1), the motif mean (m_i^1) is compared with every other motif mean in the same ABN (m_j^1) using Dynamic Time Wrapping (DTW) and the minimum distance is selected as the similarity threshold of this node τ_i (Mohammad and Nishida 2015):

$$\tau_i = \min \left(d_{DTW} \left(m_i^1, m_k^1 \right) \right), \qquad (11.14)$$

where $1 \leq k \leq n_A^1$, $k \neq i$, and n_A^1 is the number of action nodes in AN^1.

The second step is to compare the mean of node i with every other node in the second list AN^2 and a link la_{1i}^{2j} is created between an_i^1 and an_j^2 iff: $d_{DTW} \left(m_i^1, m_j^2 \right) < \tau_i$ and $\left(d_{DTW} \left(m_i^1, m_j^2 \right) - d_{DTW} \left(m_i^1, m_k^2 \right) \right) < \eta \tau_i$ for $1 \leq k \leq n_A^2$ and $k \neq j$ for some value of η greater than zero. We select η to equal 0.25 for all our experiments. If two or more nodes in AN^2 satisfy these two conditions, a link is created between an_i^1 and each of them. Each link had a value equal to the DTW distance between the means of the two nodes it links.

The final step is to remove all the conflicts in the two link lists to have at most one node in the second ABN connected to any node in the first ABN. For every link we calculate a *link competence index* (LCI) that evaluates the match between the two ABNs if this link was kept as follows:

$$LCI \left(la_{1j}^{2i} \right) = \frac{1}{v \left(la_{1j}^{2i} \right)} + \lambda_a \sum_{\substack{gn_l^2 \in Par \left(an_i^2 \right) \\ gn_k^1 \in Par \left(an_j^1 \right)}} LCI \left(lg_{1k}^{2l} \right), \qquad (11.15)$$

$$LCI \left(lg_{1j}^{2i} \right) = \frac{1}{v \left(lg_{1j}^{2i} \right)} + \lambda_g \sum_{\substack{gn_i^2 \in Par \left(an_l^2 \right) \\ gn_j^1 \in Par \left(an_k^1 \right)}} LCI \left(la_{1k}^{2l} \right), \qquad (11.16)$$

where $Par (n)$ is the set of all parents to node n. These equations constitute a set of $n_{la} + n_{ga}$ equations in the same number of variables and can be solved using a simple iterative approach similar to the value iteration for solving MDPs.

A larger LCI means that not only the primitive nodes connected by the link are similar but also their parent nodes are similar as well. After the LCI is calculated for every link, the link with highest LCI fanning out from any node is kept and the rest are discarded (Mohammad and Nishida 2010b, 2015).

After resolving all conflicts, the nodes in the two ABNs that are still linked are combined their stored pattern is re-generated from the full set of motif occurrences used when creating the two ABNs.

Combining nodes from two ABNs does not affect the edges except if it caused two nodes to be connected by more than one edge in the final ABN. In this case, the mean and variance of the delay associated with the final edge are calculated from the mean, variance, and number of occurrences in the two combined edges.

11.3.2   Deep Interaction Adaptation Algorithm

The Deep Interaction Adaptation Algorithm (DIAA) proposed here can be used to adapt the parameters of processes of interaction protocols of arbitrary depth. The main idea behind the DIAA is to monitor the difference between the current *theory* the agent has about what its partner (who plays the role to be learned) is intending at different levels of abstraction and the *simulation* of what it could have done if it was playing this role. This is simply the difference between the down-going and up-going factors described in Sect. 10.3. The DIAA then adapts the forward processes representing the target role to reduce this difference. The discussion hereafter in this section will assume a two-roles interaction to reduce notational complexity but extensions to interactions involving more than two roles is straightforward.

The goal of the DIAA is to find for every process a parameter vector ${}_i^f \hat{p}_k^l$ that minimizes the error estimate:

$$e_k^l = d \left(a_t \left({}_i P_k^l \right) - a_s \left({}_i P_k^l \right) \right), \tag{11.17}$$

where $a_t \left({}_i P_k^l \right)$ is the estimate of the activation level of process ${}_i P_k^l$ based on the down-going factor, $a_s \left({}_i P_k^l \right)$ is the estimate of the activation level of process ${}_i P_k^l$ based on the up-going factor of the simulation and $d \left(x, y \right)$ is a distance measure. Currently Euclidean distance is used.

We assume here that each process has a probability distribution that is used to set the activation level of processes in the immediate lower layer of the same role (or intentionality of intentions for the processes in the first control layer) (Mohammad and Nishida 2008).

Reverse processes are used to estimate the current understanding of what the partner is actually doing during the interaction ($a_t \left({}_i P_k^l \right)$) while forward processes represent the current predictions of the robot about how should this agent behave ($a_s \left({}_i P_k^l \right)$). If these two factors are similar, then the protocol is learned adequately and there is no need to adapt. This means that the error signal generated by comparing these two signal for all processes should be the driving factor behind the operation of DIAA.

The interactive adaptation algorithm uses these two signals to adapt the forward processes of the target roles and then executes the mirror trainer to adapt the corresponding reverse processes. The details of this algorithm are shown in Algorithm 5.

Algorithm 5 Deep Interactive Adaptation Algorithm (Mohammad and Nishida 2008)

function INTERACTIVE   ADAPTATION   ALGORITHM(η, $a_t\left(_iP_{k_l}^l\right)$, $a_s\left(_iP_{k_l}^l\right)$, $a_t\left(_iP_{1:n^{l+1}}^{l+1}\right)$,
$a_s\left(_iP_{1:n^{l+1}}^{l+1}\right)$, n^l, n^{l+1})

$\quad n \leftarrow n+1$
$\quad e_k^l \leftarrow d\left(a_t\left(_iP_k^l\right) - a_s\left(_iP_k^l\right)\right)$
$\quad$**if** $e_k^l > \tau_e$ and $_kRb > \tau_{Rb}$ **then**
$\quad\quad w_s = Ag/Ag_{max}$
$\quad\quad w_t = 1.0 - w_s$
$\quad\quad _kRb \leftarrow \frac{(n-1)\times_kRb - w_s\times e_k^l}{n}$
$\quad\quad k_{max}^{l+1} = \underset{k^{l+1}}{\arg\max}\left(w_s a_s\left(_iP_{k^{l+1}}^{l+1}\right) + w_t a_t\left(_iP_{k^{l+1}}^{l+1}\right)\right)$
$\quad\quad _iP_{k_{max}^{l+1}}^{l+1}(k_l) \leftarrow (1+\eta)\times _iP_{k_{max}^{l+1}}^{l+1}(k_l)$
$\quad\quad$**for** $j=1:k_l$ **do**
$\quad\quad\quad _iP_{k_{max}^{l+1}}^{l+1}(j) \leftarrow \frac{_iP_{k_{max}^{l+1}}^{l+1}(j)}{\sum_{m=1}^{n^l} {}_iP_{k_{max}^{l+1}}^{l+1}(m)}$
$\quad\quad$**end for**
$\quad\quad MirrorTrain\left(_iP_{k_{max}^{l+1}}^{l+1}\right)$
$\quad$**end if**
end function

The DIAA algorithm first calculates the difference between the behavior and predictions only updates the parameters of the process to be learned if this difference is above some threshold and the partner is considered robust enough to learn from him/her/it. If there is a need for adaptation:

1. The system decreases the robustness of the partner ($_kRb$) based on the age (Ag).
2. The ID process in the process in the next layer most probably responsible of this error is calculated (k_{max}^{l+1}).
3. The probability distribution of this process in the next layer ($_iP_{k_{max}^l}^l$) is updated to make this error less likely in the future.
4. The mirror trainer is executed to make $_i^r P_{k_{max}^l}^l$ compatible with $_i^f P_{k_{max}^l}^l$.

11.4   Applications

Applying the developmental learning algorithms described in this chapter to learn interaction protocols require the availability of human–human interactions from which intentions and processes can be learned. It is possible in principle to use the EICA architecture for social robotics described in the Chaps. 9 and 10 without learning by just devising the intentions and processes by hand. The gaze controllers in Sect. 9.7 and guided navigation controller of Sect. 9.8 are examples of such approach.

Nevertheless, the full power of the architecture lies in its ability to *autonomously* learn these intentions and processes using the techniques introduced in this chapter. In this section we will report case studies of the applications of these learning algorithms in the same gaze control and guided navigation tasks. In both cases, we need to a data collection experiment to collect data for the learning algorithms followed by another evaluation experiment to measure the effectiveness of the approach. Notice that the gaze control case represents an implicit protocol that are best learned as a deep architecture using the algorithms in Sects. 11.2.3 and 11.3.2 while the guided navigation scenario represents an explicit simple protocol that can be learned using a shallow architecture using the algorithms described in Sects. 11.2.1 and 11.3.1.

11.4.1   Explanation Scenario

Mohammad and Nishida (2014) reported the use of the developmental system proposed in this chapter for learning gaze and spatial body control in the explanation scenario. Twenty two sessions of human–human interactions from the H^3R interaction corpus (Mohammad et al. 2008) were used as training sessions.

Head direction and body orientation of both partners was captured using PhaseSpace motion capture system and the following interaction dimensions were calculated from it using perceptual processes (Mohammad et al. 2010; Mohammad and Nishida 2014):

1. Three absolute angles of agent head (both robot and human) (θ_y for yaw, θ_p for pitch, θ_r for roll).
2. Head alignment angle (θ_h) defined as the angle between the line connecting the forehead and back of the head sensor of the agent and the line connecting back of the head sensor and the forehead sensor of the partner.
3. Distance between listener and instructor (d).
4. Body alignment angle (θ_b) defined similar to the head alignment angle.
5. Difference between center of body coordinates in the three spatial dimensions (X, Y and Z) (d_x, d_y, d_z).
6. Salient-object alignment angle (θ_s) defined as the angle between the line connecting back of the head sensor and forehead sensor and the line connecting back of the head sensor and the location of maximum saliency according to the gaze map maintained using the algorithm described in (Mohammad and Nishida 2010a).

The interaction babbling and interaction structure learning (using DISL) stages described in Sects. 11.1 and 11.2.3 were applied to these interaction dimensions.

The system learned the following intentions Mohammad and Nishida (2014):

1. *Look@Partner* which involved continuous reduction of the absolute value of θ_h.
2. *Look@Salient* which involved continuous reduction of θ_s.
3. *align2Partner* which involved continuous reduction of θ_b.
4. *disalign2Partner* which involved continuous increase of θ_b.
5. *nod* which involved oscillation up and down of θ_p.

The DISL algorithm was able to learn two higher layers of control for the listener that involved these five intentions and did not use the two noise generated intentions that were also learned in the first stage. The second layer of control consists of four processes that corresponded to gaze toward instructor, mutual gaze, mutual attention and a process that starts nodding when the instructor looks at the listener without speech for more than 10 s. The final layer of control consisted of a single process that starts gaze toward instructor and alternates between it and the other processes based on the focus of attention of the instructor (Mohammad and Nishida 2014).

Mohammad et al. (2010) compared the performance of this autonomously learned controller and the controller developed using the base platform using the FPGA algorithm (Sect. 9.7) and showed that both approaches outperformed the expectation of participants in terms of naturalness and human-likeness which suggests that the proposed autonomous approach was compared with a strong opponent. It also showed that the autonomously learned controller received higher scores compared with the carefully designed controller in terms of human-likeness (adjusted p-value $= 0.028$, $t = 2.97$ Hedge's $g = 0.613$) naturalness (adjusted p-value $= 0.013$, $t = 3.286$, Hedge's $g = 0.605$) and comfort of the speaker (p-value $= 0.003$, $t = 3.872$, Hedge's $g = 0.674$) (Mohammad et al. 2010).

11.4.2 Guided Navigation Scenario

In a series of experiments, we applied the developmental approach proposed in this chapter for learning actor's behavior during guided navigation described in Sect. 1.4 (Mohammad et al. 2009; Mohammad and Nishida 2010b, 2015).

Eighteen subjects were recruited for this experiment (10 males and 8 females) in 6 days. The goal of the experiment was to compare the performance of the learner robot in performing the actor role in guided navigation under three settings:

- WOZ: Wizard of OZ arrangement in which the robot is remotely controlled by a hidden human operator. The hidden operator watches the gestures of the subject and issues motion commands to the robot
- Per-Participant Learner: The robot controller is developed using the first two stages of our approach from a single interaction using interaction babbling (Sect. 11.1) followed by single layer interaction structure learning (Sect. 11.2.1) in the form of ABNs and then used as the actor with the same subject.
- Accumulating Learner: The robot controller is developed using the three stages where the same algorithms are used for the first two stages and the algorithm of Sect. 11.3.1 is used for combining the ABNs of all previous interactions except with the current subject and then tested with a subject it never encountered before.

Mohammad et al. (2009) showed that the Per-participant learner could achieve the same performance as the human operator after a single training session. Moreover, the performance of the accumulating learner could achieve the same level of performance as the per-participant learner (and the WOZ operator) by the fourth day slightly outperforming both of them by the last day.

These results suggest that the proposed developmental approach was successful in learning the interaction protocol in this case in as short as 15 min for the per-participant learner and was able to generalize this learning to interactions with other participants as shown by the incremental improvement of the performance of the accumulating learner.

11.5   Summary

This chapter introduced the three stages for building a complete EICA system representing an interaction protocol by mining human–human interactions that use that protocol. The first stage involved learning social intentions as functions of percepts based on a combination of change point discovery and constrained motif discovery algorithms. The second stage builds the rest of the control architecture by learning interaction control layers form the lowest to the highest. Three different variations were proposed: single-layer learner that learns a Bayesian Network representing the causal relations in an explicit interaction protocol in a single layer. The second alternative was interaction rule induction that learns a set of probabilistic control rules representing the protocol and a session controller for deciding when to activate each of them. Finally, an algorithm for learning arbitrary deep interaction protocols was proposed. The final stage of development in the life of the social robot which starts when a completed interaction protocol is learned using one of the three algorithms proposed for interaction structure learning. The robot engages—using its learned protocol—with humans and uses the discrepancy of its expectations driven from the simulations of their roles and their actual behavior to adapt the learned protocol. We proposed two alternatives depending on the depth of the interaction protocol learned in the second stage: a single-layer and a deep interaction adaptation algorithm.

The chapter also briefly described applications of these algorithms for learning the interaction protocols in our running scenarios: explanation and guided navigation. Evaluations of these applications showed the applicability of the proposed three-stages approach for developing social robots.

References

Das K, Lin I, Mannila H, Renganathan G, Smyth P (1998) Rule discovery from time series. In: KDD'98: the 4th international conference of knowledge discovery and data mining, AAAI Press, pp 16–22

Han J, Fu Y, Wang W, Chiang J, Gong W, Koperski K, Li D, Lu Y, Rajan A, Stefanovic N, Xia B, Zaiane OR (1996) DBMiner: a system for mining knowledge in large relational databases. In: KDD'09: the international conference on data mining and knowledge discovery, AAAI Press, pp 250–255

Hinton GE, Osindero S, Teh YW (2006) A fast learning algorithm for deep belief nets. Neural Comput 18(7):1527–1554

Hoover K (1990) The logic of causal inference. Econ Philos 6:207–234

Kanda T, Kamasima M, Imai M, Ono T, Sakamoto D, Ishiguro H, Anzai Y (2007) A humanoid robot that pretends to listen to route guidance from a human. Auton Robots 22(1):87–100

Kendon A (1970) Movement coordination in social interaction: Some examples considered. Acta Pyschologica 32:101–125

Keogh E, Chu S, Hart D, Pazzani M (2001) An online algorithm for segmenting time series. In: ICDM'01: IEEE international conference on data mining, pp 289–296. doi:10.1109/ICDM.2001. 989531

Mohammad Y, Nishida T (2008) Toward agents that can learn nonverbal interactive behavior. In: IAPR workshop on cognitive information processing, pp 164–169

Mohammad Y, Nishida T (2010a) Controlling gaze with an embodied interactive control architecture. Appl Intell 32:148–163

Mohammad Y, Nishida T (2010b) Learning interaction protocols using augmented baysian networks applied to guided navigation. In: IROS'10: IEEE/RSJ international conference on intelligent robots and systems, IEEE, pp 4119–4126. doi:10.1109/IROS.2010.5651719

Mohammad Y, Nishida T (2014) Learning where to look: autonomous development of gaze behavior for natural human-robot interaction. Interac Stud 14(3):419–450

Mohammad Y, Nishida T (2015) Learning interaction protocols by mimicking: understanding and reproducing human interactive behavior. Pattern Recogn Lett 66(15):62–70

Mohammad Y, Nishida T, Okada S (2009) Unsupervised simultaneous learning of gestures, actions and their associations for human–robot interaction. In: IROS'09: IEEE/RSJ international conference on intelligent robots and systems, IEEE Press, NJ, IROS'09, pp 2537–2544

Mohammad Y, Nishida T, Okada S (2010) Autonomous development of gaze control for natural human–robot interaction. In: The IUI 2010 workshop on eye gaze in intelligent human machine interaction, IUI'10, pp 63–70

Mohammad Y, Xu Y, Matsumura K, Nishida T (2008) The H^3R explanation corpus:human-human and base human–robot interaction dataset. In: ISSNIP'08: the 4th international conference on intelligent sensors, sensor networks and information processing, pp 201–206

Sugar CA, James GM (2003) Finding the number of clusters in a dataset: an information-theoretic approach. J Am Stat Assoc 98:750–763

Chapter 12
Fluid Imitation

Chapter 13 will review several algorithms for learning from demonstration ranging from inverse optimal control to symbolic modeling. What all of these algorithms share is the assumption that demonstrations are segmented from the continuous behavioral stream of the model (i.e. the demonstrator). Chapter 7 discussed how infants learn and one thing that is clear is that they do not need this pre-segmented demonstrations to acquire a model of the behaviors they perceive. They seem to be able to imitate even when there are no clear signs about what should be imitated and they are able to produce their newly learned skills at appropriate occasions (at least most of the time). This form of fluid imitation has several advantages over standard learning from demonstration. Firstly, it frees the model from having to clearly indicate the beginnings and ends of the behavior to be demonstrated which leads to more intuitive interaction. Secondly, by allowing the model to behave normally, the learner can see the behaviors as they are actually performed in the real world including how they are eased into and eased out from to other behaviors. This can be useful in discovering not only the basic behaviors to be learned but their relations and appropriate execution times. Finally, this mode of fluid imitation allows the learner to learn from unintended demonstrations. Every interaction (even with unwilling partners) becomes an opportunity for learning. For all of these advantages, we believe that fluid imitation will be a key technology in creating the adaptive social robots of tomorrow.

This chapter introduces our efforts towards realizing a fluid imitation engine to augment traditional learning from demonstrations engines (of the kind discussed in Chap. 13). The proposed engine casts the problem as a well-defined constrained motif discovery problem (See Sect. 4.6) subject to constraints that are driven from object and behavior saliency, as well as behavior relevance to the learner's goals and abilities. Relation between perceived behaviors of the demonstrator and changes in objects in the environment is quantified using a change-causality test (Chap. 5) that is shown to provide better results compared to traditional g-causality tests. The main

© Springer International Publishing Switzerland 2015
Y. Mohammad and T. Nishida, *Data Mining for Social Robotics*,
Advanced Information and Knowledge Processing,
DOI 10.1007/978-3-319-25232-2_12

advantage of the proposed system is that it can naturally combine information from all available sources including low-level saliency measures and high-level goal-driven relevance constraints. The chapter also reports a series of experiments to evaluate the utility of the proposed engine in learning navigation tasks with increasing complexity using both simulated and real world robots.

12.1   Introduction

Learning from demonstration is an important research area in robotics (Aleotti and Caselli 2008; Argall et al. 2009; Abbeel et al. 2010) because it allows the robot to acquire new skills without explicit programming. There are two main directions in robotic imitation research. The first direction tries to utilize imitation as an easy way to *program* robots without explicit programming (Nagai 2005). This use usually goes by other names like *learning from demonstration* (Billing 2010), *programming by demonstration* (Aleotti and Caselli 2008) and *apprenticeship learning* (Abbeel et al. 2010). Researchers here focus on task learning and some examples of the results of such work were reported in Chap. 13. The second direction tries to use imitation to bootstrap social learning by providing a basis for mutual attention and social feedback (Nagai 2005; Iacoboni 2009). Researchers here focus on interaction learning and some results from this direction of research are reported in Chap. 7.

We can say that, roughly, in the first case, imitation is treated as a *programming mode* while in the second, it is treated as a *social phenomenon*. In this chapter, we focus on the task-learning aspect but we try to extend it to allow robots to learn from unaware teachers and from continuous streams of data without predefined action boundaries. As discussed in Chap. 7, in some animals, including humans, imitation is a social phenomenon (Nagai 2005) that was studied intensively by ethologists and developmental psychologists. Social psychology studies have demonstrated that imitation and mimicry are pervasive, automatic, and facilitate empathy. Neuroscience investigations have demonstrated physiological mechanisms of mirroring at single-cell and neural-system levels that support the cognitive and social psychology constructs (Iacoboni 2009). Neural mirroring and imitation solves the "problem of other minds" and makes inter-subjectivity possible, thus facilitating social behavior. The ideomotor framework of human actions assumes a common representational format for action and perception that facilitates imitation (Iacoboni 2009). Furthermore, the associative sequence learning model of imitation proposes that experience-based Hebbian learning forms links between sensory processing of the actions of others and motor plans (Iacoboni 2009).

For a robot to be able to learn from a demonstration, it must solve many problems. Most important of these problems are the following seven challenges (Mohammad and Nishida 2012) discussed in details in Chap. 7:

- *Action Segmentation*: Where are the boundaries of different *elementary behaviors* in the perceived motion stream of the demonstrator?
- *Behavior Significance for Imitation*: What are the interesting behaviors and features of behavior that should be imitated? This combines the *what* and *who* problems identified by Nehaniv and Dautenhahn (1998).
- *Perspective Taking*: How is the situation perceived in the eyes (or sensors) of the demonstrator?
- *Demonstrator modeling*: What are the primitive actions (or actuation commands) that the demonstrator is executing to achieve this behavior? What is the relation between these actions and the sensory input of the demonstrator?
- *Correspondence Problem*: How can actions and motions of the demonstrator be mapped to the learner's body and frame of reference?
- *Evaluation Problem*: How can the learner know that it succeeded in imitating the demonstrator and how to measure the quality of the imitation in order to improve it? This evaluation would usually require feedback from the demonstrator or other agents and can utilize social cues (Scassellati 1999).
- *Quality Improvement Problem*: How can the learner improve the quality of its imitative behavior over time either by adapting to new situations or by modifying learned motions to better represent the underlying goals and intentions of perceived demonstrations?

Most of the research in imitation learning has focused on the perspective taking, demonstrator modeling and the correspondence problems above (Argall et al. 2009). In most cases, the action segmentation problem was ignored and it is assumed that the demonstrator (*teacher*) will provide segmented examples of the action or behavior to be learned. In this chapter, we rely on constrained motif discovery to solve this problem (See Sect. 4.6).

Behavior significance for imitation is also usually ignored under the assumption that the demonstrator will only demonstrate *significant* or important behavior to the learner. This may be the case in laboratory-controlled experiments used to test LfD schemes, yet it is far from being true for realistic situations in which robots are sharing our spaces and looking to learn from our actions. It is certainly not true for how humans learn.

There is a distinction between two aspects of the behavior significance problem. The first aspect is deciding the importance of a specific feature of the behavior perceived (e.g. a specific limb motion) for the action being learned. This problem can usually be handled well with some LfD systems like GMM/GMR and SAXImitate (See Chap. 13). The problem that is usually ignored is the second aspect of behavior significance. Whether or not the *whole* behavior is worth imitating or learning.

There are many factors that affect the significance of a behavior for the learner. Some of these are behavior intrinsic features that may make it interesting (e.g. novelty, and repetition). Others are object intrinsic features related to the objects affected by the behavior (e.g. color, motion pattern) that can make that behavior interesting. These two kinds of features determine what we call the *saliency* of the behavior and its calculation is clearly bottom-up.

Another factor that affects the significance of behavior is its relevance to the goal(s) of the agent executing it. This factor is called *relevance* of the behavior and its calculation is clearly top-down.

The context at which the behavior is executed can also affect its significance for imitation. This factor is dubbed *sensory context* of the behavior.

A final factor that affects significance is the capabilities of the learner. It makes little sense to try to imitate a behavior that is clearly outside the capabilities of the learner.

A solution to the significance problem needs to smoothly combine all of these factors taking into account the fact that not all of them will be available all the time. The ability to solve the action segmentation and significance problems allows the learner to learn not only from explicit demonstrations but from watching the behavior of other agents around it and their interactions. We call this kind of imitative learning fluid imitation and believe that it provides an important building block for achieving real *autonomous sociality*.

The problem of fluid imitation can be defined as (Mohammad and Nishida 2012): *Given a continuous stream of actions from the demonstrator, find the boundaries of significant behaviors for imitation.*

12.2   Example Scenarios

It is useful to have concrete running examples to introduce our fluid imitation system. In this chapter, we use two running examples: navigation and cooking. By the end of the chapter, we will report briefly on an experimental evaluation of the first scenario drawn from Mohammad and Nishida (2012).

Navigation learning scenario happens in a 2D space filled with different kinds of objects. One agent (the model) knows how to navigate this environment and react to different objects in it and another agent (the learner) is a naive agent who just watches the model and tries to learn how to navigate and react in this environment. Different objects will elicit different kinds of behaviors from the model. For example, when it perceives a cylindrical obstacle it rotates around it in a circle while it will rotate around any robot in the environment in a square. It may also spontaneously generate some motion primitives like triangles. Moreover, the model may have a goal and tries to achieve it by navigational motions. For example, it may know that rotating twice around another robot will disable this robot and its goal can be to disable all robots in the arena.

This kind of environment provides a simplified toy example for the fluid imitation problem by allowing us to selectively select the level of complexity of the factors affecting behavior significance for imitation.

The second scenario is more realistic and involves a humanoid robot watching people cocking different meals and learning through fluid imitation various actions related to cooking. This scenario raises some issues regarding what to do with learned behaviors and how to learn complex plans of them (See the introduction of Chap. 13) that will be discussed later in this chapter.

12.3 The Fluid Imitation Engine (FIE)

Figure 12.1 shows the main components of the fluid imitation engine (FIE). There are four main components: perspective taking, significance estimator, self initiation engine and the imitation engine.

Perspective taking is needed to convert perceived environmental state and model behavior into the frame of reference of the model then mapping the whole thing into the frame of reference of the learner. This gives the learner an idea about how would the environment appear from its point of view *if it were in the shoes of the model* and how are the motions of the model transformed into its own actuation space. This component is task specific, yet we will consider it in two difference contexts in Sect. 12.4. This component generates a transformed version of environmental state and model's behavior that is passed to the self initiation engine.

The second basic component is the significance estimator which calculates the significance of perceived motions/behaviors based on their top-down relevance to the current set of goals of the learner and bottom-up saliency of the behavior or environmental context and objects involved in it. We will discuss the component in greater details in Sect. 12.5.

The self initiation engine is the main focus of this chapter and we will elaborate on it further later. Briefly, it receives the transformed environmental and behavior signals and discovers in the behavior signals significant segmented actions that are then passed to the imitation engine. We will discuss it in details in Sect. 12.6.

The final component shown in Fig. 12.1 is the imitation engine which simply receives segmented action demonstrations and learns the corresponding action models. This engine can rely on any LfD system from the ones introduced in Chap. 13. In our implementations we use SAXImitate (Sect. 13.4) in combination with GMM/GMR (Sect. 13.3.2).

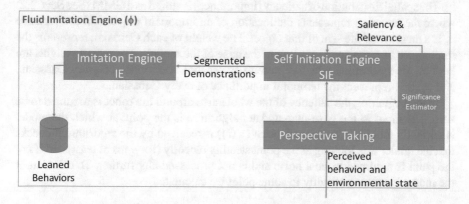

Fig. 12.1 Overview of the proposed fluid imitation engine and its main components

12.4   Perspective Taking

The perspective taking module has two goals. The first is to transform the environmental state into streams representing the objects in the environment to be useful for significance estimation and model-relative perceptions. The second is to transform the behavior of the model to the embodiment of the learner (in effect solving the correspondence problem).

12.4.1   Transforming Environmental State

To achieve the first goal, an object detection module needs to be implemented to discover objects in the environment that may be affected by the behavior of the model or that can affect that behavior. Object recognition is a traditional research subject in machine vision and many mature systems exist that can be directly employed for implementing this component when the input is provided through cameras (Belongie et al. 2002; Bo et al. 2013; Guo et al. 2014). For the navigation scenario, the learner and model both have infrared sensors for detecting nearby objects that can be used as the basis for this component. For the cooking scenario, either environmental mounted cameras or robot-mounted cameras can be used.

One interesting approach to object discovery that requires no prior knowledge of the objects involved and utilizes social cues is the use of gaze maps (Mohammad and Nishida 2010). A gaze map is a representation of the saliency of different points in space calculated based on the gazing direction of the model superimposed on an environmental map. The main advantage of a gaze-map in our context is that it provides built-in saliency calculation by focusing on the objects attended to by the model which simplifies the job of the significance estimator.

The spatial distribution of saliency (importance) is stored as a GMM (See Sect. 2.2) where the mean μ_i represents the location of an important object and the variance σ_i is a measure of the size of that object. The weight of each Gaussian represents the importance of the place according to the gaze of the model. Two sets of weights are kept by the system: w_i which represents the spatial importance of every Gaussian, and w_{r_i} that represent the temporal importance of every Gaussian.

In the beginning the saliency of the whole area around the robot is assumed to be zero. The input to the gaze-map updater algorithm is the point at which the model is currently looking. Once a new point $(g\,(i))$ is received by the algorithm it checks that the model was looking at the point steadily recently (for ε ms at least) and if not the point is assumed to be a noise and is not processed any further. To calculate if the model is looking steadily to some point we calculate:

$$\left\| g\,(i) - \sum_{j=0}^{i-1} (j+1)\,g\,(j) / \sum_{j=0}^{i-1} (j+1) \right\|.$$

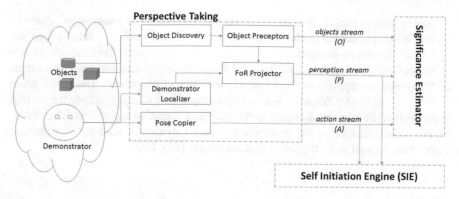

Fig. 12.2 The building blocks for perspective taking showing its internal structure and relation to other components of the fluid imitation engine

This formula calculates the difference between the instantaneous gaze estimation and the gaze direction estimated as a weighted average of previous instantaneous gaze directions.

Now if $g(i)$ is not near to any one of the available Gaussians ($\{G\}$) then a new Gaussian is added at this point otherwise it is combined with the Gaussian component nearest to it updating both the mean and variance of this component. For the detailed algorithm of this process refer to (Mohammad and Nishida 2010). The weights of the component in the spatial and time domains (w_i, w_{i_r} respectively) are then updated to increase the weight of the area near the new point.

w_i is increased proportional to the distance between the new point and the center of the winning Gaussian under the assumption that the model will usually tend to look near the *center* of the object (s)he is interested in at any point of time. w_i is continuously decreasing at a constant rate to implement forgetting into the system. w_{t_i} is calculated as the reciprocal of the summation of the time of last access to every Gaussian in the mixture and a fixed term ε_t. By controlling this parameter it is possible to control how *forgetful* is saliency calculation. More details and evaluations of the accuracy of this technique are given by Mohammad and Nishida (2010).

Discovering objects is just the first step in the operation of environmental state transformation. A set of object perception modules are then implemented that capture different features of the objects perceived including their states, locations, geometrical relations, etc. These are depicted in Fig. 12.2 as *object preceptors* and they are application dependent. For example, consider the navigation scenario. Important features of objects may be their type (cylinder, cube, another robot, etc.) and distance from the learner. For the cooking example, these may include different *accordances* provided by the objects and whether they were moved compared with the last evaluation cycle, etc. The output of object preceptors are combined together to form a multidimensional stream called the *objects stream* (O) hereafter.

The *objects stream* represents important features of the environment as perceived by the learner but not the model. They will only be used for saliency calculation (See Fig. 12.2). To imitate the model, the learner needs to put itself in the its shoes. This is achieved through a Frame of Reference (FoR) projector that transforms all the outputs of the object perceptions to the frame of reference of the model. The justification of this process comes from the *like-me* hypothesis (See Chap. 7) where the learner assumes that the model has similar perceptual and cognitive repertoires. The results of this transformation is the *perception stream* (P) that is used for significance calculation as well as by the self-initiation engine (Fig. 12.2).

12.4.2   Calculating Correspondence Mapping

The second task of the perspective taking module is to map the behavior of the model to the learner's actuation space. This is known as the correspondence problem in literature (Nehaniv and Dautenhahn 2001).

The same name is used in two different senses in robotics and developmental psychology. In robotics, it is usually used to indicate the problem faced by the *designer* in matching the actions of the model agent (usually a human) and a specific robot that may not have the same body form (i.e. a non-humanoid) or may share the general form of the model (i.e. a humanoid) but with different relative lengths of limbs and body compared with the model (Nehaniv and Dautenhahn 1998). The focus here is on different embodiments. In developmental psychology, the correspondence problem refers to the problem faced by the infant in determining the correspondence between the motions it perceives from other people and its own repertoire of motion. Here the focus is not in the difference in embodiment but in autonomously learning the correspondence between body parts and motion of these parts and their counterparts in the infant's own body (Meltzoff 2005). Chapter 7 discussed briefly the problem from the viewpoint of developmental psychology and this section will focus on the problem from the viewpoint of robotics.

The difficulty of the correspondence problem depends on the type of behavior being learned and the degree of dissimilarity between the embodiments of the model and learner. For example, learning words from spoken language of people (a possible application of fluid imitation), requires no correspondence mapping as the robot is assumed to be able to generate the same vocalizations as the model and there is no spatial viewpoint to map. For the navigation scenario, the learner and model are both having the same embodiment and this simplifies the problem to become trivial copying of behavior. For the cooking scenario, things are more complicated as the robot—even being a humanoid—will in general have different limb length ratios and body form from the human model. More complex situations can be envisioned in which the learner has completely different body form from the model and in this case it is usually possible to achieve emulation (e.g. copying of goals) instead of imitation (See Sect. 7.1 for the difference between imitation and emulation).

We will focus on the situation when the model is a human, the learner is a humanoid robot with rigid body parts and the motion to be copied is arm motion. This problem is solved enough to give closed form solutions, yet realistic enough to be of practical value (e.g. in the cooking scenario).

The following terminology will be used (Mohammad and Nishida 2013; Mohammad et al. 2013): U is the length of the upper arm and L is the lower arm length, A_i^j stands for the point-vector A_i expressed in frame j. T_i^j stands for a homogeneous transformation matrix converting vectors in frame i to corresponding vectors in frame j. R_i^j is the corresponding rotational matrix and D_i^j is the origin of frame i in frame j. T, S, E, W, and H (case insensitive) stand for the torso, shoulder, elbow, wrist and hand of the learner respectively, while $\hat{T}$, $\hat{S}$, $\hat{E}$, $\hat{W}$, and $\hat{H}$ (case insensitive) stand for the torso, shoulder, elbow, wrist and hand of the model. A represents configuration dependent vectors and Λ represents vectors fixed by design.

Our goal here is to copy the limb configuration of the human body. This can be achieved by preserving the unit vectors pointing from the shoulder to the elbow and from the elbow to the wrist. This can be formalized as: Let $\hat{a}_i^{i-1} = A_i^{i-1}/\|A_i^{i-1}\|$ be a unit vector in the direction of joint i as defined in the frame attached to the previous joint $(i-1)$ for the three upper-body kinematic chains of a humanoid robot. Given $(\Lambda_s^{\hat{i}}, A_e^{\hat{i}}, A_w^{\hat{i}})$, find the set of all possible vectors of joint angles for the robot that preserve $\hat{a}_i^{i-1}$ for all joints.

Pose copying can be achieved in two stages. The first stage (called retargeting) converts all joint locations of the model from the model's own torso frame to the learner's torso frame. This is a simple homogeneous transformation that can be achieved using:

$$A_e^t = \Lambda_s^t + \frac{U}{\left\| A_e^{\hat{i}} - A_s^{\hat{i}} \right\|} \left(A_e^{\hat{i}} - A_s^{\hat{i}} \right), \tag{12.1}$$

$$A_h^t = A_e^t + \frac{L}{\left\| A_h^{\hat{i}} - A_e^{\hat{i}} \right\|} \left(A_h^{\hat{i}} - A_e^{\hat{i}} \right). \tag{12.2}$$

After this transformation is applied, we need to solve a problem similar to the standard inverse kinematics problem but rather than being given the homogeneous transformation of the end effector (the wrist in our case), we are given the elbow and wrist positions. Reliance on positions only has two justifications. Firstly, the position estimates of most off-body motion capture systems are more accurate than orientation estimates. Secondly, the standard inverse kinematics solutions may give multiple solutions if the problem is underspecified and in this case, multiple possible elbow positions can be generated and only one of them will correspond to the sought after location corresponding to the demonstration.

The two vectors found from the retargeting step provide 4 degrees of freedom (they are normalized) and they can be used to calculate 4 joint angles. Most humanoids available today have one of two general configuration of joints in the hands: either a 3DoF spherical shoulder with a single DoF at the elbow (similar to the human body)

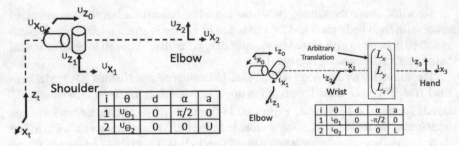

Fig. 12.3 Kinematics models used for pose copying. Reproduced with permission from Mohammad et al. (2013). **a** Upper arm model. **b** Lower arm model

or a 2DoF pan-tilt shoulder with another 2DoFs at the elbow. The method developed by Mohammad et al. (2013) can be used with both of these types of robots.

Under fairly general assumptions the two common configurations of the arm discussed above will have the same forward kinematics which means that a general solution can be devised for solving the pose copying problem for both of them.

The main idea behind our is to divide the problem of arm pose copying into three steps: elbow positioning, wrist positioning and hand inverse orientation.

Hand inverse orientation can be achieved using classical inverse kinematics (Paul et al. 1981) and is not considered a part of the pose copying problem.

Elbow positions requires a model of the upper arm and wrist positioning requires a model of the lower arm. An appropriate transformation will also be needed to convert the output from the elbow positioning step to the same frame of reference required for the wrist positioning step.

Using standard kinematics reasoning from the upper arm model shown in Fig. 12.3a, it can be shown that:

$$\theta_1^U = \tan^{-1}\left(\frac{A_e^s(y)}{A_e^s(x)}\right),$$

$$\theta_2^U = \tan^{-1}\left(\frac{A_e^s(z)}{\sqrt{A_e^s(x)^2 + X_e^s(y)^2}}\right),$$

where θ_1^U and θ_2^U are the first DoFs of the upper arm model that are shared between the two possible arm configurations considered here (spherical and pan-tilt shoulder types).

To solve the wrist orientation problem, we need to convert the wrist location to the frame of reference of the elbow. This is achieved in two steps. Firstly, we use the same upper model used above for projection using:

$$^U A_w^e = \left(T_e^s\right)^{-1} A_w^s. \tag{12.3}$$

This transformation is done using the frame of reference associated with the upper model and needs to be mapped to the lower model. This mapping depends on the robot and whether it has pan-tilt or spherical shoulder configurations. This mapping for the spherical should type is:

$$T_L^U = \begin{bmatrix} 0 & 0 & 1 & -U \\ 1 & 0 & 0 & 0 \\ 0 & 1 & 0 & 0 \\ 0 & 0 & 0 & 1 \end{bmatrix}.$$

In the pan-tilt-shoulder case, it becomes

$$T_L^U = \begin{bmatrix} 0 & 0 & 1 & 0 \\ 1 & 0 & 0 & 0 \\ 0 & 1 & 0 & 0 \\ 0 & 0 & 0 & 1 \end{bmatrix}.$$

Given this mapping we can find the wrist location in the elbow frame of reference using:

$$A_w^e = T_L^U * {}^U A_w^e. \tag{12.4}$$

Using the lower arm model of Fig. 12.3b and assuming that L_y and L_z are not zero, we can show that

$$\begin{aligned} A_w^e(x) &= Lc_2 - L_y s_2, \\ A_w^e(y) &= L_y c_1 c_2 - L_z s_1 + Lc_1 s_2, \\ A_w^e(z) &= L_z c_1 + L_y c_2 s_1 + Ls_1 s_2, \end{aligned} \tag{12.5}$$

where $c_i = cos(\theta_i^L)$ and $s_i = sin(\theta_i^L)$.

Few intermediate variables need to be defined:

$$\begin{aligned} \gamma &= A_w^e(y)^2 + A_w^e(z)^2 - L^2 - L_z^2, \\ \alpha &= Lc_2, \\ \beta &= L_y c_2, \\ a &= L_y c_2 + Ls_2, \\ b &= L_z. \end{aligned}$$

After some manipulations, it is possible to prove the following relations.

$$\begin{aligned} c_2 &= \frac{A_w^e(x)L \pm \sqrt{A_w^e(x)^2 L^2 + \gamma(L^2 + L_y^2)}}{L^2 + L_y^2}, \\ s_2 &= \frac{Lc_2 - A_w^e(x)}{L_y}, \\ c_1 &= \frac{a A_w^e(y) + b A_w^e(z)}{a^2 + b^2}, \\ s_1 &= \frac{A_w^e(z) - bc_1}{a} = \frac{ac_1 - A_w^e(y)}{b}. \end{aligned} \tag{12.6}$$

Finally the 2DoFs of the lower model can be found as:

$$\theta_i^L = \tan^{-1}\left(s_i/c_i\right). \tag{12.7}$$

The cases when L_y or L_z is zero are handled as special cases using the same approach but result in slightly different equations. Due to lack of space the details of these cases are not presented here.

The final step in the solution for elbow and wrist positioning problem is to convert the angles θ_i^L and θ_i^U to corresponding angles in the forward full model (θ_i). Mohammad et al. (2013) have shown that this final transformation can be found as:

$$\theta_1 = \theta_1^U, \theta_2 = \theta_2^U, \theta_3 = \theta_1^L, \theta_4 = \theta_2^L + \pi/2. \tag{12.8}$$

The output of the pose copier (See Fig. 12.2) will constitute the action stream (A) that is then passed to the significance estimator and self-initiation engine (SIE).

12.5   Significance Estimator

The significance estimator receives three continuous streams of data from the perspective taking module (A, O, P) and generates estimations of the significance of behaviors in the A stream without segmenting them.

Figure 12.4 shows the building components of the significance estimator. There are two independent computational stacks. The first is a top-down stack for calculating the relevance of perceived behavior to the goals of the agent. The second is a bottom-

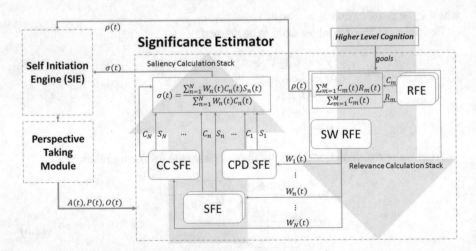

Fig. 12.4 The building blocks for the significance estimator showing its internal structure and relation to other components of the fluid imitation engine. All rounded rectangles receive a copy of A, P, O but the arrows are not shown to reduce clutter

up stack for calculating the saliency of the behavior or associated objects in the perceptual stream P. This combination of bottom-up and top-down processing is one common feature between the fluid imitation engine and L_iEICA (See Chap. 10).

Saliency calculation processes $\{A(t), O(t), P(t)\}$ streams by applying a set of N saliency feature extractors that receive A and P and generate an output saliency score $S_n(t)$ ranging from zero to one and a confidence score $C_n(t)$ of the same range. All saliency feature extractors run in parallel.

Currently, we implemented few saliency feature extractors (SFEs) but the architecture of FIE allows other saliency feature extractors to be plugged into the saliency evaluation stack directly.

The simplest SFE is designed to discover beginning and ending of motion in all inputs that are marked as position data. It simply calculates the difference of each input channel from its previous value (d). When $S_{t-1} = 0$ (i.e. when no saliency was announced), it outputs a linearly increasing value from 0 to 1 in τ steps when this difference d is greater a global threshold ε and then saturates at the value 1. When $S_{t-1} > 0$, then the output linearly decreases to 0 in S_{t-1}/τ steps when $d \leq \varepsilon$.

Another slightly more complicated SFE (CPD SFE in Fig. 12.4) applies a change point discovery algorithm to every dimension of all its inputs. In our current implementations we use a variant of SSA based CPD introduced in Sect. 3.5.

The main assumption we make here is that *if it is changing, then it is salient*. We assume that this is true both for behavior and environmental state. For the perceptual stream P, it is clear that objects that do not change their state, are not very interesting for the learner (at least in themselves). For the action stream A, we base our hypothesis onto two points. Firstly, from the computational point of view, what we are searching for are important behaviors that may appear recurrently in the stream (otherwise, there is no need to learn them) and this means that they should be different from whatever follows and precedes them leading to change points around them. Secondly, it is well-known that care-givers use very specific types of motion when teaching their children that they do not use when interacting with adults. These special motion patterns (called motionese (Nagai and Rohlfing 2007)) where shown to increase the saliency of objects and behavior by moving the first and exaggerating the later (Nagai and Rohlfing 2007).

A third SFE that is always available is the Change Causality SFE (CC SFE in Fig. 12.4). This SFE calculates the change causality score between dimensions in A and both O and P using the system proposed in Sect. 5.5. The main idea here is that behaviors of the model that *cause* changes in the environmental state should be considered important and targeted for possible imitation.

Other notions of saliency that may be application and context dependent can be added easily. For example, if we have a gaze-map (See Sect. 12.4), then an SFE for extracting changes in gaze direction or points of sustained gaze can easily be added.

The SFEs output $2N$ values at every time-step $\{S_1, S_2, \ldots, S_N, C_1, C_2, \ldots, C_N\}$. Each SFE has an associated weight $W_n(t)$ that specifies how important is this SFE for the discovery of motions to be imitated. The weights are calculates through the relevance calculation stack as will be explained later. The final saliency score is then calculated as:

$$\sigma(t) = \frac{\sum_{n=1}^{N} W_n(t) C_n(t) S_n(t)}{\sum_{n=1}^{N} W_n(t) C_n(t)}. \tag{12.9}$$

The relevance calculation stack uses the goals provided by higher level cognition modules of the robot (e.g. a BDI agent) to calculate a relevance level for behaviors and environmental state changes found in A, O, and P. The stack consists of a set of M relevance feature extractors (RFEs) that generate relevance R_m and confidence scores C_m in the same way that SFEs work. These scores are combined linearly similar to Eq. 12.9 to generate the final relevance estimate $\rho(t)$.

A special RFE is the Saliency Weighing RFE (SW SFE in Fig. 12.4) which calculates the weights of SFEs. In the current implementation, we add a weight for objects in O and P that decays exponentially with the distance between the object and the model under the assumption that only objects accessible to the model are interesting. The weights for model's behavior (A) is fixed to 1 for upper-body joints and 0 for lower body joints because the current imitation engine cannot learn lower-body motions and keep the balance of the robot in the same time. This limitation can be removed in the future and the SW RFE modified accordingly.

The output of the significance estimator is the tuple $\{\sigma, \rho\}$ that is then passed to the self initiation engine.

12.6 Self Initiation Engine

The self initiation engine is the core of FIE. It receives the outputs of the perspective taking module (the action stream A, the perception stream P, and the objects stream O) and significance measures (saliency σ and relevance ρ) from the significance estimator (Sects. 12.4 and 12.5) and uses them to segment the action stream A (or the perception stream P if object-directed motions are to be learned). Segmented motions are then combined into sets that are fed to the imitation engine.

The self initiation engine in the current implementation consists of a constrained motif discovery algorithm that uses the significance and relevance measures to bias the search for motifs (See Sect. 4.6).

12.7 Application to the Navigation Scenario

In this chapter we report a series of proof-of-concept experiments with increasing complexity aiming at showing the applicability of the proposed system to the navigation scenario that were first reported by Mohammad and Nishida (2010). The task used for all of the experiments involves robot navigation (See Sect. 12.2). All simulations were done using the V-REP simulator with realistic physics based on the Bullet Physics engine. A simulated e-puck robot was used as a model and a learner in all experiments reported in this section.

In the first experiment, the simulator moved randomly in the arena executing one of three geometrical shape randomly every while (a triangle, a circle and a square). In this case, the significance estimator could only report change points in the motion behavior which rendered the SIE a simple application of constrained motif discovery as described in Sect. 4.6.

The *perspective taking* module received a 3D time-series representing the location and orientation of the model over time $(x\,(t), y\,(t), \theta\,(t))$ and outputs a 1D time series $(\hat{l}\,(t))$ representing the location of the model in the arena.

The pose copying module simply applied inverse kinematics and there was no need to solve the correspondence problem as the learner and model had the same exact embodiment. Applying this inverse kinematics to $(\hat{x}\,(t), \hat{y}\,(t), \hat{\theta}\,(t))$ to generate the corresponding motor commands $(\hat{m}_l\,(t), \hat{m}_r\,(t))$.

The imitation engine simply calculated the mean of all demonstrations reported to belong to the same motion.

Mohammad and Nishida (2010) used the log of a single session (of 120–400 seconds each) sampled at 20 Hz as the input to our system containing four occurrences of each of the three behaviors. The experiment was repeated 100 times and for each of them artificial Gaussian noises of zero mean and a standard deviation of zero up to the maximum value of each dimension was added (generating 8 trials for each experiment). This gives a total of 800 trials.

The average occurrence accuracy of the system over the 800 trials was 93.8% (93.18% for circles, 94.72% for squares, and 93.51% for triangles) (Mohammad and Nishida 2010).

The first experiment tested mostly the solution to the action segmentation problem as significance calculation was trivial. Nevertheless, even in this case, significance calculation using CPD proved effective in reducing the computational cost by allowing the motif discovery algorithm at the heart of the SIE to focus on important parts of the input.

To test the applicability of the Fluid Imitation Engine for solving the significance estimation problem (alongside action segmentation), we need to have different kinds of objects with different associated behaviors. Mohammad and Nishida (2010) used two kinds of objects: Cylinders and K-Junior robots. The model rotated in a square around every cylinder it encountered and in a circle around every K-Junior robot.

In this second experiment, significance calculation was completely determined from the bottom-up. In other words, relevance was not relevant, only saliency mattered. We used the log of 1000 trials similar to the ones used for the first experiment. The learner, learned only circle and square primitives as expected and with detection mean occurrence accuracy of 93.95% (93.18% for circles, 94.72% for squares) This suggests that the causality estimator was able to remove the unnecessary triangle pattern from consideration (Mohammad and Nishida 2010).

The last experiment reported by Mohammad and Nishida (2010) in the simulated environment involved making the K-Junior robots rotate continuously in place until the model rotates around them and giving the learner the *goal* of stopping all K-Junior robots.

This case tested the relevance component of the system and we used the log of 1000 trials similar to the ones used in the previous two experiments. The learner, learned only circle primitive as expected and with detection mean accuracy of 93.18 % (Mohammad and Nishida 2010).

12.8   Summary

This chapter introduced our fluid imitation engine. FIE is a first step toward increasing the fluency of robotic imitation by making it more natural. The main difference between fluid imitation and traditional learning from demonstration is that the action streams perceived from the model is not segmented. The model is not even fixed as the robot can learn from any human (or other robot). The core step added to standard LfD approaches is a solution to the action segmentation challenge. We used a combination of change point discovery, motif discovery and causality analysis to segment interesting motions from the continuous streams of actions perceived by the robot. Interesting actions in this case are those that are repeated or that happen around major changes in environmental state related to the robot goals. The chapter also reported an application of the engine by a miniature robot learning to navigate an arena and execute appropriate navigation patterns of actions around different kinds of objects.

References

Abbeel P, Coates A, Ng AY (2010) Autonomous helicopter aerobatics through apprenticeship learning. The Int J Robot Res 29(13):1608–1639

Aleotti J, Caselli S (2008) Grasp programming by demonstration: a task-based quality measure. In: RO-MAN'08: IEEE international symposium on robot and human interactive communication, IEEE, pp 383–388

Argall BD, Chernova S, Veloso M, Browning B (2009) A survey of robot learning from demonstration. Robot Auton Syst 57(5):469–483

Belongie S, Malik J, Puzicha J (2002) Shape matching and object recognition using shape contexts. IEEE Trans Pattern Anal Mach Intell 24(4):509–522

Billing E (2010) A formalism for learning from demonstration. Paladyn 1(1):73–102. doi:10.2478/s13230-010-0001-5

Bo L, Ren X, Fox D (2013) Unsupervised feature learning for RGB-D based object recognition. In: Experimental robotics. Springer, pp 387–402

Guo Y, Bennamoun M, Sohel F, Lu M, Wan J (2014) 3D object recognition in cluttered scenes with local surface features: a survey. IEEE Trans Pattern Anal Mach Intell 36(11):2270–2287

Iacoboni M (2009) Imitation, empathy, and mirror neurons. Annu Rev Psychol 60:653–670. doi:10.1146/annurev.psych.60.110707.163604

Meltzoff AN (2005) Imitation and other minds: the "like me" hypothesis. Perspect Imitation Neurosci Soc Sci 2:55–77

Mohammad Y, Nishida T (2010) Controlling gaze with an embodied interactive control architecture. Appl Intell 32:148–163

Mohammad Y, Nishida T (2012) Fluid imitation: discovering what to imitate. Int J Soc Robot 4(4):369–382

Mohammad Y, Nishida T (2013) Tackling the correspondence problem: closed-form solution for gesture imitation by a humanoid's upper body. In: AMT'13: international conference on active media technology. Springer, pp 84–95

Mohammad Y, Nishida T, Nakazawa A (2013) Arm pose copying for humanoid robots. In: RO-BIO'13: IEEE international conference on robotics and biomimetics, IEEE, pp 897–904

Nagai Y (2005) Joint attention development in infant-like robot based on head movement imitation. In: The 3rd international symposium on imitation in animals and artifacts, April, pp 87–96

Nagai Y, Rohlfing KJ (2007) Can motionese tell infants and robots. What to imitate? In: 4th international symposium on imitation in animals and artifacts, pp 299–306

Nehaniv C, Dautenhahn K (1998) Mapping between dissimilar bodies: affordances and the algebraic foundations of imitation. In: Demiris J, Birk A (eds) EWLR'98: the 7th European workshop on learning robots, pp 64–72

Nehaniv CL, Dautenhahn K (2001) Like me?-measures of correspondence and imitation. Cybern Syst 32(1–2):11–51

Paul RP, Shimano B, Mayer GE (1981) Kinematic control equations for simple manipulators. IEEE Trans Syst, Man, Cybern 11:445–449

Scassellati B (1999) Knowing what to imitate and knowing when you succeed. In: The AISB symposium on imitation in animals and artifacts, pp 105–113

Chapter 13
Learning from Demonstration

Creating robots that can easily learn new skills as effectively as humans (or dogs or ants) is the holly grail of intelligent robotics. Several approaches to achieve this goal have appeared over the years. Many of the approaches focused on using standard machine learning methods in an effort to develop the most intelligent *autonomous* robot which handles learning new skills without the need of human intervention.

This is asking too much from the robot. To see that, consider how humans learn new skills. Even though autonomous exploration is an important feature of human learning, we learn most of what we know by capitalizing of the knowledge of other humans. We learn swimming not by just jumping into water and trying, nor by studying swimming books but by watching others swimming and then being taught explicitly how to improve our style. After having enough tutelage to achieve minimal skill, we can continue to improve our skill on our own by exploring local variations of the learned motions or even by trying completely new motions but only after we have enough confidence based on the initial mimicking and tutelage phase.

This suggests that intelligent robots may be advised to take a similar route and capitalize on human knowledge to learn new skills. This is the idea behind learning from demonstrations. This chapter explores the history and current state of the art in learning from demonstration as alternative techniques for implementing the *imitation engine* part of the fluid imitation engine (FIE) discussed in Chap. 12. The same techniques can be used for controller generation in the first stage of development of EICA agents discussed in Sect. 11.1.2.

Learning from demonstration (LfD) is a crucial component in imitation learning as it answers the *how to imitate* and—to some extent—the *what to imitate* questions. We can divide current methods for LfD into two main categories. Primitive learning LfD methods focuses on learning basic motions from a single or multiple demonstrations. This is the lowest level LfD possible and can be used to learn motion primitives to be employed by the second category. The second category is plan learning LfD methods. These focus on learning a plan (in most cases a serialization of motion primitives) to achieve some goal or execute some task. Both categories can be grouped

© Springer International Publishing Switzerland 2015
Y. Mohammad and T. Nishida, *Data Mining for Social Robotics*,
Advanced Information and Knowledge Processing,
DOI 10.1007/978-3-319-25232-2_13

together under a general formulation of policy learning in which the robot learns a policy from the demonstrations. A policy is a mapping from state to actions. For primitive learning, the state may be time or current configuration and the action is a specific configuration. For plan learning the state is perceived situation including both environmental conditions and robot's internal state and the actions are primitives to be executed. The treatment of this chapter is biased toward primitive learning as this is the main component we need for our social developmental approach and fluid imitation.

13.1 Early Approaches

Learning from Demonstrations has a long history in robotics. Earliest systems that utilize some form of LfD appeared in 1980s (Dufay and Latombe 1984). Motion replay is one of the earliest methods for teaching robots by demonstrating tasks. In its simplest form, the robot is teleoperated and instructed to perform some task while saving the complete commands sent to its actuators. This exact sequence of motor commands can then be played back whenever the task is to be executed. This approach is not expected to work well in practice because any small variation in the environment or any noise or measurement inaccuracies during recording may cause a failure in reproducing the task. This is specially correct when the task involves manipulation of external objects.

Early approaches to LfD tried to avoid this problem by encoding the demonstrated task as a logical program and using inductive reasoning to produce a logical representation that can then be used for reproducing the task even within a variable environment. Teleoperation was again used to guide the robot into completing the task while recording both its state (e.g. the position and orientation of the end effector) in the environment in relation to other important objects (e.g. the motion goal) as well as the motor commands sent to the robot. The recording was then segmented into episodes that achieve subgoals. These subgoals were encoded appropriately for industrial robots. For example a subgoal may be to achieve a certain orientation in relation to the object being manipulated. The information of the recording was then converted with the help of these subgoals into a graph of states connected by actions or a set of *if-then* rules encoded using symbolic relationships like "In contact with", "over", "near". The rules needed to convert numeric readings of the state into these symbolic relationships were set by hand prior to the learning session. Once this graph of state–action associations or set of rules is learned, inductive inference can then be used to execute the task even when small variations in the environment or sensory noise is present (Dufay and Latombe 1984).

This symbolic approach suffers from the standard problems of deliberative processing in robotics: it is difficult to scale to complex situations, and requires adhoc decisions regarding the conversion from numeric values to logical constructs. Soon, the generalization problem to other robotic platforms and situations led to utilization of machine learning techniques.

13.2 Optimal Demonstration Methods

Given the shortcomings of classical deliberative approaches to learning from demonstration in cases of trajectory or primitive learning; a host of other approaches were proposed over the years. This section explores optimization and dynamical approaches to the problem. Given an input trajectory, these algorithms assume that this trajectory carries information about the optimal policy utilized by the demonstrator to achieve the required task or goal. The goal of the LfD system is then to learn a *generalized form* of this motion that does not only allow the robot to achieve the same goal but provides a basis for the robot to achieve other goals and adapt to environmental change.

13.2.1 Inverse Optimal Control

One of the early approaches to LfD that appeared in 1990s was inverse optimal control (IOC). In this case, the robot learns within an optimal control framework. To understand the system, we first introduce the main structure of optimal control Problems.

The optimal control framework casts the control problem (how to act on a dynamical system) into an optimization problem of some optimization criterion $J(x, u, t)$. A full specification of an optimal control problem consists of the following:

- A specification of the state variables $x(t)$ and controls $u(t)$.
- A specification of the system model usually in the form $\dot{x}(t) = f(x(t), u(t), t)$.
- A specification of the performance index to be optimized $J(x(t), u(t), t)$ which can depend on both states and controls or directly on time. For example, in the minimum time problem, the performance index to be optimized (minimized) is time itself.
- A specification of constraints on states and controls including boundary conditions. For example, starting and final states can be considered as state constraints and maximum allowable torque can be considered as a control constraint. Mixed constraints that depend on states and controls can also be specified.

Given the above mentioned ingredients, a fairly general optimal control problem can be specified as:

$$\arg\min_{u} J(x(t), u(t), t),$$

$$s.t.$$
$$\dot{x}(t) = f(x(t), u(t), t),$$
$$and$$
$$h_k(x(t), u(t)) < 0 \quad 1 \leqslant k \leqslant K_h,$$
$$g_l(x(t), u(t)) = 0 \quad 1 \leqslant l \leqslant K_g.$$

In normal optimal control problems, J is chosen by the engineer to achieve some predefined goal. In inverse optimal control problems, J itself is learned which is where demonstrations fit.

The simplest approach to utilize LfD in inverse optimal control is to define J as the distance between some function of the state $g\left(x\left(t\right)\right)$ and some function of the human demonstration $\zeta\left(d\left(t\right)\right)$ where $d\left(t\right)$ is the demonstration. Standard optimal control methods like dynamic programming or Pontryagen's Maximum Principle can then be used to recover the *optimal* control input $u\left(t\right)$.

This method was one of the earliest LfD approaches and was employed as early as 1997 by Atkeson and Schaal (1997b) where a single demonstration was used to teach a manipulator (SARCOS arm) to swing up a pendulum (initially hanging downwards) and balance it as an inverted pendulum. To simplify the task, only horizontal motion was allowed both in the demonstration and learned motion. Human demonstrations were captured by a stereo vision system with markers attached to the pendulum to capture its orientation over time.

It is important to understand the details of this early system. Firstly, the learning problem can be divided into two parts: learning to swing the pendulum up to approach the desired state (at the top) and learning to balance the pendulum once it is near the top. The output of the solution to the first problem is an open-loop controller while the output of the second is a closed-loop controller. Learning from demonstration appears only in the solution to the first problem. Nevertheless, we will consider a simplified version of the balancing problem for completeness. Through out this discussion the state of the system will be called $s = \left[\theta, x, \dot{\theta}, \dot{x}\right]^{T}$ where θ is the angle of the pendulum measured couterclockwise from the top position and x is the horizontal position of the hand (or the end effector of the robot arm).

Balancing at the top can be achieved using a linear quadratic regulator (LQR) working on a linearized model of system dynamics at the top. Atkeson and Schaal (1997b) showed that system dynamics can be modeled as:

$$\dot{\theta}_{k+1} = (1 - \alpha_1)\,\dot{\theta}_k + \alpha_2\left(sin\left(\theta_k\right) + \ddot{x}_k cos\left(\theta_k\right)/g\right), \qquad (13.1)$$

where α_1 is the drag coefficient, $\alpha_2 = \frac{g}{fl}$, $f = 60$ is the sampling frequency, $l = 0.35$ is the pendulum length, and $g = 9.81$ is the gravitational acceleration. Notice that this is an idealized model assuming that the mass of the pendulum is concentrated completely on its tip.

To design the balancing controller, this model is linearized near the top ($\theta = 0$). This can be achieved by putting the system into states near the top and running a single step of system dynamics using Eq. 13.1 to obtain training data for a linear regressor. Using MATLAB's standard *mvnregress* function, we can obtain a linear model to be used with LQR. Applying this technique we get to the following linearized system:

$$s_{k+1} = As_k + Bu_k, \qquad (13.2)$$

where the input u_k is the acceleration $\ddot{x}$ and

$$A = \begin{bmatrix} 1.000 & 0.000 & 1/f & 0.000 \\ 0.456 & -0.000 & 0.991 & -0.000 \\ 0.000 & 1.000 & 0.000 & 1/f \\ 0.000 & 0.000 & 0.000 & 1.000 \end{bmatrix},$$

and

$$B = \begin{bmatrix} 0.000 \\ 0.046 \\ 0.000 \\ 1/f \end{bmatrix}.$$

Using standard LQR (e.g. through MATLAB's *dlqr* function), we can then obtain a linear controller $u = Ks$ that can be used for balancing. For our case, this controller will be:

$$\ddot{x} = u = [-48.963, -3.057, -10.795, 2.632]\, s.$$

Now implementing this controller can be shown to balance the pendulum near the top as shown in Fig. 13.1. As the figure shows, the balancing controller can successfully balance the pendulum at the top. No information from the demonstration was used to solve this problem. One reason for that is that it is solvable using standard

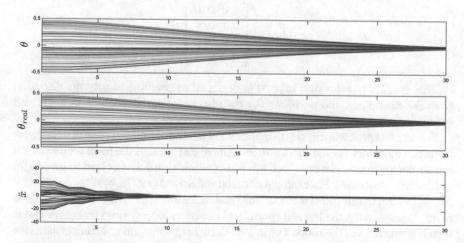

Fig. 13.1 The results of applying the balancing controller learned using LQR on the linearized model for 50 different initial conditions showing the first half second. The *top panel* shows the angle of the pendulum according to the linearized model. *Middle panel* shows the angle according to the nonlinear idealized model. *Bottom panel* shows the input acceleration applied to the end effector position

optimal control once we have an appropriate linear model. Another reason is that the hand motion used for balancing is usually much smaller than that used for swinging up. This means that accurate sensing of this motion is much harder. Now, we turn our attention to the second problem of swinging up the pendulum to a state that allows this balancing controller to take over.

The demonstration (called d hereafter) is used to teach the robot how to do the swing up of the pendulum. This formulated as another optimal control problem with a quadratic cost function that penalizes deviation from the demonstrated motion. More formally, the swing up problem can be formulated as the following optimal control problem:

$$\arg\min_{\ddot{x}} J\left(s\left(t\right), \ddot{x}\left(t\right)\right),$$

$$s.t.\ s\left(t+1\right) = f\left(s\left(t\right), \ddot{x}\left(t\right)\right),$$

where the dynamics $f\,()$ is given by Eq. 13.1.

$$J\left(s\left(t\right), \ddot{x}\left(t\right)\right) = \sum_{t=1}^{T} \left(s\left(t\right) - d\left(t\right)\right)^{T} Q_{lfd}\left(s\left(t\right) - d\left(t\right)\right) + R_{lfd}\ddot{x}(t)^{2},$$

and the quadratic costs are defined as $R_{lfd} = 1$ and

$$Q_{lfd} = \begin{bmatrix} 1 & 0 & 0 & 0 \\ 0 & 1 & 0 & 0 \\ 0 & 0 & 10 & 0 \\ 0 & 0 & 0 & 1 \end{bmatrix}.$$

This specific penalization scheme is designed to penalize mostly the deviation from demonstrated hand positions (x) but also penalizes deviations in angle and speeds (Atkeson and Schaal 1997b).

Solving this problem directly using standard nonlinear optimal control techniques leads to a trajectory for the robot arm movement that mimics the hand motion of the human demonstrator. Nevertheless, this direct mimicking does not lead to successful imitation in most cases. For example Atkeson and Schaal (1997a, b) show only examples of failures of this direct method. The reasons for this failure can be traced to three main causes. Firstly, the learned controller is an open-loop desired trajectory that is then implemented in the robot. Deviations during real-time are not compensated for or—if a feedback controller is employed on top of the LfD system—are compensated for assuming that the error resides in deviations from the learned hand trajectory not from the desired trajectory of the pendulum itself. Secondly, the controller is not perfect which means that even if implementing the exact trajectory would have succeeded, the controller will not follow the exact learned trajectory. Finally, the task is not exactly the same because the grip of the human and the response to the pendulum to motion may be different than the case with the robot arm.

To resolve this problem, we need to *improve* the quality of the learned policy. One possible approach is to use data from the failed attempt by the robot to improve the model (as specified by Eq. 13.1). This is done by learning the coefficients α_1 and α_2 using the data from the failed attempt. A second trial is then conducted with the newly learned model which usually fails but comes closer to a full swing up. This process is repeated until there is more possible improvement of the model (remember that it is an idealized model anyway and cannot capture the full dynamics of the pendulum). At this point, Atkeson and Schaal (1997b) suggest increasing α_1 gradually because doing so takes energy from the system which is then compensated by the controller leading to a higher swing. This approach succeeded from the third trial (Atkeson and Schaal 1997b).

This example gives the flavor of IOC approaches to LfD but is only one possible approach. Extensions and novel adaptations of IOC for learning from demonstrations tasks are abound (see for example Park and Levine 2013; Byravan et al. 2014; Doerr et al. 2015; Chen and Ziebart 2015; Huang et al. 2015; Byravan et al. 2015).

The IOC approach presented in this section combines modeling of the plant (for the balancing problem), learning from demonstration (for getting the initial policy or trajectory) and iterative improvement of the model and motion. It also requires some understanding of system dynamics to achieve the final iterative process (e.g. the effect of the dragging coefficient α_1 on system dynamics) and is not completely autonomous. As presented, noise and measurement inaccuracies were not considered but they can be taken into account—at a great computational cost—by defining rewards in the form of expectations over the stochastic parameters of the system. The following section presents a related approach that takes the stochastic nature of the system into consideration from the beginning.

13.2.2 Inverse Reinforcement Learning

Inverse reinforcement learning (IRL) is related to IOC. LfD is sometimes called apprenticeship learning in this context (Abbeel et al. 2010). Rather than casting the problem as a control problem of a dynamical system, IRL casts the problem as a Marcov Decision Process (MDP) where we are interested in learning a policy that maximize the expected accumulated reward. Again, the problem is inverted in the sense that the reward function is not given but is learned from the demonstrations.

This method was applied successfully to learning helicopter acrobat maneuvers from expert demonstrations (Abbeel and Ng 2004; Abbeel et al. 2010).

As an example of this approach, we present the method proposed by Abbeel and Ng (2004). Here, the system is modeled as a Markov Decision Process (MDP). An MDP is a tuple $\mathcal{M} = (S, A, T, S_0, \gamma, R)$, where, S is the set of states, A is the set of possible actions, $T : S \times A \rightarrow S$ is the transition probability from current state and action to next state, $\gamma \in [0, 1)$ is a discount factor used to calculate infinite horizon rewards and

$R : S \times A \to (0, 1)$ is the reward function. We also define $\mathcal{M} \backslash R = (S, A, T, \gamma, S_0)$. In IRL, we assume that we are given $\mathcal{M} \backslash R$ and some demonstrations sampled from a policy $\pi^D : S \to A$ that is usually (but not necessarily) assumed to be optimal. IRL can then be used to recover an approximation of the underlying reward function $\bar{R}$ completing the model $\bar{\mathcal{M}} = (S, A, T, S_0, \gamma, \bar{R})$. The completed model $\bar{\mathcal{M}}$ can then be used with a standard RL solver to recover an optimal policy for the learner $\bar{p}i$ that can achieve performance similar to π^D on the original MDP $\mathcal{M}$. One major advantage of this approach is that it provides the clearest definition of the *intention* of the demonstrator in the form of the reward function learned. In fact, RL can be used as a unifying framework for modeling learning from demonstrations (Argall et al. 2009) and imitation-like behaviors (Melo et al. 2007). It can also support interleaved stages of learning from demonstration and autonomous explorations (Oudeyer et al. 2014).

To simplify the problem, the underlying reward function to be learned R is assumed to be a linear combination of features. These features are defined as mappings to the unit hypercube ($\phi : S \to [0, 1]^k$).

$$R(s, a) = w\phi(s),\tag{13.3}$$

where $w \in \mathbb{R}^k$ and $\|w\|_1 \le 1$. We assumed following Abbeel and Ng (2004) that features depend only on the state leading to a reward function that depends only on the state. Extension to the more general case of dependence on both state and action is straightforward.

This limitation of the reward function form makes sense in practice. In realistic RL applications, we usually have a set of goals to achieve and these can be encoded as sub-rewards (features) that are then combined using appropriate weights. For example, the reward function of driving is usually defined as a linear combination of rewards for keeping safe distance, taking the appropriate lane, not exceeding the speed limit, having minimal jerk and reducing accelerations, etc.

The value of any policy $\pi : S \to A$ is given by:

$$E_{s_0 \sim S_0} \left[V^\pi(s_0) \right] = E \left[\sum_{t=0}^\infty \gamma^t R(s_t) | \pi \right].\tag{13.4}$$

Given Eqs. 13.3 and 13.4 we get:

$$E_{s_0 \sim S_0} \left[V^\pi(s_0) \right] = wE \left[\sum_{t=0}^\infty \gamma^t \phi(s_t) | \pi \right].\tag{13.5}$$

We can then define a shorthand version of feature expectations as:

$$\mu(\pi) \equiv E_{s_0 S_0} \left[V^\pi(s_0) \right],$$

which leads to the simplified expression:

$$E_{s_0 \sim S_0}\left[V^\pi\left(s_0\right)\right] = w\mu\left(\pi\right). \tag{13.6}$$

Equation 13.6 shows that, under our assumption, the value of any policy is a linear combination of the values of this policy using features as rewards and the weights are the same as these used to combine the features in reward definition (See Eq. 13.3).

Given a set of policies $\pi^1, pi^2, \ldots, pi^n$, we can define a probability distribution over them $\left(p^1, p^2, \ldots, p^n\right)$ that sums to one. We can define the policy π^p defined by selecting one policy π^i with probability p^i then executing it. It can be shown that:

$$E_{s_0 \sim S_0}\left[V^{\pi^p}\left(s_0\right)\right] = \sum_{i=1}^{n} p^i E_{s_0 \sim S_0}\left[V^{\pi^i}\left(s_0\right)\right].$$

With these preliminaries, we are ready to explain how to use IRL in LfD under the aforementioned assumption. Notice that the final goal for LfD is not the discovery of R but the discovery of some policy $\bar\pi$ that approximates and generalizes the performance of the demonstrator. The approach presented here (due to Abbeel and Ng 2004) uses the linearity assumptions stated earlier to find an appropriate policy $\bar\pi$ directly without the need to estimate R.

We assume that we have N demonstrations $\{D^n\}$ each is a sequence of states of lengths T^n ($D^n = \left\{s_0^n, s_1^n, \ldots, s_{T_n-1}^n\right\}$). We firstly estimate the value of this policy using:

$$\hat\mu^D = \frac{1}{N}\sum_{n=1}^{N}\sum_{t=0}^{T_n}\gamma^t\phi\left(s_t^n\right). \tag{13.7}$$

A random policy π^0 is then selected and its value evaluated as $\mu^0 \equiv \mu\left(\pi^0\right)$. The weights of the features are then estimated as:

$$w^1 = \bar\mu^D - \mu^0,$$

and we set the first estimate of the approximate policy to $\bar\mu^0 = \mu^0$.

We then enter a loop until convergence with values of $i = 1, 2, \ldots$ iterating the following steps:

1. $\bar\mu^{i-1} \leftarrow \bar\mu^{i-2} + \frac{\left(\mu^{i-1}-\bar\mu^{i-2}\right)^T\left(\bar\mu^D-\bar\mu^{i-2}\right)}{\left(\mu^{i-1}-\bar\mu^{i-2}\right)^T\left(\mu^{i-1}-\bar\mu^{i-2}\right)}\left(\mu^{i-1} - \bar\mu^{i-2}\right).$
2. $w^i \leftarrow \bar\mu^D - \mu^{\bar i-1}.$
3. $t^i \leftarrow \|\bar\mu^D - \bar\mu^{i-1}\|_2.$
4. Terminate iff $t^i \le \varepsilon$, where ε is a small allowable error in $\bar\mu$.
5. $\bar{\mathcal{M}} \leftarrow \left(S, A, T, S_0, \gamma, \bar R^i\right)$, where $\bar R^i = w^i\phi.$
6. Solve the MDP $\mathcal{M}$ using any standard RL approach (e.g. value or policy iteration) to get an optimal policy. Call this optimal policy π^i and its value μ^i.

Abbeel and Ng (2004) showed that this algorithm always terminates in at most $O\left(\frac{k}{(1-\gamma)^2\varepsilon^2}\log\frac{k}{(1-\gamma\varepsilon)}\right)$ iterations. When it terminates, we return all the policies π^i

and a final policy $\bar{\mu}$ can be found by solving a quadratic programming problem to minimize the difference between it and $\bar{\mu}^D$ by solving:

$$\min_{\mu} \|\bar{\mu}^D - \mu\|_2$$

$$s.t. \mu = \sum_i \lambda_i \mu^i, \sum_i \lambda_i = 1.$$

Consider a small problem with 10 features, $\gamma = 0.9$, and $\varepsilon = 0.1$, The worst case number of iterations is 2×10^5 iterations each involves the solution of an MDP. Despite this high computational cost, the approach proved appropriate in teaching robots to navigate in a simulated environment (Abbeel and Ng 2004) and a related approach was used to advance the state of the art in autonomous helicopter flight aerobatics (Abbeel et al. 2010). A related approach could even provide performance better than all demonstrations (Coates et al. 2008).

13.2.3 Dynamic Movement Primitives

A third approach that is gaining increased support in the research community is modeling the task to be learned by a regulated dynamical system (Ijspeert et al. 2013; Schaal 2006). In this approach, we start by a dynamical system with guaranteed convergence to the goal state of the motion. We then regulate the resulting motion based on the demonstrations to achieve demonstration-like trajectories. This approach has several advantages: firstly, it is adequate to describe both linear and cyclic motions (e.g. drumming motions). Secondly, the resulting model is parameterized by the goal state which can be changed to generalize the motion. Thirdly, the speed at which the motion is generated is controllable using a single parameter.

The most unique feature of DMP is how it represents control policies. Traditionally control policies could be represented as functions of the state space or as a desired trajectory with an implicit linear PD controller. DMP, on the other hand, represents the controller by a dynamical system.

Consider the following dynamical system:

$$\ddot{x} = \zeta_\alpha (x_d - x) - \alpha_\alpha \dot{x}. \tag{13.8}$$

This equation describes a dynamical system in which the acceleration is directly proportional to the difference between the current state x and some desired state x_d plus another factor proportional to the velocity $\dot{x}$ and tending to reduce it (assuming that α is positive). This system has a single attractor at $x = x_d$ and will converge to it. This is simply a PD control signal. If we assume that the controlled degrees of freedom of the robot are encoded by x, this simple system gives us a way to guarantee the convergence to any desired goal given that it is accessible from the starting state but without any further control on the path used by the system to achieve that goal. For now we assume that the demonstration is a single-dimensional time-series.

To make the robot follow a specified demonstration (encoded as a time-series X), Eq. 13.8 needs to be modified in order to allow us to control the path used to approach the goal $x_d = x(T)$ where T is the length of the time-series X. One possible approach is to add another term (called the forcing term) to the Equation that controls how the acceleration is to change as shown in Eq. 13.9.

$$\ddot{x} = \zeta_\alpha (x_d - x) - \alpha_\alpha \dot{x} + \Gamma. \tag{13.9}$$

This additional Γ term is a nonlinear function of the state x and is calculated based on another dynamical system called the canonical system which has the following simple form:

$$\dot{z} = -\alpha_z z. \tag{13.10}$$

Given that α_z is positive, this dynamical system converges to zero from any starting point. By having Γ being linear in z, we make sure that its effect is reduced over time leading the system of Eq. 13.9 to behave approximately the same as the system described by Eq. 13.8 near the end of the motion which keeps the guarantee to converge to the goal. Nevertheless, in the beginning and middle of the motion, z can have appreciable value leading to a different path than the simple approach taken by Eq. 13.8. This aforementioned discussion suggests that $\Gamma \propto z$.

Moreover, we can further ensure that the nonlinear dynamical system of Eq. 13.9 approaches the goal x_d by having $\Gamma \propto (x_d - x_0)$ where x_d is the desired state (goal) and x_0 is the initial state.

The aforementioned discussion suggests the following form of Γ:

$$\Gamma = \Xi z (x_d - x_0),$$

where Ξ is the nonlinear term that will be used to control the path taken from x_0 to x_d. This nonlinear component (Ξ) is what will be learned based on the demonstration. One possible choice is to build this term as a weighted combination of Gaussian kernels. By controlling the weights of this mixture of Gaussians, it is possible to achieve the desired smooth path known from the demonstration. This entails the following definition of the Γ term:

$$\Gamma = \frac{\sum_{k=1}^{K} w_k \psi_k}{\sum_{k=1}^{K} w_k} z (x_d - x_0),$$

$$\psi_k = \exp\left(-\frac{(z - \mu_k)^2}{\sigma_k^2}\right). \tag{13.11}$$

Equations 13.9 and 13.11 provide a model of the motion (e.g. acceleration) as a dynamical system that is still guaranteed to converge to the goal x_d from the starting point x_0. A main advantage of this technique of modeling is that we can change the desired goal on the fly and the system will adapt its motion without the need to modify every point in the acceleration time-series. This allows us to handle changing environmental conditions effectively. It is possible to extend this spatial flexibility to the temporal domain by allowing arbitrary scaling of the trajectory with a controllable parameter τ as shown in Eq. 13.12.

$$\ddot{x} = \tau^2 \left(\zeta_\alpha \left(x_d - x \right) - \alpha_\alpha \dot{x} + \Gamma \right),$$
$$\dot{z} = -\tau \alpha_z z. \tag{13.12}$$

These simple modifications of the main and canonical dynamical systems (compare Eq. 13.12 with Eqs. 13.9 and 13.10) enables us to control the speed of execution of the motion by setting τ to a value between 0 and 1 to slow it down or to a value larger than 1 to speed it up.

In the most general case, we can learn the sets μ_k, σ_k and w_k that comprise the nonlinear factor. We may even try to learn the number of kernels to use (K). In most cases though, it is possible to get good performance by fixing μ_k, σ_k and K and learning only the set of weights w_k.

When setting $= mu_k$, we should try to make them equally spaced along the execution time of the motion. The problem is that these centers are defined in terms of the variable z which is monotonically decreasing in time but not linearly. According to Eq. 13.10, the canonical dynamical system will converge to zero exponentially fast not linearly. For this reason, μ_k are usually selected to be equally spaced in the logarithmic space $log\,(z)$.

Given this logarithmic approach to setting μ_k, we also need to adapt σ_k to provide larger variance for the kernels that are activated for longer times. This means that later kernel Gaussians should have smaller variances. A rule of thumb that usually works is to set the variances as

$$\sigma_k = \sqrt{\frac{\mu_k}{K}}.$$

Now the problem of learning a dynamical system to represent a given demonstration $(x\,[t])$ is translated to the problem of estimating the set of weights (w_k) given an appropriate number of kernels and their positions μ_k and variances σ_k^2. Several standard methods can be used to solve this problem.

As a simple example, consider that we are given a demonstration as a T-points time-series $X_d = x_d\,(1 : T)$. We first start by differentiating it twice to get a time-series representing the acceleration according to:

$$\ddot{x}_d = \frac{\partial^2 x_d}{\partial t^2}. \tag{13.13}$$

Given some value for dt, this can be approximated by double differencing. Substituting from Eq. 13.13 into Eq. 13.9, we can find the desired value of Γ as

$$\Gamma(t) = \ddot{x}_d(t) - \left(\zeta_\alpha \left(x_d(T) - x_d(t) \right) - \alpha_\alpha \dot{x}_d(t) \right). \tag{13.14}$$

Now we try to find the values of w_k that fits $\Gamma(t)$ as defined in Eq. 13.14 for $1 \leq t \leq T$. This is a standard machine learning problem. An effective approach for solving this problem is Locally Weighted Regression (LWR) (Schaal et al. 2002). LWR assumes that Γ is of the form given in 13.11 and tries to minimize the summation of squared errors weighted by the value of the kernel function at every point. This leads to K optimization problems of the form:

$$\arg\min_{w_k} \sum_{t=1}^{T} \psi_k(t) \left(\Gamma_k(t) - w_k \left(z(t) \left(x_d(T) - x_d(0) \right) \right) \right)^2. \tag{13.15}$$

A solution for this problem can be found in (Schaal et al. 2002) and it takes the form:

$$w_k = \frac{\upsilon^T \Psi_k \Gamma}{\upsilon^T \Psi_k \upsilon}, \tag{13.16}$$

where,

$$\upsilon = \left(z(1) \left(x_d(T) - x_d(0) \right) \ldots \ldots z(T) \left(x_d(T) - x_d(0) \right) \right),$$

$$\Psi_k = \begin{pmatrix} \psi_i(1) & \ldots & 0 \\ 0 & \ddots & 0 \\ 0 & \ldots & \psi_i(T) \end{pmatrix}. \tag{13.17}$$

The values of w_k learned from Eq. 13.16 are then used to drive the DMP during motion execution.

Several other approaches, for solving this or related optimization problems, have been proposed over the years. For example, Locally Weighted Projection Regression (LWPR) sets the locations of the kernels μ_k and their standard deviation σ_k based on the complexity of the motion so that approximately linear parts of the time-series are modeled by few large kernels while complicated parts are modeled by many smaller ones (Vijayakumar and Schaal 2000). Recently, Gaussian Mixture Modeling was proposed to solve this optimization problem in multidimensional LfD (Calinon et al. 2012).

All of the aforementioned discussion assumes that the input is single dimensional. To model multidimensional motions, that are ubiquitous in robotics, we can assign a DMP to every degree of freedom. Notice that the output of the DMP is an open-loop controller that gives the desired trajectory to be followed by the learned degrees of freedom.

If the input to the DMP are joint angles of the robot during the demonstration, the output of the DMP can directly be used to set the desired acceleration of the robot. It is a simple matter to run the DMP for a short time and then correct the current state based on sensors attached to the robot leading to a closed loop controller that can adjust immediately to environmental conditions.

It is also possible to train the DMP on the task-space and then use an underlying inverse kinematic model (e.g. operational space controller) to derive the robot in real time. Again extension to closed loop control is easy to achieve.

The DMP framework was extended in several ways since its appearance in early 2000s. For example, Pastor et al. (2009) proposed a method to integrate collision avoidance in the DMP framework using another dynamical system that moves the robot away from obstacles. DMPs can be coupled with measured and predicted sensor traces to learn goal-directed behaviors. Recently, the applications of DMPs have been extended beyond modeling robot-object interactions to modeling Human Robot Interaction.

The DMP framework is very flexible and allows for real-time modification of the goal state and motion speed. This also translates to easy modulation of the motion based on error in case of closed loop control. For example, if the output of the DMP was too fast for the controller to follow or if a disturbance changed the current state of the robot, we can easily *temporarily slow down* DMP output by modifying the *dt* value used for driving the canonical system as follows:

$$\tau \dot{z} = -\alpha_z z \frac{1}{1 + \alpha_e e^2},$$

where e is the difference between the desired state as generated from the DMP and the actual state of the robot and α_e is a modulation parameter that controls how much the DMP slows down. Other kinds of modulation of the time evolution of the canonical system are also possible like coupling joints to achieve synchronization (Nakanishi et al. 2004) or resetting it on certain events like heel strike during walking.

All of the aforementioned discussions assume that the main dynamical system is linear. DMPs can easily be used with cyclic motion (which is one of its main strengths) by appropriately modifying the equations for the main and canonical dynamical systems. In this case the main dynamical system will not be designed to have a simple attractor but to have a limit cycle for which it converges.

Two of the most challenging problems of DMPs are smooth serialization of multiple DMPs which is usually not easy to achieve (Nemec and Ude 2012) and selection of the number of kernel functions to be used for modeling Γ. Statistical approaches based on Gaussian Mixture Modeling and Regression do not suffer from the first problem and symbolization can be used to alleviate the second one as will be discussed later in this chapter.

13.3 Statistical Methods

Statistical methods have seen wide spreed in all kinds of machine learning tasks in the recent decades. Learning from Demonstration is not an exception and several LfD systems utilize statistical methods for learning as well as motion generation. This section describes some of the main techniques used for this purpose focusing on Gaussian Mixture Modeling.

13.3.1 Hidden Markov Models

Hidden Markov Models (HMM) are probabilistic graphical models that can encode time-variant signals and were discussed in Sect. 2.2.8. Given a demonstration (or a set of demonstrations) a HMM can be learned using the Baum–Welch algorithm as discussed in Sect. 2.4.3 and can then be used for generating similar behavior.

Using the Baum–Welch algorithm for learning a HMM representation of the motion, generates a model that can adapt to timing variations and the covariance matrices of the observation distribution encode the internal correlations between degrees for freedom in a scalable way which was not possible with DMPs. The problem though is that the encoding provides only a single trajectory output which cannot be parameterized by for example a goal as in DMPs.

There are several ways to alleviate this trajectory specificness of the HMM and one of the most promising is Parametric Hidden Markov Models (PHMM) proposed by Herzog et al. (2008). PHMM assumes that D demonstrations are available ($x_d^1 : x_d^D$) with some parameter (to be generalized over) given for each demonstration ($p^1 : p^D$). This parameter can be for example the goal of the motion in a 2D surface or 3D space. The main idea of PHMM is to learn a HMM for each demonstration associated with a parameter value and then to generate a new HMM on the fly by interpolating these learned HMMs whenever a new parameter value (e.g. goal) is given.

For this to be successful, the states of different HMMs learned from the demonstrations must be representing the same portion of the motion. If this condition is not satisfied, interpolating the HMMs will be meaningless. A global HMM that is learned from the complete set of demonstrations is used to achieve this state-equivalence between the specific HMMs learned from individual demonstrations.

13.3.2 GMM/GMR

Gaussian Mixture Modeling combined with Gaussian Mixture Regression is one of he most widely used approach to robot learning from demonstration (Calinon et al. 2010). The approach has spawned several new directions including Quantum GMM (Chatzis et al. 2012), and variational methods (See Sect. 13.3.2.1).

The input in this case can be joint angles of the robot or end effector configuration. This input is a set of time-series $\{X^d\}$ for $1 \le d \le D$ and D is the total number of demonstrations which will have lengths T_d and N dimensions. The basic approach is straightforward. The input time-series are first aligned using Dynamic Time Wrapping (DTW) to be of the same length. This is necessary to make the values at the same time-step directly comparable for different demonstrations. Now we have a set of times-series all of the same length (T). We will continue to use X^d to indicate member time-series of this list.

The second step is to augment each X^d in one of two ways to generate an augmented time-series $\tilde{X}^d$ of the same length but with one extra dimension. The first approach is to simply add one dimension representing time which is proportional to the sequence $0 : T - 1$. The second approach is to use the derivative of the input dX^d $(dx_i^d = x_{i+1}^d - x_i^d)$ if a dynamical system is thought. In the first case, the resulting time-series will have the dimensionality $\tilde{N} = N + 1$, and in the second case $\tilde{N} = 2 \times N$. We will focus on the first approach in this first treatment of the subject.

All of the augmented time-series are then concatenated to give a $\tilde{N}$ dimensional time series X of length $D \times T$. We will call the original demonstration time-series dimensions (e.g. joint angles or end effector configuration when augmented with time) α, and the dimensions augmenting it (e.g. time) β. A GMM is then fitted to this time-series using—for example—the Expectation Maximization algorithm discussed in Sect. 2.4.4. To complete this step, we need to specify a number of Gaussians to use K. This is a standard model selection problem and we will propose some solutions to it later in this chapter.

This gives us a mixture model that represents the joint probability $p(\alpha, \beta)$. This joint distribution carries information about not only the expected values for joint angles or end effect configuration at every time-step but also the correlations of these values.

Generating desired motions from this learned demonstration can simply be achieved using Gaussian Mixture Regression as discussed in Sect. 2.2.9. This procedure is repeated here for concreteness. Motion generation amounts to a conditioning operation of the form shown in Eq. 13.18.

$$p(\alpha|\beta) = \frac{p(\alpha, \beta)}{\sum_\alpha p(\alpha, \beta)}. \tag{13.18}$$

Our model is a GMM model which leads to the following form of the joint distribution:

$$p(\alpha, \beta) = \sum_{k=1}^{K} p(k) \, p_k(\alpha, \beta),$$

$$p_k(\alpha, \beta) \equiv \mathcal{N}(\mu_k, \Sigma_k). \tag{13.19}$$

Given this joint distribution, it is possible to generate time-series points by conditioning on the augmenting variable β (time). The use of a GMM to model the joint distribution ensures that the conditional distribution is also a GMM.

Equation 13.18 simplifies to the following in this case:

$$p\left(\alpha|\beta\right) = \sum_{k=1}^{K} \pi_k^c p_k^c\left(\alpha|\beta\right),$$

$$p_k^c\left(\alpha|\beta\right) \equiv \mathcal{N}\left((\alpha, \beta)^T ; \mu_k^c, \Sigma_k^c\right). \tag{13.20}$$

Now assume that we make the following definitions:

$$\mu_k \equiv \left(\mu_k^\alpha, \mu_k^\beta\right)^T, \tag{13.21}$$

$$\Sigma_k \equiv \begin{pmatrix} \Sigma_k^\alpha & \Sigma_k^{\alpha\beta} \\ \left(\Sigma_k^{\alpha\beta}\right)^T & \Sigma_k^\beta \end{pmatrix}. \tag{13.22}$$

Notice that $\mu_k^\beta \in \mathbb{R}$ and $\Sigma_k^\beta \in \mathbb{R}$ while $\mu_k^\alpha \in \mathbb{R}^N$ and $\Sigma_k^\alpha \in \mathbb{R}^N \times \mathbb{R}^N$.

Given these definitions, the new parameters of the conditional GMM are given by the following set of equations:

$$\pi_k^c = \frac{p\left(k\right)\mathcal{N}\left(\beta; \mu_k^\beta, \Sigma_k^\beta\right)}{\sum \pi_k^c}, \tag{13.23}$$

$$\mu_k^c = \mu_k^\alpha + \Sigma_k^{\alpha\beta}\left(\Sigma_k^\beta\right)^{-1}\left(\beta - \mu_k^\beta\right), \tag{13.24}$$

$$\Sigma_k^c = \Sigma_k^\alpha - \Sigma_k^{\alpha\beta}\left(\Sigma_k^\beta\right)^{-1}\left(\Sigma_k^{\alpha\beta}\right)^T. \tag{13.25}$$

Furthermore, using properties of Gaussians, Eq. 13.20 can be further simplified by approximating the GMM with a single Gaussian:

$$p\left(\alpha|\beta\right) \cong \mathcal{N}\left(\alpha; \bar{\mu}_\alpha, \bar{\Sigma}_\alpha\right), \tag{13.26}$$

where $\bar{\mu}_\alpha = \sum_{k=1}^{K} \pi_k^c \mu_k^c$ and $\bar{\Sigma}_\alpha = \sum_{k=1}^{K} \left(\pi_k^c\right)^2 \Sigma_k^c$.

Equation 13.26 defines a Gaussian at every time-step. The mean $\bar{\mu}_{alpha}$ can be used as *desired* output (e.g. joint angle or end effector configuration) at time β for $0 \leq \beta T - 1$ and a feedback controller (e.g. a PD controller or a LQR) can be used to stabilize this path. Moreover, the variance $\bar{\Sigma}_\alpha$ can be used to inform the controller about how *accurate* it needs to follow the desired path at every time-step. The availability of this extra information about how accurate should the path

be followed is a major advantage of statistical modeling in general compared with
dynamical approaches.

The GMM/GMR approach provide several advantages for real-world LfD problems. Firstly, it is very easy to utilize multiple demonstrations by simply appending
them together with the appropriate time augmentation (that is in contrast with DMP).
We have already pointed out to the availability of information about the required accuracy in following every point in the learned path. Moreover, the same GMR approach
can be used not only to condition on time but on any other dimension (or group of
dimensions) of the input time-series. For example, we can estimate what the left hand
of the robot should be doing given what the right hand is doing if both were available in the demonstrated data. Moreover, it is very simple to combine learned basic
motions together to generate more complex behaviors by simply adding together the
GMM models (contrast this with DMP for which serialization is a research subject).

This flexibility of the GMM/GMR resulted in several extensions and improve-
ments that we will consider later but there are some limitations to the approach.
Firstly, we need to solve the model selection problem (i.e. the value of K). Secondly, even though we have correlation information at every time-step, we have no
such information about variables at different time-steps. More importantly, in its
basic form, GMM/GMR learns a single path from the demonstrations (encoded in
the mean of the conditioned distribution), and this makes it challenging to deal with
variable goals (a major advantage of DMP), or to handle changing environmental
conditions. Partial solutions to some of these limitations will be discussed later in
this section.

This basic algorithm has been extended in several ways since it was proposed by
Calinon et al. (2006). The supporting website of this book provides an implementation
of this approach in the function $doGMM()$ based on the toolbox provided by Calinon.

Explicit dependence on time may appear as a restricting factor in our formulation
but it can be handled in several way. The simplest approach is to use the sequence
$0 : T - 1$ as the augmenting dimension and for β but then to scale it when generating
the motion using GMR which leads to variable speed generation of the same behavior.
A more interesting approach is to use a dynamical systems approach in the form:

$$\dot{x} = f(x).$$

In this case, α will correspond to $\dot{x}$ and β will correspond to x. Notice that in this
case, the state β may not be just joint angles but may include as well their derivatives
and the variable α will correspond to desired acceleration at the state α. This gives a
time-independent dynamical system that can be used for motion generation without
any major modification to the GMM/GMR framework.

13.3.2.1 Variational Inference

The standard GMM/GMR approach presented in Sect. 13.3.2 requires the specification of a number of Gaussian (K) to be used by the EM algorithm to learn the GMM

model (See Sect. 2.4). BIC can be used to select an appropriate number of Gaussian but it requires training multiple times using all $k \in 1, 2, \ldots, K$ and estimating the likelihood for each case.

This section introduces a method that allows us to learn the GMM model in a single EM-like application with the maximum number of Gaussians K_{max} only needing to be specified. The approach is based on variational inference and was first utilized by Corduneanu and Bishop (2001) and used for LfD by Hussein et al. (2015).

Equation 13.19 defines model. We assume that the data is generating by selecting a Gaussian component based on the probability $p(k)$ and then using it to sample a point from the corresponding Gaussian. For convenience we define $\pi_k \equiv p(k)$ and $\pi = \{p_k\}$. At every time-step t, a hidden variable z^t can be introduced to represent this Gaussian selection process (we will drop the superscript when not necessary for understanding). Following (Bishop 2006) we will define it as K binary random variables where only one of them equals 1 (representing the selected model) and the rest of them equaling zero.

The value of z_k then satisfies $z_k \in \{0, 1\}$ and $\sum_k z_k = 1$ so we can write the conditional distribution of z, given the mixing coefficient π, in the form

$$p(Z|\pi) = \prod_{n=1}^{\mathcal{N}} \prod_{k=1}^{K_{max}} \pi_k^{z_k^t}, \tag{13.27}$$

and the conditional distribution of the observed data, given Z, will be:

$$p(X|Z, \mu, \Sigma^{-1}) = \prod_{t=1}^{T} \prod_{k=1}^{K_{max}} \mathcal{N}(x_t|\mu_k, \Sigma_k)^{z_k^t}. \tag{13.28}$$

To evaluate $p(x|\pi)$ we have to marginalize Eq. 13.28 with respect to z, μ, Σ which is analytically intractable and variational inference is used to approximate the solution to this problem.

The main difference between this approach and standard EM is that it is a Bayesian approach and suffers from no overfitting problems. This means that only computation time limits the maximum number of Gaussians we can use (K_{max}) and usually many of them will have mixing coefficients π_k indistinguishable from zero. This built-in model selection is what makes this approach attractive compared with standard BIC.

The VB algorithm has two steps similar to the Expectation and Maximization steps of the EM algorithm (and with only slightly higher computational overhead) (Corduneanu and Bishop 2001). In the Expectation (E) step, the current distribution is used to calculate $\mathbb{E}[z_k^t] = r_k^t$ using:

$$\mathbb{E}_{\mu_k, \Sigma^{-1}_k}[(X_t - \mu_k)^T \Sigma^{-1}_k (X_t - \mu_k)] = D\beta_k^{-1} + \upsilon_k(X_t - m_k)^T W(X_t - m_t), \tag{13.29}$$

$$\ln \tilde{\Sigma}^{-1}_k = \mathbb{E}[\ln|\Sigma^{-1}_k|] = \sum_{i=1}^{D} \phi(\frac{\upsilon_k + 1 - i}{2}) + D\ln 2 + \ln|W_k|, \tag{13.30}$$

$$ln\tilde{\pi}_k = \mathbb{E}[ln\pi_k] = \phi(\alpha_k) - \phi(\widehat{\alpha}),\tag{13.31}$$

$$ln\rho_{nk} = \mathbb{E}[ln\pi_k] + \frac{1}{2}\mathbb{E}[ln|\Sigma^{-1}{}_k|] - \frac{D}{2}ln(2\pi) - \frac{1}{2}\mathbb{E}_{\mu_k,\Sigma^{-1}{}_k}[(X_t - \mu_k)^T\Sigma^{-1}{}_k(X_t - \mu_k)],\tag{13.32}$$

$$r_k^t = \frac{\rho_k^t}{\sum_{j=1}^{K_{max}}\rho_j^t},\tag{13.33}$$

where β_0 and W_0 are the parameters of a Gaussian-Wishart distribution which represents the mean and precision of the Gaussian component and α_0 is the parameter of Dirichlet process which represent π.

The second step (Maximization) recalculates the distribution parameters form the new r_k^t as follows:

$$N_k = \sum_{t=1}^{T} r_k^t,\tag{13.34}$$

$$\alpha_k = \alpha_0 + N_k,\tag{13.35}$$

$$\beta_k = \beta_0 + N_k,\tag{13.36}$$

$$\bar{X}_k = \frac{1}{N_k}\sum_{t=1}^{T} r_k^t X_t,\tag{13.37}$$

$$m_k = \frac{1}{\beta_k}(\beta_0 m_0 + N_k \bar{X}_k),\tag{13.38}$$

$$v_k = v_0 + N_k,\tag{13.39}$$

$$S_k = \frac{1}{N_k}\sum_{t=1}^{T} r_k^t (X_t - \bar{X}_k)(X_t - \bar{X}_k)^T,\tag{13.40}$$

$$W_k^{-1} = W_0^{-1} + N_k S_k + \frac{\beta_0 N_k}{\beta_0 + N_k}(\bar{X}_k - m_0)(\bar{X}_k - m_0)^T.\tag{13.41}$$

Iterating these two steps until the likelihood no longer improves gives an estimate for the mixture parameters (π, μ and Σ^{-1}) that represent the Gaussian components. These can be further refined using standard EM.

13.4 Symbolization Approaches

Another approach to learning from demonstration consists of converting real-valued demonstration trajectories into strings of symbols from a predefined vocabulary and generating as an output a string representing the learned behavior which can then converted back to real-valued desired trajectories (after smoothing). Mohammad and Nishida (2014) used this approach in a system called SAXImitate. As the name implies, the symbolization is done using the SAX algorithm (See Sect. 2.3.2) after appropriately extending it to multi-dimensional time-series.

The input time-series representing the demonstrations X_n are equalized in length using Dynamic Time Wrapping (a common step with GMM/GMR systems). The resulting time-series are then transformed using the MSAX algorithm (See Sect. 2.3.2) to a set of K strings called S_i hereafter.

Given The set of strings S_i, we need to generate a model of the demonstration that is usable in reproducing it. This is achieved on two steps. Firstly, we combine these strings to generate a single model string S which is then used to create a $D \times T$ model time-series that is used for re-generating the motion and a T-points vector corresponding to the variability at every timestep.

The main data-structure used in the first step is a $K \times N$ matrix called the *Z-Matrix* where rows (r) correspond to different strings (S_r) and columns (c) correspond to different symbols within these strings S_i (c). The matrix is used to find the relative importance/confidence to be assigned to every input demonstration (now encoded as a string) for every T/N steps (now represented as individual symbols) of the original task.

The main insight here is that the importance of a symbol depends on the relative similarity between it and symbols in other strings at the same location but with the constraint that the dominant string(s) is not changed frequently (to avoid creating discontinuities in the real-valued model to be generated). This rule takes care of both confusions (existence of demonstrations that are outright wrong and do not correspond to the task) and distortions (segments of demonstrations that are extra-noisy). This means that we can initialize the *Z-Msatrix* using:

$$ Z(r, c) = \sum_{d=0}^{M-1} e^{-\tau d} \, |\{i \,|d = |S_i(c) - S_r(c)|\}|. \tag{13.42} $$

This equation multiplies exponentially decreasing weight to the number of strings that are d characters different from the current character (character c in string r) and adds all of these factors together. In Eq. 13.42, τ determines how fast is the exponential decay and in all our experiments we just set it to 1.

The resulting *Z-Matrix* gives an initial evaluation of the relative confidence we can assign to each symbol in each string. This confidence value is not enough though to select a good representation of the motion because in many cases it may have ties. For example, if we have five demonstrations that have very similar behavior in the beginning of the motion, the resulting strings will likely all have the same symbol in the first location and Eq. 13.42 will have exactly the same value in all members of the first few (say four) columns. Now how would we create a good model of the motion in this case: one option is to use the mean of all motions and another is to use the mean of any subset of them (even a single demonstration). The information in the four columns cannot help us make a decision here. If we just use the mean of all the signals and the fifth column had only two strings with high score compared with the rest, it would have been beneficial to use these two strings only for the first four columns to avoid having an unnecessary discontinuity in the final model.

To break these kinds of ties while preserving the continuity of the final model as much as possible, we update the $Z\text{-}Matrix$ by having the score at every cell be added to its two (or one) adjacent cells using Eq. 13.43 except at the boundaries where we have a single neighbor in the string.

$$Z(r, c) \leftarrow Z(r, c) + \lambda (Z(r, c-1) + Z(r, c+1)). \qquad (13.43)$$

This operation is continued until ties at all columns are resolved, no changes in the increment beyond a simple scale is happening or a predefined number of iterations is executed.

The maximum of every column in the $Z\text{-}Matrix$ is then found forming a N-dimensions vector (called Z_m). A weight matrix W of the same size as the $Z\text{-}Matrix$ is then calculated as:

$$W(r, c) = \begin{cases} \frac{e^{Z(r,c)/Z_m(c)-1}}{\sum_i W(i,c)} & e^{Z(r,c)/Z_m(c)-1} > \beta \\ 0 & otherwise \end{cases}. \qquad (13.44)$$

The weight matrix W is then used to calculate an initial model of the motion by concatenating T/N points from different demonstrations now weighted by the corresponding W entries. The final model of the motion is then generated by smoothing this initial model by applying local regression using weighted least squares. A string representation of the model can also be found by concatenating the symbols with the maximum weight in W at every position.

The initial value of the $Z\text{-}Matrix$ (called Z^0) before applying Eq. 13.43 and its column-wise max (Z_m^0) can be used to calculate a *variability* score (V) to each T/N segment in the final model (corresponding to a single symbol in the K corresponding strings):

$$V\left((c-1)\frac{T}{n} + 1 : \frac{cT}{n}\right)$$
$$= \sum_{k=1}^{K} \left(\begin{cases} 1 - e^{Z^0(k,c)/Z^0_m(c)-1} & e^{Z^0(k,c)/Z^0_m(c)-1} > \beta \\ 1 & otherwise \end{cases} \right). \qquad (13.45)$$

This variability score can be used during motion execution to control how faithful should the controller follow the learned model at various parts of the task. This variability representation can be enough for some situations but in other cases it would be beneficial to have a full-fledged covariance matrix for various parts of the motion similar to the GMM/GMR system.

It was shown by Mohammad and Nishida (2014) that SAXImitate can be combined with GMM/GMR to produce full GMM models with full covariance matrices instead of the variability score generated by SAXImitate. Moreover, the SAXImitate step can be utilized to get an appropriate number of Gaussians for the GMM step removing the need for complex algorithms like variational inference or repeated calculations when BIC is used. The resulting number of Gaussians though can be shown to be suboptimal (Hussein et al. 2015).

The main advantage of SAXImitate over other LfD systems is its high resistance to two types of problems with demonstrations: confusing and distorted demonstrations. Distorted demonstrations are ones with short bursts of distortions due to sensor problem. These distorted demonstrations are common when external devices are used to capture the motion of the demonstrator. Confusing demonstrations are ones that do not even belong to the motion being learned. Both of these kinds of demonstrations can easily be avoided when demonstrations are collected in controlled environments by a careful designer. Nevertheless, when robots rely on unsupervised techniques for discovering demonstrations (as in the fluid imitation case presented in Chap. 12), both confusing and distorted demonstrations become unavoidable.

Even though SAXImitate can be used to estimate the number of Gaussians for a GMM model or the number of kernels for a DMP model (Mohammad and Nishida 2014), a better approach would be to use piecewise linear approximation. The set of Equiprobable points of a 2D Gaussian is always an ellipse in which the transverse diameter is a line that rotates with changes in the covariance matrix and the conjugate diameter is another line that is affected by the determinant of this matrix. The location of the ellipse depends on the mean vector. From this, it is clear that a single Gaussian cannot faithfully represent more than a single line in the original distribution. We can utilize this fact to base the number of Gaussians used on the number of straight lines in the input trajectory. This allows us to specify an acceptable model complexity without the need to use more complex model selection approaches. Moreover, using this technique, we can add Gaussians to the mixture whenever a new line is found in the trajectory.

This focused on 2D Gaussians but the idea presented can be extended to arbitrary dimensionality by noticing that a Multidimensional Gaussian can always be projected into a 2D Gaussian between any two of its dimensions. In the case of motion trajectory, we can always use time as one of these two dimensions which means that we can use piecewise linear fitting at every other dimension (with time as the dependent variable) exactly as discussed for the 2D case. An algorithm for estimating the number of Gaussians for a GMM based on this idea was proposed by Mohammad and Nishida (2015) to effectively estimate the number of Gaussians in a motion copying task.

13.5   Summary

This chapter discussed learning from demonstration (LfD) in robotics. LfD systems were divided into low level trajectory learning and high level plan learning systems. The chapter then focused on the trajectory learning literature as this is the approach used intensively in our systems. We covered the four major approaches in modern LfD: inverse optimal control, inverse reinforcement learning, dynamical motor primitives, statistical methods including HMM and GMM/GMR, and symbolization approaches based on multidimensional extensions of the SAX transform and piecewise linear approximations.

References

Abbeel P, Coates A, Ng AY (2010) Autonomous helicopter aerobatics through apprenticeship learning. Int J Robot Res 29(13):1608–1639

Abbeel P, Ng AY (2004) Apprenticeship learning via inverse reinforcement learning. In: ICML'04: the twenty-first international conference on machine learning, ACM. http://www.machinelearning.org/proceedings/icml2004/papers/335.pdf

Argall BD, Chernova S, Veloso M, Browning B (2009) A survey of robot learning from demonstration. Robot Auton Syst 57(5):469–483

Atkeson C, Schaal S (1997a) Robot learning from demonstration. In: ICML'97: International conference on machine learning, pp 12–20

Atkeson CG, Schaal S (1997b) Learning tasks from a single demonstration. In: ICRA'97: IEEE International conference on robotics and automation, vol 2. IEEE, pp 1706–1712

Bishop CM (2006) Pattern recognition and machine learning. Springer

Byravan A, Montfort M, Ziebart B, Boots B, Fox D (2014) Layered hybrid inverse optimal control for learning robot manipulation from demonstration. In: NIPS workshop on autonomous learning robots. http://www.ias.informatik.tu-darmstadt.de/uploads/ALR2014/Byravan_ALR2014.pdf

Byravan A, Monfort M, Ziebart B, Boots B, Fox D (2015) Graph-based inverse optimal control for robot manipulation. In: IJCAI'15: the 24th international joint conference on artificial intelligence. http://mobilerobotics.cs.washington.edu/papers/graph-based-IOC-ijcai-2015.pdf

Calinon S, Guenter F, Billard A (2006) On learning the statistical representation of a task and generalizing it to various contexts. In: ICRA'06: IEEE international conference on robotics and automation, IEEE, pp 2978–2983

Calinon S, D'halluin F, Sauser EL, Caldwell DG, Billard AG (2010) Learning and reproduction of gestures by imitation. IEEE Robot Autom Mag 17(2):44–54

Calinon S, Li Z, Alizadeh T, Tsagarakis NG, Caldwell DG (2012) Statistical dynamical systems for skills acquisition in humanoids. In: Humanoids'12: IEEE international conference on humanoid robots, IEEE, pp 323–329

Chatzis SP, Korkinof D, Demiris Y (2012) A quantum-statistical approach toward robot learning by demonstration. IEEE Trans Robot 28(6):1371–1381

Chen X, Ziebart BD (2015) Predictive inverse optimal control for linear-quadratic-gaussian systems. In: The 80th international conference on artificial intelligence and statistics, pp 165–173

Coates A, Abbeel P, Ng AY (2008) Learning for control from multiple demonstrations. In: ICML'08: the 25th international conference on machine learning, ACM, pp 144–151

Corduneanu A, Bishop CM (2001) Variational Bayesian model selection for mixture distributions. In: Artificial intelligence and statistics, vol 2001. Morgan Kaufmann Waltham, MA, pp 27–34

Doerr A, Ratliff N, Bohg J, Toussaint M, Schaal S (2015) Direct loss minimization inverse optimal control. In: RSS'15: robotics science and systems. http://www.roboticsproceedings.org/rss11/p13.pdf

Dufay B, Latombe JC (1984) An approach to automatic robot programming based on inductive learning. Int J Robot Res 3(4):3–20. doi:10.1177/027836498400300401

Herzog D, Krüger V, Grest D (2008) Parametric hidden markov models for recognition and synthesis of movements. In: BMVC'08: British machine vision conference, British machine vision association, pp 163–172

Huang DA, Farahmand Am, Kitani KM, Bagnell JA (2015) Approximate maxent inverse optimal control and its application for mental simulation of human interactions. In: AAAI'15: 29th AAAI conference on artificial intelligence, pp 2673–2679

Hussein M, Mohammed Y, Ali SA (2015) Learning from demonstration using variational Bayesian inference. In: Current approaches in applied artificial intelligence. Springer, pp 371–381

Ijspeert AJ, Nakanishi J, Hoffmann H, Pastor P, Schaal S (2013) Dynamical movement primitives: learning attractor models for motor behaviors. Neural Comput 25(2):328–373

Melo F, Lopes M, Santos-Victor J, Ribeiro MI (2007) A unified framework for imitation-like behaviors. In: 4th international symposium on imitation in animals and artifacts. http://www.eucognition.org/network_actions/NA141-1_outcome.pdf#page=32

Mohammad Y, Nishida T (2014) Robust learning from demonstrations using multidimensional SAX. In: ICCAS'14: 14th international conference on control, automation and cystems, IEEE, pp 64–71

Mohammad Y, Nishida T (2015) Simple incremental GMM modeling using multidimensional piecewise linear segmentation for learning from demonstration. In: ICIAE'15: the 3rd international conference on industrial application engineering, The Institute of Industrial Applications Engineers, pp 471–478

Nakanishi J, Morimoto J, Endo G, Cheng G, Schaal S, Kawato M (2004) A framework for learning biped locomotion with dynamical movement primitives. In: Humanoids'04: the 4th IEEE/RAS international conference on humanoid robots, IEEE, vol 2, pp 925–940

Nemec B, Ude A (2012) Action sequencing using dynamic movement primitives. Robotica 30(05):837–846

Oudeyer PY et al. (2014) Socially guided intrinsic motivation for robot learning of motor skills. Auton Robots 36(3):273–294

Park T, Levine S (2013) Inverse optimal control for humanoid locomotion. In: Robotics science and systems workshop on inverse optimal control and robotic learning from demonstration. http://www.cs.berkeley.edu/~svlevine/papers/humanioc.pdf

Pastor P, Hoffmann H, Asfour T, Schaal S (2009) Learning and generalization of motor skills by learning from demonstration. In: ICRA'09: international conference on robotics and automation, pp 763–768

Schaal S (2006) Dynamic movement primitives-a framework for motor control in humans and humanoid robotics. In: Adaptive motion of animals and machines. Springer, pp 261–280

Schaal S, Atkeson CG, Vijayakumar S (2002) Scalable techniques from nonparametric statistics for real time robot learning. Appl Intell 17(1):49–60

Vijayakumar S, Schaal S (2000) Locally weighted projection regression: an o(n) algorithm for incremental real time learning in high dimensional spaces. In: ICML'00: the 17th international conference on machine learning, pp 288–293

Chapter 14
Conclusion

This book provided a systematic introduction to the engineering of autonomous sociality in robots using techniques from time-series analysis and data mining. This is the first book sized treatment of this young field of inquiry. Rather than providing a comprehensive list of contributions and results related to autonomous sociality, the book started by detailing the foundations of the subject in time-series analysis, and data mining in the first part of the book and built upon this foundation in the second part providing two computational architectures for achieving autonomous sociality and fluid imitation and several case studies in the second part of the book. To learn and model human's social behavior, it is modeled as the result of executing several context dependent interaction protocols. Learning and executing these interaction protocols, thus, was the focus of this book.

Most of the algorithms discussed in the book are implemented in two open-source MATLAB toolboxes which enables the reader to experiment with the ideas introduced in the book and can provide a basis for her own research in autonomous sociality and related robotics and agent applications.

The introductory chapter defined the central problem of autonomous sociality as learning interaction protocols (either implicit or explicit) using unsupervised learning techniques and detailed the relation between this goal and research in several fields including interaction studies, robotics, neuroscience, experimental psychology, machine learning and data mining. The chapter also introduced two interaction scenarios that will serve as running examples throughout the book: natural listening to an explanation and guided navigation. Given the focus on nonverbal interaction protocols in this book, the introduction also presented an overview of nonverbal communication modalities in human–human and human–robot interactions. Since the proposed approach to autonomous sociality will be implemented as an HRI specific architecture, we also discussed briefly the history of behavioral robotic architectures introducing both reactive and hybrid architectures.

Autonomous sociality in our approach is achieved through unsupervised processing of human–human interaction records to discover patterns using which the interaction protocols are modeled. In many cases, these interaction records take the form

© Springer International Publishing Switzerland 2015
Y. Mohammad and T. Nishida, *Data Mining for Social Robotics*,
Advanced Information and Knowledge Processing,
DOI 10.1007/978-3-319-25232-2_14

of multidimensional real-valued time-series. The foundational technology for our approach to autonomous sociality is, thus, time-series analysis. The first part of the book introduces this area of research for social scientists and roboticists. We did not assume any prior knowledge in time-series analysis or signal processing and introduced all of the basic building blocks upon which later chapters of the book will heavily rely. Several generation models of time-series were introduced including linear additive time-series model, random walks, moving average processes, auto-regressive processes, ARMA and ARIMA processes, state-space models, Markov Chains, HMMs, GMMs, and Gaussian Processes.

After covering the generation processes commonly used in time-series analysis (specially for our purposes of learning interaction protocols) we discussed several alternative representation and transformation techniques that can be used to modify the form of the time-series to make it more amenable to analysis. The techniques covered in the book were piecewise aggregate approximation, symbolic aggregate approximation, discrete Fourier and wavelet transforms and singular spectrum analysis. The book also discussed simple preprocessing techniques that are usually employed to clean-up and condition time-series before applying mining algorithms to them including smoothing, thinning, normalization, de-trending, and dimensionality reduction. After that, we discussed learning techniques that can be employed to recover the parameters of the underlying generation process given example time-series and the related model selection problem. These kinds of inverse problems are encountered repeatedly in the second part of the book when learning interaction protocols.

Given this basic knowledge in time-series analysis, the book moved on to discuss the three building blocks of time-series mining that will be used to build our computational methods in the second part of the book. These are change point discovery, motif discovery and causality analysis. Several approaches for each of these three basic problems were discussed in the first part of the book. Other than presenting the main algorithms and results in each of these three areas, the book tried to provide objective approaches to comparing different algorithms and applications from the world of social robotics that can be implemented using each of these three technologies.

The second part of the book presented in details our approach to social robotics based on data mining techniques specially the three main technologies: change point discovery, motif discovery and causality analysis. The book started by introducing social robotics focusing on human's response to social robots emphasizing the importance of expectation in this context. Three specific HRI architectures are then discussed: C4, situated modules and HAMMER. These architectures were shown to be based on complimentary principles with C4 emphasizing the cognitive aspect of the theory of mind embodied by the robot, situated modules emphasizing the ability of the designer to combine and match different context specific behaviors to generate appropriate social actions from the robot, and HAMMER emphasizing on learning through a mirroring mechanism. These three concepts informed the design of the Embodied Interactive Robotic Architecture (EICA) proposed in this book.

The architecture advocated in this book is based on two main theoretical foundations: intention modeling through a dynamical view that avoids the problems with traditional intention as a plan approach and a theory of mind approach that combines aspects of the simulation theory and theory of theory. A description of each of these principles with discussion of their relation to autonomy, sociality and embodiment is presented and lessons from them are then used to motivate the design of EICA.

While simulation is the core of EICA's behavior generation mechanism, imitation is the core of its learning system which is our way to achieve autonomous sociality in a robot. To motivate this learning methodology, we discuss different definitions of imitation and its importance in animal and human behavior then present two studies of the social aspects of imitation in robotics. The first used imitation to bootstrap social understanding on an animal-like robot and the second analyzed the effect of back-imitation on the perception of imitative skill, human-likeness and naturalness of robot behavior.

After presenting these theoretical foundations, the book presents the details of EICA progressively. Firstly, the basic behavioral architecture is discussed which implements a hybrid action integration mechanism combining combinative and selective approaches under the control of higher dynamical processes. The intention function discussed theoretically earlier is implemented in this architecture through a set of interacting intentions each with its own level of activation, attentionality and intentionality. This intention functions evolves under the control of higher processes that can range from purely reactive to purely deliberative components and interact with each other using a variety of data and control channels. This basic behavioral architecture is not specific to social robots or HRI applications and can be used to implement any kind of behavior. This generality comes with the price of having too many knobs to adjust and parameters to select. A floating point genetic algorithm (FPGA) for selecting appropriate values of these parameters is then presented and the whole system is utilized to develop applications for our running example scenarios: gaze control and collaborative navigation.

These applications provide a proof of concept that human-like behavior in simple social situations is achievable through careful engineering of different intentions and processes of a behavioral system. This is not though the approach advocated in this book because of its reliance on careful analysis of human–human interaction records leading to high development costs. Moreover, this approach means that the resulting robot is not independent from its designer toward learning interaction protocols. The book then discusses the proposed approach to behavior generation using the down-up-down mechanism and mirror training. This approach to behavior generation is inspired by our earlier discussions of the simulation theory of mind.

The book then presents the main learning mechanism proposed to achieve the goal of autonomous sociality in the form of a three stages developmental system. The first stage, interaction babbling, takes as input the records of human–human interactions exemplifying the interaction protocol to be learned and uses change point discovery and motif discovery to learn the forward intentions corresponding to the basic interaction acts found in these records. A controller is then generated for each of these intentions using a suitable controller generation mechanism. The second stage, interaction structure learning, builds progressively more complex interaction protocols based on the basic acts (intentions) learned in the first stage. Three alternative approaches are discussed: single-layer interaction structure learning, induction of rules and deep interaction protocol learning. Depending on the complexity of the interaction protocol to be learned, one or the other of these approaches will be most appropriate. The final stage of the proposed developmental system adapts the learned protocol through analysis of the differences between forward and inverse directions of behavior generation during interactions with humans. This stage capitalizes on the simulations carried out as a part of the DUD behavior generation mechanism of EICA. Again both single layer protocols and deep interaction protocols are discussed. The book then presents two applications of the complete EICA system to the explanation and guided navigation scenarios showing the superiority of this autonomous learning technique over the engineering approach presented earlier using the underlying behavioral platform of EICA combined with the FPGA algorithm.

EICA provides a technique that allows the robot to learn interaction protocols by watching human–human interactions. Yet, robots are usually not designed just for interaction. They have tasks to do and interacting with people provides useful data for them to acquire the skill required to complete these tasks and enhance their performance of them. The second architecture discussed in the book was the fluid imitation engine which is designed to achieve this skill transfer from humans to robots without relying on pre-segmented out-of-context demonstrations that are usually employed in learning from demonstration research. The basic building blocks of the fluid imitation engine are the same as those of EICA: change point discovery, motif discovery and causality analysis. The book discusses the perspective taking process necessary for all forms of imitation learning presenting algorithms for transforming environmental state to the frame-of-reference of the learner as well as solutions to the correspondence problem mapping actions of the imitatee to the imitator. Change point discovery and causality analysis are then used for significance estimation of perceived behaviors of humans or other *expert* robots and drive the self-initiation engine at the heart of our fluid imitation system. The book then discussed the application of fluid imitation to a navigation problem.

Both EICA and the fluid imitation engine rely on the robot's ability to generate accurate controllers that can execute a task given a set of examples (learning from motif discovery). This is the standard learning from demonstration in robotics. The final chapter of the book briefly discusses this problem in historical context. Early approaches that relied on encoding the behavior to be learned as a logical program are discussed and their limitations are used to motivate modern approaches to learning from demonstration. The book then introduces and compares modern learning from demonstration approaches including inverse optimal control, inverse reinforcement learning, dynamic movement primitives and statistical modeling methods including HMMs and GMM/GMRs. Recent algorithms based on symbolization are then introduced to complete the coverage of learning from demonstration.

The approach advocated in this book for autonomous sociality in robots is but one possible route for this ambitious goal and a first step in that road. Future directions of research suggested by the work presented in this book are diverse and are being pursued by our research groups. One such direction is integrating the interaction protocol mechanism of EICA into fluid imitation to enable learning of the appropriate time for producing newly learned behaviors by watching how infants and children succeed in that task. Another possible direction of future research is developing an incremental interaction structure learning mechanism for the second stage of development in EICA that combines interaction structure learning and adaptation. This will allow the robot to learn interaction protocols without the need of offline processing of large interaction corpora brining it even closer to how human children learn to socialize. We are sure that the reader can envision more directions for extending the work presented in this book and hope that the provided theoretical tools and practical toolboxes can be of help in this task.

The introduction of this book started with the following question: *How to create a social robot that people do not only* operate *but relate to?* This book is a detailed long answer based on fundamental ideas from neuroscience, experimental psychology, machine learning and data mining. Other answers are certainly possible and the approach proposed here is but one small step toward this ambitious goal.

Index

A

Acive intramodal mapping, 201
Action integration, 232, 234
    combinative, 234
    hybrid coordination, 235
    selective, 234
Action segmentation, 25, 277
Action theory, 215
Adaptive resonance theory (ART), 22, 246
Affordance learning, 195
APCA, 51
Appearances theory, 213
Apprenticeship learning, 24
ARIMA process, 40
ARMA process, 40
Authocorrelation coefficient, 69
Auto-regressive process, 40, 91
Autonomous socility, 209
Autonomy, 174, 207, 208

B

Back imitation, 201, 202
Bahavior symbols, 121
Baum–Welch algorithm, 73
Bayesian information criteria (BIC), 47, 77, 153
Bayesian network, 188, 214
BDI, 22, 220
Behavior copying
    part-oriented, 196
    process-oriented, 196
    product-oriented, 196
Behavioral robotics, 21

C

C4, 234
Catalano's algorithm, 134
Causality, 149, 150
    common cause, 149
    cycles, 149
    granger, 150, 153
Causality discovery, 150
    change causality, 162
    convergent cross mapping, 153
    delay consistency-discrete, 164
    granger, 151
Chameleon effect, 199
Change point discovery (CPD), 85
Clique, 117
Clustering, 53, 116
Common ground, 8
Conrrespondence problem, 25, 277
Contagion, 195
Contribution theory, 8
Controller generation, 257
Convergent cross mapping, 153
Correlation, 150
Correspondence problem, 282
CPD localization, 85, 98
CPD scoring, 85

© Springer International Publishing Switzerland 2015
Y. Mohammad and T. Nishida, *Data Mining for Social Robotics*,
Advanced Information and Knowledge Processing,
DOI 10.1007/978-3-319-25232-2

Cross validation, 77
Cumsum, 93
Cyclic time-series, 37

D
Data mining, 11
Development, 3
Developmental robotics, 173
Direct location inference, 86
Discrete fourier transform (DFT), 54
Distance
    Euclidean, 125, 129, 131
    Hamming, 110
    Levinstein, 110
    normalized Euclidean, 125
    scale normalized, 131
Do–As–I–Do studies, 196
DUD, 249
Dynamic, 233
Dynamic movement primitives (DMP), 302
Dynamic plan, 223
Dynamic time wrapping, 313
Dynamical coherence, 223

E
Earth mover's distance, 102
EICA, 4, 224, 229
    action integration, 234
    design procedure, 237
    features, 233
Eigen triple, 59
Embodiment, 172
    historical, 210
    historical social, 211, 224, 225, 245
    physical, 210
    social, 210
    structural, 209
Empathy, 24, 173, 199, 200
Emulation, 195
Equal sampling rate, 104
Euclidean distance, 54
Expectation maximization, 73
Experimental psychology, 12
Explanation scenario, 13, 270
Exploration scenario, 240
Extended LAT, 37

F
Fluid imitation, 275
Fourier transform, 37
FPGA, 238

Frobenius norm, 60

G
Gaussian mixture regression (GMR), 45
Gaussian process, 47
Gaze control, 17
Gemoda algorithm, 115
Generalized likelihood ratio, 91
Gesture, 17
GHMM, 44
GMM, 45
    GMR, 307
Grounding, 8
Guided navigation scenario, 12, 242, 271

H
Haar wavelet, 55
HAMMER, 185, 214, 248
Hankel matrix, 95
Heteronomous, 208
Hidden marokov model (HMM), 19, 43, 307

I
Imitation, 24, 193, 195
    affordance learning, 195, 197
    contagion, 195, 218
    context, 197
    developmental, 198
    emulation, 195, 197
    fluid, 275
    goal-oriented, 198
    human, 198
    moral development, 199
    nativist, 198
    part-oriented, 196, 197
    process-oriented, 196, 198
    product-oriented, 196, 198
    social facilitation, 195
    stimulus facilitation, 197
Imitation challenges, 25
Intention
    EICA, 231
    goal, 222
    implementation, 222
    intention in action, 222
    mutual, 223, 247
    prior, 222
Intention function, 233, 245
Intention modeling, 218
    expression, 221
    joint attention theory, 221

mtual, 223
    planning theory, 220
    recognition, 221
Intention through interaction, 223–225, 233
Interaction adaptation, 7
    deep, 268
    single-layer, 266
Interaction babbling, 6, 255
Interaction protocol, 4, 225, 229, 245
Interaction structure learner, 259
    deep, 264
    rule induction, 261
    single-layer, 259
Interaction structure learning (ISL), 6
Interaction studies, 7
Inverse optimal control, 295
Inverse reinforcement learning, 299

J
Joint attention theory, 221

K
Kinds theory, 213
KL divergence, 99, 102

L
Learning from demonstration, 2, 45
    dynamic movement primitives, 302
    GMM/GMR, 307, 315
    hidden marvok models, 307
    inverse optimal control, 295
    inverse reinforcment learning, 299
    SAXImitate, 313
Least squares, 67, 69, 151
Like-me hypothesis, 218
Linear additive time-series model (LAT), 36

M
Machine learning, 11
MAP, 90
Markov chain, 42
Maximal extensibility, 116
Maximal occurrence coverage, 116
Maximum likelihood, 72, 73, 94
Mentalizing, 211
Mimicry, 24
Mind reading, 211
Minimum description length, 120
Mirror neuron, 248
Mirror neurons, 197

Mirror training, 252
MK
    MK+, 127, 132
    MK++, 129, 130, 132, 145
    MN, 131, 133
    naive, 127
MK algorithm, 125
Model selection, 77
Mofid discovery
    minimum description length, 120, 124
Motif
    exact, 124
    gemoda, 116, 128, 136, 141
    K, 119
    pair, 124, 127
    range, 123
    variable length, 116
Motif discovery, 109
    catalano, 134, 137
    common, 111
    constrained, 136
    discretization, 118
    exact, 124
    gemoda, 116, 123, 128, 135, 138
    gemoda (real valued), 138
    GSteX, 129, 138, 145
    MCFull, 136, 137
    minimum description length, 130
    MK, 125
    MK+, 127, 132
    MK++, 129, 132, 145
    MN, 131, 133
    pattern first, 111
    planted(n,l), 114
    repeated, 112
    sequence first, 111
    shift density, 140, 145, 146
    stochastic, 134
    variable length, 116
Motor babbling, 200, 201, 257
Moving average process, 38
MSAX, 54
Mutual intention, 223

N
Naive MK, 127
Naturalness, 4, 105
Nonverbal, 14
Normalization, 78

O
Organ identification, 201

P
Page-hinkley stopping rule, 93
Perspective taking, 25, 277, 280
Piecwise aggregate approximation (PAA), 51
Planning theory of intention, 220
Priming, 197
Principle component analysis (PCA), 53, 80, 163
Programming by demonstration, 24
Projections algorithm, 114

Q
Q-Learning, 19

R
Random walk, 37
Reactive architectures, 21
Risedual, 85
Robot, 3
Robovie, 20
Robust singular spectrum transform (RSST), 95, 96

S
Self determination theory, 207
Shakey, 10
Significance, 25, 277
Simulation theory, 186, 200, 217
Singular spectrum analysis (SSA), 56, 95
Singular value decomposition (SVD), 53, 57, 95
Situated modules, 173, 181, 234
Smoothing, 77
Sociability, 208
Social facilitation, 195, 197
Sociality, 209
Social referencing, 202
Social robot, 172
Social robotics, 171, 174
Speech act theory, 7
Squared exponential kernel, 48
State space model, 41
Stigmergy, 172
Structure equation model, 149
Subsequence, 36

Subsumption, 21, 182
Supervised learning, 25
Symbolic aggregate approximation (SAX), 52, 313

T
Teleoperation, 5
Theory of mind (ToM), 200, 211
  bayesian, 214
  child-scientist, 213–215
  copy theorist, 215
  Enactive, 211
  modular, 211, 215
  simulation, 186, 200, 217, 246
  theory, 211
Theory theory, 211
Thinning, 78
Time-series, 36
Tren, 37
Triangular inequality, 125, 127, 131
Two–actions studies, 197
Two-Models CPD, 90
Two-stages regression, 70

U
Uncanny valley, 175

V
Viterbi algorithm, 90

W
Wavelet, 55
Winnower algorithm, 113, 114

X
XLAT, 37

Y
Yule–Walker equations, 69

Z
Z-score, 52, 53, 125